Three-noded Triangular Elements

Shape Functions

$$T_1(\xi,\eta) = -\frac{1}{2}(\xi+\eta) \qquad T_2(\xi,\eta) = \frac{1}{2}(1+\xi) \qquad T_2(\xi,\eta) = \frac{1}{2}(1+\eta)$$

Coordinate Transformation $\quad x = \sum_{i=1}^{3} x_i T_1(\xi,\eta) \qquad y = \sum_{i=1}^{3} y_i T_i(\xi,\eta)$

In solving Poisson's equation $\dfrac{\partial^2 u}{\partial x^2} + \dfrac{\partial^2 u}{\partial y^2} + h(x,y) = 0$ the element stiffness matrix and force vectors are, for constant h and area A:

$$[K] = \frac{1}{4A}\begin{bmatrix} b_1^2+c_1^2 & b_1b_2+c_1c_2 & b_1b_3+c_1c_3 \\ b_1b_2+c_1c_2 & b_2^2+c_2^2 & b_2b_3+c_2c_3 \\ b_1b_3+c_1c_3 & b_2b_3+c_2c_3 & b_3^2+c_3^2 \end{bmatrix} \qquad \{F\} = \frac{hA}{3}\begin{Bmatrix} 1 \\ 1 \\ 1 \end{Bmatrix}$$

$$b_1 = y_2 - y_3 \quad b_2 = y_3 - y_1 \quad b_3 = y_1 - y_2$$

$$c_1 = x_3 - x_2 \quad c_2 = x_1 - x_3 \quad c_3 = x_2 - x_1 \quad A = \frac{1}{2}(b_2c_3 - b_3c_2)$$

Four-noded Quadrilateral Elements

Shape Functions

$$L_1 = \frac{1}{4}(1-\eta)(1-\xi) \qquad L_2 = \frac{1}{4}(1-\eta)(1+\xi)$$

$$L_3 = \frac{1}{4}(1+\eta)(1+\xi) \qquad L_4 = \frac{1}{4}(1+\eta)(1-\xi)$$

Coordinate Transformation $\quad x = \sum_{i=1}^{4} x_i L_i(\xi,\eta) \quad y = \sum_{i=1}^{4} y_i L_i(\xi,\eta)$

For a rectangular element $x_1 \le x \le (x_1+2b)$, $\quad y_1 \le y \le (y_1+2a)$ the element stiffness matrix and force vector arising from Poisson's equation are

$$[K] = \frac{a}{6b}\begin{bmatrix} 2 & -2 & -1 & 1 \\ -2 & 2 & 1 & -1 \\ -1 & 1 & 2 & -2 \\ 1 & -1 & -2 & 2 \end{bmatrix} + \frac{b}{6a}\begin{bmatrix} 2 & 1 & -1 & -2 \\ 1 & 2 & -2 & -1 \\ -1 & -2 & 2 & 1 \\ -2 & -1 & 1 & 2 \end{bmatrix} \qquad \{F\} = \frac{hA}{4}\begin{Bmatrix} 1 \\ 1 \\ 1 \\ 1 \end{Bmatrix}$$

The Finite Element Method

Principles and Applications

P. E. Lewis and J. P. Ward

Loughborough University of Technology

 ADDISON-WESLEY PUBLISHING COMPANY

Wokingham, England · Reading, Massachusetts · Menlo Park, California
New York · Don Mills, Ontario · Amsterdam · Bonn · Sydney · Singapore
Tokyo · Madrid · San Juan · Milan · Paris · Mexico City · Seoul · Taipei

Cover designed by Hybert Design and Type, Maidenhead and printed by
The Riverside Printing Co. (Reading) Ltd.
Typeset by the authors using TEX.
Printed in Great Britain by T.J. Press (Padstow) Ltd, Cornwall.

First printed 1991.

British Library Cataloguing in Publication Data
Lewis, P. E. (Peter Edward), 1943–
The finite element method: principles and applications
I. Title II. Ward, J. P. (Joseph Patrick), 1949–
624.101515353

ISBN 0–201–54415–6

Library of Congress Cataloging in Publication Data
Lewis, P. E., 1943–
The finite element method: principles and applications by P.E. Lewis and J.P. Ward.
p. cm.
Includes bibliographical references and index.

ISBN 0–201–54415–6
1. Boundary value problems — Numerical solutions. 2. Finite element method. I. Ward, J.P., 1949– . II. Title.
QA379.L49 1991
515′.35—dc20

91-18840
CIP

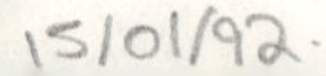

Preface

This book is intended as an introductory text on the finite element method. The emphasis is on finite elements as an approximate numerical technique for the solution of boundary value problems. Because such problems arise in many areas of engineering, science and applicable mathematics it is our hope that the book will be useful to a wide variety of practitioners and students of these subjects.

We, and our colleagues, have long felt that there is a need for a finite element text such as this which attempts to steer a middle ground between books on the subject which are orientated almost entirely to structural mechanics and those which are highly mathematical. The level of mathematics that is used and the prerequisites (which are mainly basic differential and integral calculus and matrix manipulation) should be well within the compass of the target audience. An elementary introduction to elasticity theory is provided to help readers to appreciate some finite element applications in the solid mechanics area, this of course being the field in which the technique was first applied. We have also used specific boundary value problems from other application areas but a detailed knowledge of the derivation of the governing equations (although included for completeness) is not needed in order to understand the solution technique.

Over the past thirty years or so the finite element method has become a very popular numerical technique: indeed it is the computational basis of many CAD systems. The method is implemented on a computer and many commercial packages are available for this purpose. However this book is addressed to those who seek a straightforward, concise explanation of the principles of finite elements. We do not discuss the practical implementation in any great detail although we have included a few simple programs for illustrative purposes.

The basic theory of the finite element method may be developed using the calculus of variations or using the concept of weighted residuals. We have concentrated on the latter approach because we feel it is more likely to be understandable and acceptable to our intended readers than the variational approach. However, for completeness and for the benefit of those readers who will wish to move on to other more advanced texts we have provided an introductory account of the variational basis of the finite element method in Chapter 8.

The layout of the book is straightforward. After introducing the concept of boundary value problems and the 'natural occurrence' of finite elements in simple structures in Chapter 1 we discuss in Chapter 2 the Galerkin weighted residual method as an approximate technique for the solution of boundary value problems in one dimension. The concept of division of the domain of the problem into 'elements' is introduced in Chapter 3 where much of the basic 'jargon' of the subject – shape functions, nodes, linear elements, quadratic elements, local coordinate systems – is explained. We can then, in Chapter 4, present a full discussion of the finite element formulation of one-dimensional, second-order linear boundary value problems. Chapter 5 introduces linear elasticity theory. The concepts of strain energy and potential energy, which play a basic role in the formulation of finite elements in this important area, are described fully, particularly in two-dimensional (plane strain) applications. Also in this chapter we consider the 'archetypal' finite element – the pin-jointed strut.

In Chapters 6 and 7 we move on to two-dimensional problems, Chapter 6 containing the necessary numerical groundwork on triangular and quadrilateral elements which, in Chapter 7, is then applied to the solution of linear second-order problems but now in two dimensions.

Throughout these basic chapters we have tried, via the medium of worked examples to explain carefully the systematic steps involved in finite element analysis.

Chapters 8 and 9 are independent and need not be read in order. Chapter 8 is the introduction to the variational formulation already referred to while Chapter 9 is intended as useful background for those interested in understanding, or perhaps even developing, computer programs which implement the finite element method.

In the final chapter, we extend finite element concepts to the solution of fourth-order boundary value problems.

As well as detailed worked examples throughout the text there are many end-of-chapter examples for readers to test their understanding of the material. The solution of these problems does not require access to a computer. We have provided answers to all odd-numbered examples and outline solutions to many of these.

Our grateful thanks are due to Karen Syme who, with skill and patience, typed the original draft of this text. We are of course responsible for any errors that arise. The comments of readers and suggestions for improvement are welcome.

P.E. Lewis
J.P. Ward
Loughborough, March 1991

Contents

To my wife and children
P.E.L.

To my wife Rosemarie and my children Joanne and Frances
J.P.W.

1

Preliminaries

PREVIEW

This chapter discusses the need for studying the finite element method and briefly outlines the type of problem to which the method will be applied. In particular we examine one-dimensional linear second-order boundary value problems. We also describe eigenvalue problems and problems in two-dimensions as well as approximate techniques for their solution. Finally a brief introduction to the direct approach to finite elements is given.

1.1 Introduction

The finite element method is essentially a numerical method for the approximate solution of practical problems arising in engineering and scientific analysis. Although the distinction is not completely sharp we can divide applications of the finite element method into two broad categories:

(a) Solution of problems arising in structural mechanics. Historically this was the area where the method had its foundations, the motivation being to solve complex structural problems arising in the aerospace industry. A number of now classical research papers in this field were published in the late 1950s and early 1960s.

(b) Solution of differential equations particularly equations arising in engineering analysis for which exact solutions are not available.

It is with this second broad set of applications that we are mainly concerned in this text. Typical subject areas of interest are Heat Transfer, Acoustics, Electromagnetism, Fluid Mechanics and Solid Mechanics. The application of finite elements to solving problems in these (and other) areas probably began in the mid-1960s when the link between the finite element technique and variational methods for solving differential equations became recognized. Solutions of two of the more common differential equations – those of Laplace and Poisson – were early applications of the finite element method and useful results in Heat Transfer, Fluid Mechanics and, later, Electromagnetics were readily obtained.

The scope of the finite element method was further extended by the realization in the late 1960s that the basic equations necessary for finite element analysis could be obtained by a method of weighted residuals. This method could be applied to problems where variational techniques were not available and, arguably, provided for easier analysis than variational methods even when both techniques were applicable.

The finite element method involves utilizing mathematical methods to transform differential equations into a form amenable to numerical solution. The end-point is normally a system of linear or non-linear algebraic equations, perhaps thousands of such equations in large scale engineering problems. Clearly, therefore, the availability of a digital computer is an essential adjunct to the finite element method both for the setting up and the solution of the system of equations. Indeed many standard computer packages are available for implementing the method.

The engineer or scientist to whom this text is mainly directed might at first question the need for a detailed study of a technique for which there appear to be well developed computer packages for solving many problems. The authors feel that this is a somewhat short-sighted point of view since, although many of the presently available finite element computer packages have the appearance of great sophistication and applicability, many of them are in fact somewhat limited. In order to extend the applicability of these packages or, more importantly, to develop finite element programs for **specific** engineering problems not treated by currently available packages, a thorough understanding of the fundamentals of the finite element approach is advisable.

Also, part of the task of the modern engineer is to interpret the 'output' from computer packages. He/she should be able to assign what might be termed a 'level of credibility' to approximate solutions generated by a 'black-box' computer package whether the data is in graphical or numerical form. It is all too easy to view an impressive picture on a computer terminal and conclude that the information being displayed is without error and faithfully represents reality. The finite element method, in common with any other numerical method applied to an engineering problem, produces a solution which **approximates** the desired (but unknown) true solution. To understand the quality of the approximation, as one certainly should with the application of any numerical technique, a basic understanding of the underlying principles is desirable, if not essential.

This text covers the fundamentals of the finite element method and will, we hope, be sufficient reading for the great majority of engineers and scientists who encounter the technique in practice or who study it as part of a university or college course. It is also an adequate and readable introduction to more advanced texts.

1.2 Boundary value problems

Differential equations take on many different forms. Perhaps the simplest type is when the equation involves one independent and one dependent variable, for example the simple harmonic motion equation:

$$\frac{d^2u}{dx^2} = -\lambda^2 u$$

where λ^2 is a positive constant. This equation may be readily solved analytically to give

$$u = A\cos\lambda x + B\sin\lambda x$$

where A and B are arbitrary constants. The determination of A and B can be achieved if two conditions are known about the solution u.

If the values of u and of its derivative are known at some particular value, say x_0, of the independent variable then we are said to have an **initial-value problem**.

On the other hand, if the values of u at two distinct values of x, say $x = a$ and $x = b$, are known then we are said to have a **boundary value problem**. Various other boundary conditions are possible; for example we might know the value of u at $x = a$ and the value of du/dx at $x = b$. The points $x = a$ and $x = b$ are called boundary points.

Mathematically we could state a typical simple boundary value problem as

$$\frac{d^2u}{dx^2} = -\lambda^2 u \qquad a \le x \le b \tag{1.1a}$$

$$u(a) = \alpha \tag{1.1b}$$

$$u'(b) = \beta \tag{1.1c}$$

where α and β are given. The term $u'(b)$ means du/dx evaluated at $x = b$, that is (1.1c) is a **derivative** boundary condition.

To be more precise (1.1) defines a one-dimensional linear second-order boundary value problem. The term one-dimensional arises because the domain of the problem can be looked on as a line segment $a \le x \le b$ between the boundary points. The term 'second-order' is because the governing differential equation (1.1a) is of second order i.e. the highest derivative present is a second derivative. The term 'linear' refers to the fact that in (1.1) the **degree** of each term involving the dependent variable u is 1, the degree being the number of times that u and/or its derivatives appear in that term.

The equations

$$u\frac{d^2u}{dx^2} = 1 \quad \text{and} \quad \left(\frac{d^2u}{dx^2}\right)^2 = -\lambda^2 u$$

are examples of non-linear differential equations.

Although the finite element method can be applied more widely we will, in this introductory text, be mainly concerned with its application to linear boundary value problems, particularly second-order problems. However in the final chapter we will discuss fourth-order boundary value problems which are particularly important in solid mechanics applications.

An example of a fourth-order boundary value problem is

$$\frac{d^2}{dx^2}\left(EI(x)\frac{d^2u}{dx^2}\right) = F(x) \qquad 0 \le x \le \ell \tag{1.2a}$$

$$u(0) = \frac{du}{dx}(0) = 0 \tag{1.2b}$$

$$u(\ell) = \frac{du}{dx}(\ell) = 0 \tag{1.2c}$$

Note the **four** boundary conditions here, two at each boundary point. The boundary value problem (1.2) governs the deflection u of a beam deformed by a given distributed transverse force $F(x)$. The quantities E and $I(x)$ are given properties of the material and geometry of the beam.

It can be shown that the most general linear second-order boundary value problem may be written in the form

$$-\frac{d}{dx}\left(p(x)\frac{du}{dx}\right) + q(x)u = r(x) \qquad a \le x \le b \tag{1.3a}$$

$$\alpha_0 u(a) + \alpha_1 u'(a) = \gamma_1 \tag{1.3b}$$

$$\beta_0 u(b) + \beta_1 u'(b) = \gamma_2 \tag{1.3c}$$

in which $p(x)$, $q(x)$ and $r(x)$ are given functions and α_0, β_0, α_1, β_1, γ_1, γ_2 are given constants.

If $r(x) \equiv 0$ then the differential equation (1.3a) is said to be **homogeneous**. Similarly if $\gamma_1 \equiv 0$ or if $\gamma_2 \equiv 0$ then the corresponding boundary condition is said to be homogeneous. Not surprisingly if $r(x) \equiv 0$, $\gamma_1 \equiv 0$ and $\gamma_2 \equiv 0$ the boundary value problem is said to be homogeneous. Normally, the only solution to a homogeneous boundary value problem is the so-called trivial solution $u(x) \equiv 0$. However, a parameter λ may be present either in the differential equation or in the boundary conditions and particular values of this parameter may permit a non-trivial solution. In this case we say that the boundary value problem is an **eigenvalue problem**, the special values of λ being called **eigenvalues.**

Eigenvalue problems arise most frequently in the analysis of vibrations and wave motion. For example, if we consider the homogeneous boundary value problem

$$\frac{d^2u}{dx^2} = -\lambda^2 u \qquad 0 \leq x \leq 1$$

$$u(0) = 0 \quad u'(1) = 0$$

then the solution to the differential equation is, as already mentioned,

$$u = A\cos\lambda x + B\sin\lambda x$$

Imposing the first boundary condition implies

$$0 = A$$

while the second boundary condition implies

$$0 = B\lambda\cos\lambda$$

Now if $B = 0$ (or indeed if $\lambda = 0$) we obtain the usual trivial solution $u = 0$ which is of no interest. However, if we choose λ so that

$$\cos\lambda = 0 \qquad \text{i.e.} \quad \lambda = (2n+1)\pi/2 \quad n = \text{integer or zero}$$

then both the differential equation and the boundary conditions are satisfied. Thus we have, for these special values of λ (the eigenvalues), an infinite number of non-trivial solutions

$$u_n = B_n \sin \frac{(2n+1)\pi x}{2} \qquad n = \text{integer or zero}$$

these solutions being called **eigenfunctions**.

The main problem which arises in trying to determine exact solutions to one-dimensional second-order boundary value problems is finding the solution to the given differential equation. The imposition of the boundary conditions is relatively straightforward because they are only imposed at two points.

By contrast the difficulty with analytical solutions of two- and three-dimensional problems is often the imposition of the boundary conditions. For example a two-dimensional boundary value problem (one where two independent variables are present) has as its essential ingredients

- a **partial differential equation** defined over a region D
- given boundary conditions defined on the boundary of D, a curve Γ.

In this case the boundary conditions have to be imposed at all points on the boundary, clearly a more difficult task than in one-dimensional problems. Indeed analytical solutions of two-dimensional boundary value problems are only possible if the domain D is of a simple geometric form, such as rectangular or circular. Even in such cases the exact solution may have a very complicated form. For example, the boundary value problem

$$\frac{\partial^2 u}{\partial x^2} + \frac{\partial^2 u}{\partial y^2} = 0 \qquad 0 \le x \le \ell \quad 0 \le y \le \ell \tag{1.4a}$$

$$u(x,0) = 0 \qquad u(x,l) = u_0 \qquad 0 \le x \le \ell \tag{1.4b}$$

$$u(0,y) = 0 \qquad u(l,y) = 0 \qquad 0 \le y \le \ell \tag{1.4c}$$

(which is Laplace's equation in a square) has, as its exact solution, the infinite series

$$u(x,y) = \frac{4u_0}{\pi} \sum_{\substack{n=1 \\ n \text{ odd}}}^{\infty} \frac{1}{n} \frac{\sin(n\pi x/\ell)\sinh(n\pi y/\ell)}{\sinh n\pi} \tag{1.5}$$

A second example of a linear two-dimensional boundary value problem is

$$\frac{\partial^2 u}{\partial x^2} + \frac{\partial^2 u}{\partial x^2} = -2 \qquad (x, y) \text{ in } D \tag{1.6a}$$

$$u = 0 \quad (x, y) \text{ on } \Gamma \tag{1.6b}$$

where D may be an irregular shaped region bounded by the curve Γ.

This boundary value problem arises in the analysis of the torsion of a beam of uniform cross section (the region D) whose axis is aligned along the z-axis. However, the governing equation (1.6a) also arises in a variety of other contexts and is known as Poisson's Equation.

An example of a fourth-order boundary value problem in two-dimensions is

$$\frac{\partial^2}{\partial x^2}\left(\frac{\partial^2 u}{\partial x^2} + \frac{\partial^2 u}{\partial y^2}\right) + \frac{\partial^2}{\partial y^2}\left(\frac{\partial^2 u}{\partial x^2} + \frac{\partial^2 u}{\partial y^2}\right) = F(x, y) \quad (x, y) \text{ in } D \tag{1.7a}$$

$$u = 0 \quad \text{and} \quad \frac{\partial^2 u}{\partial x^2} + \frac{\partial^2 u}{\partial y^2} = 0 \quad \text{on } \Gamma \tag{1.7b}$$

This problem arises in the analysis of the deflection of a plate (the region D) which is simply supported all along its edge (the curve Γ) and which is deformed by transverse forces $F(x, y)$. Equation (1.7a) is called the Biharmonic equation when $F(x, y) \equiv 0$.

1.3 Approximate solution of boundary value problems

The fact that boundary value problems, particularly of dimension higher than one, are difficult to solve has led to the development of many approximate numerical methods. This development is further motivated by the complicated nature of those analytical solutions, such as (1.5), which are available.

We note firstly that a solution to a one-dimensional boundary value problem is a function $u(x)$ of one independent variable x and so can be represented graphically by a curve. Correspondingly, a solution $u(x, y)$ to a two-dimensional boundary value problem has a geometric interpretation as a surface. Approximate techniques for solving boundary value problems might be regarded therefore as processes of curve or surface approximation.

A well known approximate solution technique for boundary value problems is the method of **finite differences**. In this method, the derivatives arising in the governing differential equation are approximated at certain grid points by finite differences. Algorithms of varying sophistication can then be developed for finding values of the unknown function at the grid points. The finite difference method is reasonably straightforward to set up – certainly for one-dimensional problems and for two-dimensional problems involving domains with boundaries parallel to the coordinate axes. However it is less easy to implement for irregular regions, perhaps with curved boundaries.

Variational methods and weighted residual methods give rise to other approximate techniques for solving boundary value problems. These will be discussed in detail later in this text because of their use in connection with the finite element method. Both techniques predate finite elements and both involve **integral formulations** of the given boundary value problem.

The finite element method takes a quite different standpoint from the finite difference method. The domain of the problem – a line segment, a planar region, a volume according as the boundary value problem is one-, two- or three-dimensional – is divided into smaller subregions or elements. These elements are line elements in one dimension, normally triangles or quadrilaterals in two dimensions and perhaps tetrahedra and prism elements in three dimensions. (We do not in fact discuss three-dimensional problems in this text since we feel that anyone who has mastered the use of two-dimensional elements would not find the extension to three dimensions overly demanding apart from the need for increased computing power.)

Following the subdivision, the change in the unknown function with position is approximated by simple functions. The original problem is then replaced by an equivalent **integral formulation**. There are two popular methods for doing this – one based on weighted residual methods, particularly the method of Galerkin, and the other based on variational principles. We shall concentrate mainly on the first of these methods which we feel is more direct and straightforward but we will, for completeness, provide a single chapter on the variational formulation.

Clearly the details of the finite element method are more complex than those of the finite difference technique and will, of course, be discussed fully in later chapters. The end-points of both methods are essentially the same namely a system of linear or non-linear algebraic equations. The finite element method, however, has a number of advantages which justify the additional labour involved in its understanding and formulation.
Some of these advantages are

- it is well-suited to problems involving complex geometries
- it can readily handle problems where the physical parameters vary with position within the domain
- it can also be used for non-linear and/or time-varying problems although we shall be mainly concerned in this text with linear time-independent problems

- complex boundary conditions can be readily dealt with
- general computer programs to perform finite element calculations can be, and have been, developed
- the finite element method is the computational basis of many computer-aided design packages
- conventional numerical techniques can be used to solve the equations resulting from a finite element analysis.

1.4 Direct approach to finite elements in one dimension

As indicated in the previous sections our main concern in this text is the use of finite elements in connection with the solution of boundary value problems defined over a region of space. Put another way we **impose** a discretization onto a continuous domain; in one dimension we divide a line segment, the problem domain, into shorter segments or elements; in two dimensions we approximate an arbitrary region as a series of smaller elements. See Figure 1.1, for example, where the discretization is implemented using triangular elements.

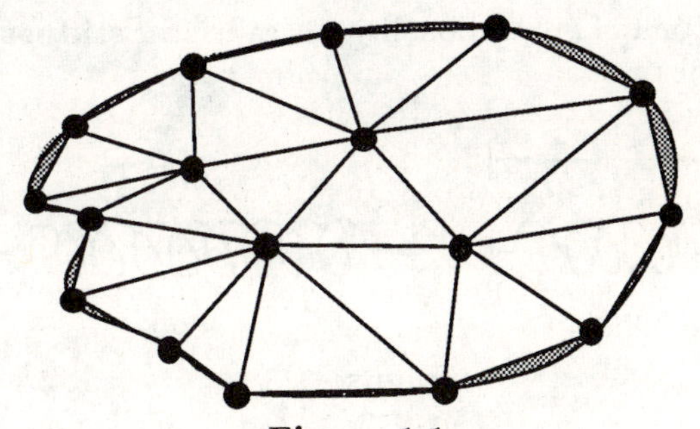

Figure 1.1

However in some situations, particularly in structural mechanics, discrete elements are already present. For example, consider the two dimensional structure shown in Figure 1.2.

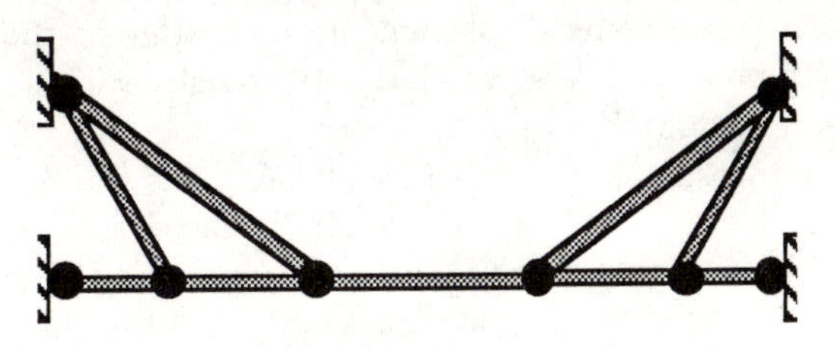

Figure 1.2

Each of the individual components of this structure can be thought of and analysed as an individual **discrete element** and the complete structure then considered as an **assembly** of such elements. Indeed, as mentioned in Section 1.1, finite element techniques were first used in the analysis of structures such as this and the 'less natural' applications to continuum problems were developed considerably later.

It is useful therefore in this first chapter to introduce finite elements in a 'direct' manner – by considering structures actually composed of discrete 'elements'. We shall consider only one-dimensional structures composed of very simple elements, viz. linear springs. A fuller discussion of structural analysis and finite elements therein will be presented in Chapter 5.

• The linear spring

A linear spring (see Figure 1.3a) is characterized by the fact that if a force is applied to one end then a displacement is produced which is proportional to that force, that is

$$F = k\,d$$

where the constant of proportionality k is called the **stiffness** of the spring (see Figure 1.3b).

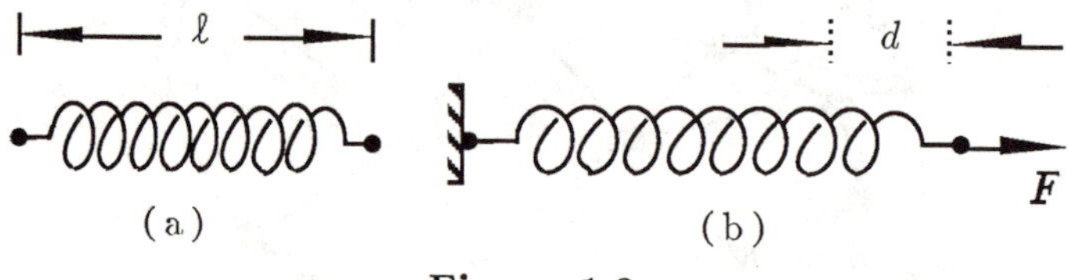

Figure 1.3

The endpoints of the spring will be called **nodes** and, in the simple situation here, one node is being held fixed and the other displaced due to the action of the force. A more general situation is one where both of the nodes may be displaced.

Suppose forces F_1 and F_2 are applied at nodes 1 and 2 respectively of a linear spring producing displacements u_1 and u_2 at these nodes (see Figure 1.4). (In fact this is a somewhat artificial situation since, when the spring is in equilibrium $F_1 = -F_2$.)

In other words, we have a single element, the spring, characterized by two nodes, for each of which nodal displacements u_1, u_2 can be measured and hence a member (or internal) force F_1, F_2 may be defined.

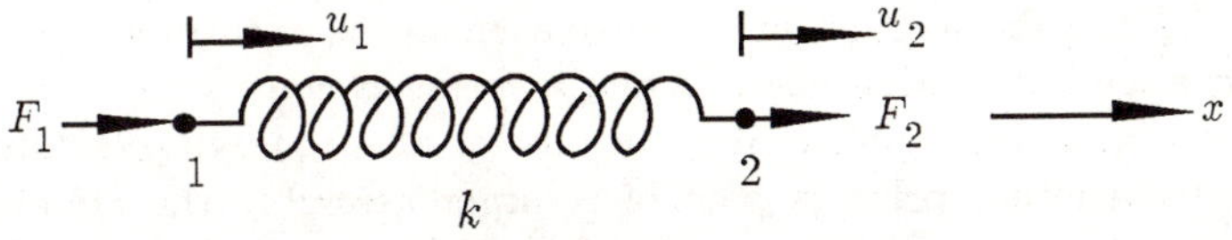

Figure 1.4

The displacement $d = u_2 - u_1$ so for node 2 we have

$$F_2 = k(u_2 - u_1)$$

and correspondingly for node 1

$$F_1 = -k(u_2 - u_1)$$

since, as noted above, F_1 must equal F_2 in magnitude if the spring is to be in equilibrium.

These two equations can be written

$$\begin{aligned} ku_1 - ku_2 &= F_1 \\ -ku_1 + ku_2 &= F_2 \end{aligned}$$

or, in matrix form,

$$\begin{bmatrix} k & -k \\ -k & k \end{bmatrix} \begin{Bmatrix} u_1 \\ u_2 \end{Bmatrix} = \begin{Bmatrix} F_1 \\ F_2 \end{Bmatrix} \tag{1.8}$$

The symmetric matrix

$$[K^{[e]}] = \begin{bmatrix} k & -k \\ -k & k \end{bmatrix}$$

is referred to as the element stiffness matrix and the column vector

$$\{F^{[e]}\} = \begin{Bmatrix} F_1 \\ F_2 \end{Bmatrix}$$

as the element force vector. (These terms are still sometimes used in applications of finite elements quite remote from the world of springs and forces.)

• A two spring assembly

Consider now a system made up of two linear springs as shown in Figure 1.5. Each spring can be thought of as an element $[e]$ where $[e]$ is $[1]$ or $[2]$ and the system has three nodes numbered 1,2,3. At each node i

a force G_i may be applied and a displacement u_i measured. Note the subtle change in notation here. When a spring is considered in isolation the nodal forces are member (internal) forces denoted by the symbol 'F'. However when the spring is part of a larger assembly the nodal forces are referred to as global forces and indicated by the symbol 'G'. For two elements with a common node the internal forces existing within each element combine (taking account of sign) to form the global force acting at that node. Alternatively we can imagine the global force applied at the common node being distributed or 'shared' within each element.

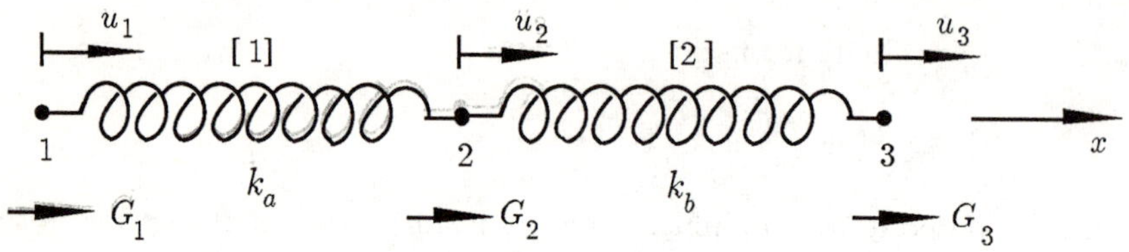

Figure 1.5

For each element we have equilibrium equations of the form (1.8):

- element [1]

$$\begin{bmatrix} k_a & -k_a \\ -k_a & k_a \end{bmatrix} \begin{Bmatrix} u_1 \\ u_2 \end{Bmatrix} = \begin{Bmatrix} F_1^{[1]} \\ F_2^{[1]} \end{Bmatrix} \tag{1.9}$$

- element [2]

$$\begin{bmatrix} k_b & -k_b \\ -k_b & k_b \end{bmatrix} \begin{Bmatrix} u_2 \\ u_3 \end{Bmatrix} = \begin{Bmatrix} F_2^{[2]} \\ F_3^{[2]} \end{Bmatrix} \tag{1.10}$$

Note that in (1.9) and (1.10) we have used superscripts [1] and [2] respectively on the internal forces to emphasize that the corresponding terms refer to distinct elements.

We wish to combine these equations to obtain equilibrium equations for the entire assembly. To do this we rewrite (1.9) as

$$\begin{bmatrix} k_a & -k_a & 0 \\ -k_a & k_a & 0 \\ 0 & 0 & 0 \end{bmatrix} \begin{Bmatrix} u_1 \\ u_2 \\ u_3 \end{Bmatrix} = \begin{Bmatrix} F_1^{[1]} \\ F_2^{[1]} \\ 0 \end{Bmatrix}$$

and (1.10) as

$$\begin{bmatrix} 0 & 0 & 0 \\ 0 & k_b & -k_b \\ 0 & -k_b & k_b \end{bmatrix} \begin{Bmatrix} u_1 \\ u_2 \\ u_3 \end{Bmatrix} = \begin{Bmatrix} 0 \\ F_2^{[2]} \\ F_3^{[2]} \end{Bmatrix}$$

It can be shown (see Chapter 5) that the overall equilibrium relations for the two-spring assembly can be obtained by simply adding (or superimposing) these relations together:

$$\begin{bmatrix} k_a & -k_a & 0 \\ -k_a & k_a + k_b & -k_b \\ 0 & -k_b & k_b \end{bmatrix} \begin{Bmatrix} u_1 \\ u_2 \\ u_3 \end{Bmatrix} = \begin{Bmatrix} F_1^{[1]} \\ F_2^{[1]} + F_2^{[2]} \\ F_3^{[2]} \end{Bmatrix}$$

The matrix

$$\begin{bmatrix} k_a & -k_a & 0 \\ -k_a & k_a + k_b & -k_b \\ 0 & -k_b & k_b \end{bmatrix}$$

is referred to as the overall or global stiffness matrix of the system and the column vector

$$\begin{Bmatrix} F_1^{[1]} \\ F_2^{[1]} + F_2^{[2]} \\ F_3^{[2]} \end{Bmatrix}$$

as the global force vector. As discussed above the latter is normally relabelled as $\{G_1, G_2, G_3\}^T$.

The discussion in this section, simple though it is, serves to introduce many of the basic ideas of the finite element method. The concepts of nodal numbers, element numbers, nodal values for physical quantities, element matrices and their superposition for a system composed of many elements will be recurring themes throughout this text. However in most cases the elements will have to be introduced artificially, as problem-solving aids rather than actual physical elements, and the actual forms of the matrices arising will have to be derived mathematically rather than by using simple physical laws.

Galerkin's Weighted Residual Method

PREVIEW

In this chapter we examine the Galerkin technique for determining approximate solutions to boundary value problems. In the basic method a trial solution satisfying all the boundary conditions of the problem is used. We also study the modified Galerkin method for which the trial solution need only satisfy some of the boundary conditions. Finally, a matrix formulation of the modified Galerkin technique is considered.

2.1 Introduction

The exact solution to a boundary value problem can rarely be found even if the governing differential equation is a linear one. The finite element method is essentially a powerful numerical method for obtaining approximate solutions to boundary value problems in which approximate trial solutions are used. If $u(x)$, in one-dimension, or $u(x, y)$ in two-dimensions denotes the unknown exact solution to a boundary value problem then we shall denote approximate trial solutions by $\tilde{u}(x)$ or $\tilde{u}(x, y)$ (read as 'u tilde').

One of the major features of the finite element method is the subdivision of the domain of the problem into smaller regions or elements and the construction of simple trial solutions within each element. However, long prior to the development of finite elements, various numerical techniques had been developed for solving boundary value problems without subdivision. Two of these methods – one using **variational principles** the other based on **weighted residual techniques** – have been incorporated into the finite element method. The variational formulation of boundary value problems will be discussed in Chapter 8. In this chapter, we give an introduction to the idea of weighted residuals, in particular to a technique known as **Galerkin's method** which we will use extensively in later chapters.

2.2 Galerkin's method for one-dimensional boundary value problems

Galerkin's method involves the concept of weighted residuals. We will immediately illustrate the concept using an example.

Example 2.1

Determine an approximate solution to the boundary value problem:

$$\frac{d^2u}{dx^2} + u = 1 \qquad 0 \le x \le 1$$

$$u(0) = 0, \;\; u(1) = 1$$

Solution

Our first step is to look for an approximate trial solution. In general we seek a trial solution which is (i) simple in form and (ii) satisfies the given boundary conditions of the problem.

We select as our trial solution here:

$$\tilde{u}(x) = x + \alpha\, x(x-1) \tag{2.1}$$

where α is a parameter to be determined. Clearly $\tilde{u}(0) = 0$ and $\tilde{u}(1) = 1$ so the boundary conditions are satisfied, regardless of the value of α. (There are in fact an infinite number of one-parameter trial solutions that could have been chosen, for example $\tilde{u}(x) = x + \alpha\, x^2(x-1)$ and $\tilde{u}(x) = x + \alpha \sin \pi x$ are possible alternatives. More generally we could select a trial solution with n parameters.)

The idea now is to select a value for α to obtain, in some sense, an optimum or 'best' solution of the form (2.1) i.e. one that is 'as close as possible' to the true solution. This selection of parameters is usually done using the concept of a residual function.

To obtain the residual function we first collect all the terms in the differential equation on the left hand side. For example we can write the given equation here as

$$\frac{d^2u}{dx^2} + u - 1 = 0$$

The exact solution will, when substituted into the left hand side, produce an answer which is identically zero for all values of x in the problem domain $0 \leq x \leq 1$. An approximate solution $\tilde{u}(x)$, such as (2.1), will not produce an identically zero function but a function, say $\varepsilon(x)$, called the residual function. In our case then

$$\varepsilon(x) = \frac{d^2\tilde{u}}{dx^2} + \tilde{u} - 1$$

(Clearly $\varepsilon(x) = 0$ if $\tilde{u}(x)$ coincides with the exact solution $u(x)$ for all x in the domain of the problem.)

A number of techniques based on the use of residual functions are available for determining the unknown parameters in trial solutions.

One such method is called the **collocation method** which involves forcing the approximate solution $\tilde{u}(x)$ to be exact at n points $x_1, x_2, \ldots, x_n$ where n is the number of unknown parameters. This implies that

$$\varepsilon(x_i) = 0 \qquad i = 1, 2, \ldots n$$

which is a set of n equations to be solved for the parameters.

A more commonly used technique involving residuals, particularly in connection with finite elements, is **Galerkin's method**. In this method we determine the n unknown parameters by selecting n **weighting functions** $W_1(x), W_2(x), \ldots W_n(x)$ and requiring that each of the n integrals

$$\int_a^b \varepsilon(x)\, W_i(x)\, dx$$

for $i = 1, 2, \ldots n$ should be zero. Here $a \le x \le b$ is the domain of the problem and again we obtain n equations to be solved.

The question immediately arises as to what n functions $W_i(x)$ should be used to weight the residual function. In Galerkin's method we use as weighting functions **those terms in the trial solution which are multiplied by unknown parameters.**

Hence, returning to the example to hand where the only unknown parameter in the trial solution (2.1) is the quantity α which multiplies the term $x(x-1)$, we select as the only weighting function

$$W_1(x) = x(x-1)$$

Now the residual function is

$$\begin{aligned}\varepsilon(x) &= \frac{d^2\tilde{u}}{dx^2} + \tilde{u} - 1 \\ &= \frac{d^2}{dx^2}(x + \alpha\, x(x-1)) + x + \alpha\, x(x-1) - 1 \\ &= 2\,\alpha + x + \alpha x(x-1) - 1\end{aligned}$$

Hence the Galerkin procedure requires α to be such that

$$\int_0^1 \varepsilon(x)\; x(x-1)\; dx = 0$$

that is,

$$\int_0^1 (2\,\alpha + x + \alpha\, x(x-1) - 1)\; x(x-1) dx = 0$$

from which, after simple integration, we find $\alpha = 5/18$. Our approximate solution (2.1) becomes

$$\tilde{u}(x) = x + \tfrac{5}{18}\, x(x-1)$$

Now the exact solution to this boundary value problem is known and is

$$u(x) = -\cos x + (\cot 1)\sin x + 1$$

Graphs of $u(x), \tilde{u}(x)$ and the residual function are shown in Figure 2.1. Clearly throughout the problem domain $0 \leq x \leq 1$ the approximation $\tilde{u}(x)$ is a good one – perhaps surprisingly so in that we have used only a one-parameter trial solution. Note, however, that despite this the residual function $\varepsilon(x)$ is not 'small' throughout the domain of the problem. This is a reflection of the fact that although $u(x)$ and $\tilde{u}(x)$ are 'close' this is not so for their second derivatives upon which $\varepsilon(x)$ depends. We will see shortly that using a two parameter approximation will improve matters in this respect.

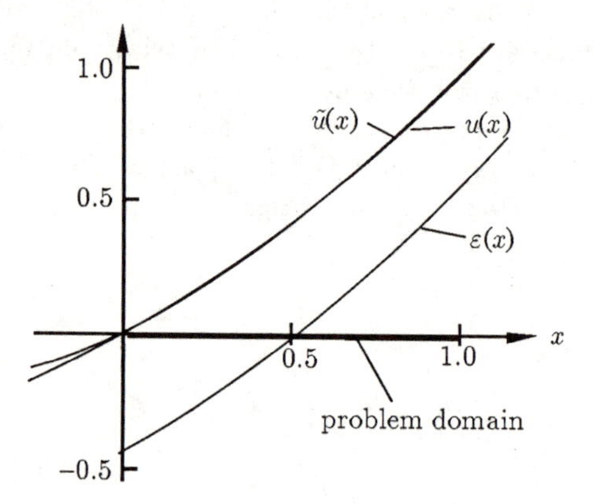

Figure 2.1

We can now make a formal statement of the Galerkin approximation procedure at least for linear second-order one-dimensional boundary value problems of the general form:

$$-\frac{d}{dx}\left(p(x)\frac{du}{dx}\right) + q(x)u = r(x) \qquad a \leq x \leq b$$

with given boundary conditions at $x = a$ and $x = b$. Approximate trial solutions are used which are of the form

$$\tilde{u}(x) = \theta_0(x) + \sum_{i=1}^{n} a_i\theta_i(x) \qquad \textbf{(2.2)}$$

where

(a) the function $\theta_0(x)$ is chosen to satisfy the given boundary conditions of the problem.

(b) the functions $\theta_1(x), \theta_2(x), \ldots \theta_n(x)$ must each satisfy the corresponding **homogeneous form** of the boundary conditions.

(The functions $\theta_i(x), i = 1, 2, \ldots n$ are sometimes called coordinate functions.)

Referring to Example 2.1 our assumed trial solution (2.1) has the form of a one-parameter version of (2.2) with

$$\begin{aligned}\theta_0(x) &= x\\ \theta_1(x) &= x(x-1)\\ a_1 &= \alpha\end{aligned}$$

Clearly $\theta_0(x)$ does indeed satisfy the given boundary conditions as $\theta_0(0) = 0$ and $\theta_0(1) = 1$. Also $\theta_1(0) = 0$ and $\theta_1(1) = 0$; i.e. θ_1 satisfies the corresponding homogeneous form of these boundary conditions.

We note, in general, that selecting $\theta_0(x)$ and $\theta_1(x), \ldots, \theta_n(x)$ as indicated in (a) and (b) above will ensure that the trial solution (2.2) satisfies the given boundary conditions regardless of the values of the parameters $a_1, a_2 \ldots a_n$.

Galerkin's method then involves determining these parameters by solving the n weighted residual equations:

$$\int_a^b \varepsilon(x)\,\theta_i(x)\,dx = 0 \qquad i = 1, 2, \ldots, n \tag{2.3}$$

Note that the coordinate functions $\theta_1(x), \ldots \theta_n(x)$ must be **linearly independent** functions. Informally this means that none of them is obtainable as a linear combination of the remainder.

We give an informal justification of Galerkin's method in the first supplement to this chapter. The interested reader may refer to that at this stage if so desired.

Example 2.2

Use the Galerkin method to obtain a two-parameter approximation to the solution of the boundary value problem

$$\frac{d^2u}{dx^2} + u = 1 \qquad 0 \le x \le 1$$
$$u(0) = 0 \qquad \frac{du}{dx} = 0 \text{ at } x = 1$$

(This is the same differential equation as in Example 2.1 but note now the derivative boundary condition at $x = 1$.)

Solution

Using (2.2) with $n = 2$ we seek a trial solution of the form

$$\tilde{u}(x) = \theta_0(x) + \sum_{i=1}^{2} a_i\, \theta_i(x)$$

As the boundary conditions are homogeneous we can put $\theta_0(x) \equiv 0$. Of the many possible choices of $\theta_1(x)$ and $\theta_2(x)$ we choose two of the simplest:

$$\theta_1(x) = x^2 - 2x \qquad \theta_2(x) = x(x-1)^2 \tag{2.4}$$

both of which satisfy the given (homogeneous) boundary conditions.

(Note that had we chosen $\theta_1(x)$ as say $x^2(x-1)^2$, which also satisfies the boundary conditions, then both θ_1 and θ_2, and hence $\tilde{u}$, would have been zero at $x = 1$. There is no reason to expect the exact solution to vanish at $x = 1$ and these choices would probably lead to a poor approximation.)

Our trial solution is thus

$$\tilde{u}(x) = a_1(x^2 - 2x) + a_2\, x(x-1)^2$$

and the corresponding residual function is

$$\begin{aligned} \varepsilon(x) &= \frac{d^2\tilde{u}}{dx^2} + \tilde{u} - 1 \\ &= 2a_1 + a_2(6x - 4) + a_1(x^2 - 2x) + a_2 x(x-1)^2 - 1 \end{aligned} \tag{2.5}$$

Following Galerkin we use $\theta_1(x)$ and $\theta_2(x)$ from (2.4) as weighting functions to obtain the equations

$$\int_0^1 \varepsilon(x)\,\theta_1(x)\,dx = 0 \qquad \text{and} \qquad \int_0^1 \varepsilon(x)\,\theta_2(x)\,dx = 0$$

Substitution of (2.4) and (2.5) into these equations and long but straightforward integration gives

$$\begin{aligned} -48\,a_1 + 7a_2 + 40 &= 0 \\ 49\,a_1 - 52a_2 - 35 &= 0 \end{aligned}$$

with solutions $a_1 = 0.8523$, $a_2 = 0.1301$. Hence we have obtained a two parameter approximation

$$\tilde{u}(x) = 0.8523\,(x^2 - 2x) + 0.1301\,x(x-1)^2$$

to the solution of the given boundary value problem. The graph of this function and of the exact solution which is

$$u(x) = -\cos x - (\tan 1)\sin x + 1$$

is shown in Figure 2.2. Note that $u(1) = -1/(\cos 1) + 1 = -0.8518$ and $\tilde{u}(1) = -0.8523$ indicating surprising accuracy in our two-parameter approximation.

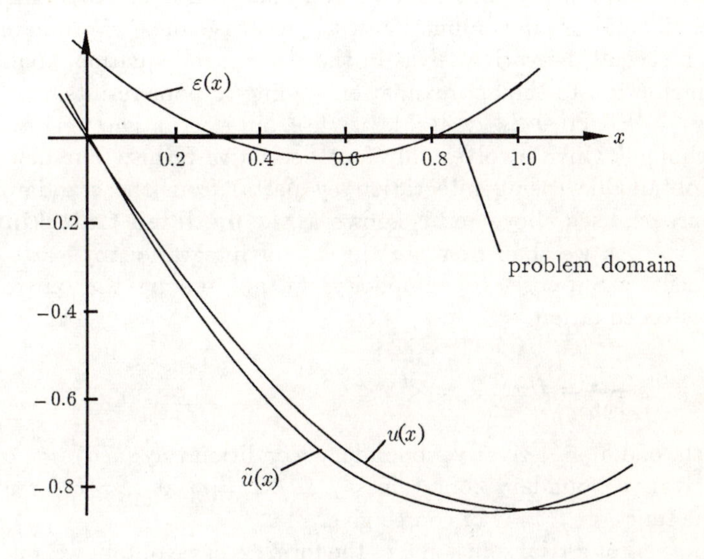

Figure 2.2

Figure 2.2 also shows the residual function $\varepsilon(x)$ and we see that, unlike Example 2.1, this function is 'small' over the domain $0 \leq x \leq 1$ of the problem. We normally find that as the order of the approximation, i.e. the number of parameters, increases then the residual function does indeed tend to decrease in magnitude through the domain.

2.3 The modified Galerkin technique

A definite integral can be assigned an unambiguous meaning (in terms of an area under a curve) if the integrand is piecewise continuous. This implies, that in order to implement the Galerkin procedure for second-order boundary value problems, the functions occurring in the integrand must have continuous derivatives (so that, at worst, the second derivative term which occurs is piecewise continuous). Thus the coordinate functions $\theta_i(x), i = 0, 1, 2, \ldots, n$, that are used in the formation (2.2) of the approximation $\tilde{u}(x)$ must have continuous derivatives.

In the finite element method we segment the domain and hence need to assume trial solutions (for second-order boundary value problems) whose derivatives are continuous only over an individual segment or element. In other words we hope to use trial solutions whose derivatives are only piecewise continuous. The simplest such trial solutions are piecewise linear functions where (obviously) the second derivatives are zero within each element. If we continue to use the Galerkin technique described above in conjunction with piecewise linear coordinate functions then second derivative terms in the differential equation would make no contribution to the approximation leading to poor results.

Hence it is desirable to use an alternative weighted residual technique which involves only first derivative terms. The new technique is obtainable (using integration by parts) from the standard Galerkin approach used above and is known as the **modified Galerkin method**.

As we shall now see the use of integration by parts has other benefits. Suppose, for simplicity, we are solving the one-dimensional Poisson equation

$$\frac{d^2u}{dx^2} = f(x) \qquad a \leq x \leq b$$

with one non-derivative boundary condition say $u(a) = \alpha$ and one derivative boundary condition say $du/dx = \beta$ at $x = b$ (or $u'(b) = \beta$). The function $f(x)$ is of course given.

For a trial solution $\tilde{u}$ of the form (2.2) we obtain weighted residual equations of the form

$$\int_a^b \left(\frac{d^2\tilde{u}}{dx^2} - f(x)\right)\theta_i(x) = 0 \tag{2.6}$$

for $i = 1, 2, \ldots, n$ where n, as usual, is the number of unknown parameters. We integrate the first term by parts:

$$\begin{aligned}\int_a^b \frac{d^2\tilde{u}}{dx^2}\theta_i(x)dx &= \left[\theta_i(x)\frac{d\tilde{u}}{dx}\right]_a^b - \int_a^b \frac{d\tilde{u}}{dx}\frac{d\theta_i}{dx}dx \\ &= \theta_i(b)\tilde{u}'(b) - \theta_i(a)\tilde{u}'(a) - \int_a^b \frac{d\tilde{u}}{dx}\frac{d\theta_i}{dx}dx \end{aligned} \tag{2.7}$$

Now on the right-hand side of (2.7) (which only involves, at worst, first derivative terms) there is the opportunity of incorporating any derivative boundary conditions that may have been imposed in the original boundary value problem. For if $u'(b)$ or $u'(a)$ are known at the outset then the corresponding approximation $\tilde{u}'(b)$, $\tilde{u}'(a)$ will be known and the approximate value can be substituted into the right-hand side of (2.7) On the other hand if a derivative boundary condition is not imposed at (say) $x = a$ then $u(a)$ must be given (as there must be **a** boundary condition at $x = a$). We conclude that at $x = a$ the coordinate functions vanish i.e. $\theta_i(a) = 0 \quad i = 1, 2 \ldots, n$ and so the term involving $\tilde{u}'(a)$ in (2.7) would be eliminated.

In our particular example the right-hand side of (2.7) becomes

$$\theta_i(b)\beta - 0 - \int_a^b \frac{d\tilde{u}}{dx}\frac{d\theta_i}{dx}dx$$

and so (2.6) becomes

$$-\int_a^b \left(\frac{d\tilde{u}}{dx}\frac{d\theta_i}{dx} + f(x)\,\theta_i(x)\right) dx \quad + \quad \beta\theta_i(b) = 0 \qquad \textbf{(2.8)}$$

We use (2.8) as the basis of a **modified Galerkin technique**. We need to select trial functions to satisfy only the non-derivative boundary condition at $x = a$ as the derivative boundary condition has been used to obtain (2.8). In this example, the derivative boundary condition assumed a particularly straightforward form. It is important to realise, at this early stage in the treatment of weighted residual methods, that boundary conditions are of two basic types, referred to as **essential** and **suppressible**. This nomenclature only applies to even-ordered boundary value problems. For second-order differential equations a boundary condition containing a derivative term is called suppressible; otherwise it is referred to as essential. For example consider a second-order differential equation over the domain $a \leq x \leq b$ with boundary conditions:

$$\alpha_0 u(a) + \alpha_1 u'(a) = \gamma_1$$

$$\beta_0 u(b) + \beta_1 u'(b) = \gamma_2$$

where $\alpha_0, \alpha_1, \beta_0, \beta_1, \gamma_1, \gamma_2$ are given constants. Here if α_1, β_1 are both non-zero then both conditions are suppressible. If $\alpha_1 = 0$ and $\beta_1 \neq 0$ then the first condition is essential, the second is suppressible and so on.

As seen in the second supplement to this chapter the information contained in a suppressible boundary condition may always be incorporated directly into the modified Galerkin procedure and consequently the constraint that the trial solution should satisfy this condition can be relaxed or 'suppressed'. On the other hand it is 'essential' that a trial solution should satisfy any essential boundary condition present as the information contained in such a condition is never incorporated fully into the modified Galerkin formulation.

Also, in the modified Galerkin technique, the order of the highest derivative term occurring will be lower than those occurring in the standard Galerkin method and consequently the 'smoothness' constraints can be relaxed accordingly.

In this text we shall normally employ the modified Galerkin technique. We shall therefore demand of the trial solution $\tilde{u}(x)$, still taken in the form (2.2), that $\theta_0(x)$ satisfies any essential boundary condition present and that $\theta_i(x)$ $i = 1, 2, \ldots, n$ should satisfy the corresponding homogeneous form of any such essential boundary condition.

We will now repeat Example 2.2 but using the modified Galerkin procedure.

Example 2.3

Use the modified Galerkin procedure with two parameters to find an approximate solution to the boundary value problem

$$\frac{d^2u}{dx^2} + u = 1 \qquad 0 \leq x \leq 1$$

with $u(0) = 0$ (an essential boundary condition) and $du/dx = 0$ at $x = 1$ (a suppressible boundary condition).

Solution

Using (2.2) we would look for an approximate solution of the form

$$\tilde{u}(x) = \theta_0(x) + \sum_{i=1}^{2} a_i \, \theta_i(x) \tag{2.9}$$

Here again we can choose $\theta_0 \equiv 0$ because the boundary condition at $x = 0$ is homogeneous. The standard Galerkin equations

$$\int_0^1 \left(\frac{d^2\tilde{u}}{dx^2} + \tilde{u} - 1 \right) \theta_i(x) dx = 0 \qquad i = 1, 2 \tag{2.10}$$

are modified by using integration by parts on the first term:

$$\int_0^1 \frac{d^2\tilde{u}}{dx^2} \theta_i(x) \, dx = \left[\frac{d\tilde{u}}{dx} \theta_i(x) \right]_0^1 - \int_0^1 \frac{d\tilde{u}}{dx} \frac{d\theta_i}{dx} \, dx$$

The first term on the right hand side disappears at $x = 0$ because we must choose both coordinate functions $\theta_i(x)$ such that $\theta_i(0) = 0$. Also, the suppressible boundary condition $d\tilde{u}/dx = 0$ at $x = 1$ ensures the disappearance of the same term at $x = 1$ regardless of the behaviour of the coordinate functions θ_i at $x = 1$. We have then from (2.10) the modified Galerkin formula

$$-\int_0^1 \frac{d\tilde{u}}{dx}\frac{d\theta_i}{dx}\,dx + \int_0^1 (\tilde{u} - 1)\,\theta_i(x)\,dx = 0 \qquad i = 1, 2 \tag{2.11}$$

which we use to determine the parameters a_1 and a_2.

We now need to select coordinate functions such that $\theta_1(0) = \theta_2(0) = 0$. Simple functions which are suitable are

$$\theta_1(x) = x \qquad \theta_2(x) = x^2 \tag{2.12}$$

Note the difference between these functions and the coordinate functions (2.4) used in the 'standard' Galerkin approach to this problem. As defined in (2.12), $\theta_1(x)$ and $\theta_2(x)$ do **not** satisfy the homogenous form of the boundary condition at $x = 1$ i.e. $d\theta_1/dx$ and $d\theta_2/dx$ are non-zero at $x = 1$.

Using (2.12) in (2.9) we obtain the trial solution

$$\tilde{u}(x) = a_1\,x + a_2\,x^2 \tag{2.13}$$

and we can readily find a_1 and a_2 using (2.11).

We have

$$\frac{d\tilde{u}}{dx} = a_1 + 2\,a_2\,x$$

Also

$$\frac{d\theta_1}{dx} = 1 \text{ and } \frac{d\theta_2}{dx} = 2x$$

For $i = 1$ (2.11) becomes

$$-\int_0^1 \{(a_1 + 2\,a_2\,x) - (a_1x + a_2\,x^2 - 1)x\}\,dx = 0$$

which simplifies to

$$a_1\left(\frac{2}{3}\right) + a_2\left(\frac{3}{4}\right) + \frac{1}{2} = 0 \tag{2.14}$$

Similarly for $i = 2$ (2.11) becomes

$$-\int_0^1 \{(a_1 + 2\,a_2\,x)2x - (a_1\,x + a_2\,x^2 - 1)x^2\}\,dx = 0$$

which simplifies to

$$a_1\left(\frac{3}{4}\right) + a_2\left(\frac{17}{15}\right) + \frac{1}{3} = 0 \tag{2.15}$$

Solving (2.14) and (2.15) and substituting in (2.13) we obtain

$$\tilde{u} = -1.6403x + 0.7913x^2$$

This is a very good approximation to the exact solution as is shown in Figure 2.3.

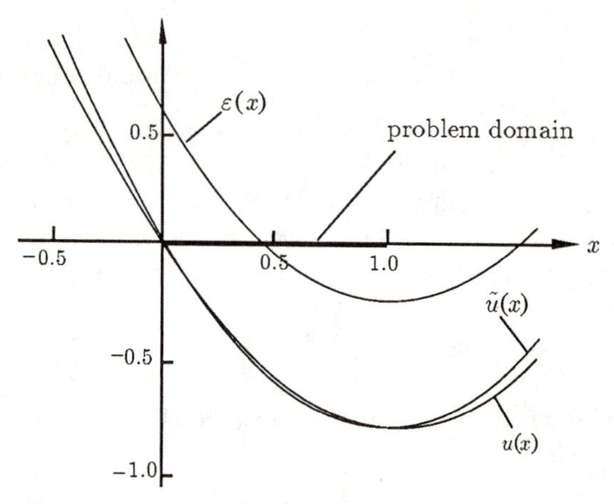

Figure 2.3

In this section the modified Galerkin method has been applied to determine approximate solutions to a one-dimensional Poisson equation with simple boundary conditions. In the second supplement to this chapter we explore the applicability of this technique to general one-dimensional linear boundary value problems of second order.

2.4 Matrix formulation of the modified Galerkin method

In applying the modified Galerkin method we obtain a system of linear equations for the unknown parameters $a_1, a_2, \ldots a_n$. We now demonstrate a general formulation for this system of equations which is independent of the value of n and which will be particularly convenient when we apply the modified Galerkin method to finite elements proper.

Consider, for the sake of example, the boundary value problem:

$$\frac{d}{dx}\left\{(x+1)\frac{du}{dx}\right\} = 0 \qquad 1 < x < 2$$

$$u(1) = 1\,, \quad \frac{du}{dx} = \frac{1}{3} \text{ at } x = 2$$

Assume an n parameter Galerkin approximation (2.2)

$$\tilde{u}(x) = \theta_0(x) + \sum_{j=1}^{n} a_j\, \theta_j\,(x)$$

where we must choose $\theta_0(1) = 1$ and $\theta_j(1) = 0$, $j = 1, 2, \ldots, n$. From the discussion in the previous section we need not choose functions to satisfy the (suppressible) condition at $x = 2$. The residual function is

$$\varepsilon(x) = \frac{d}{dx}\left\{(x+1)\frac{d\tilde{u}}{dx}\right\}$$

and the standard Galerkin technique would yield a system of n equations

$$\int_1^2 \frac{d}{dx}\left\{(x+1)\frac{d\tilde{u}}{dx}\right\}\theta_i\,(x)\,dx = 0 \qquad i = 1, 2, \ldots, n$$

However, we must integrate by parts to convert to the modified Galerkin technique giving

$$\left[\theta_i(x)(x+1)\frac{d\tilde{u}}{dx}\right]_1^2 - \int_1^2 (x+1)\frac{d\tilde{u}}{dx}\,\frac{d\theta_i}{dx}\,dx = 0 \qquad i = 1, 2, \ldots, n$$

or, using the property $\theta_i(1) = 0$ and the suppressible boundary condition at $x = 2$,

$$\int_1^2 (x+1)\frac{d\tilde{u}}{dx}\,\frac{d\theta_i}{dx}\,dx = \theta_i\,(2) \qquad i = 1, 2, \ldots, n$$

which is the modified Galerkin formula for this problem. Substituting for $\tilde{u}$ and transposing the term in $\theta_0(x)$ so that only the terms involving the unknown coefficients a_j appear on the left hand side we obtain

$$\int_1^2 (x+1)\left[\sum_{j=1}^{n} a_j\,\frac{d\theta_j}{dx}\right]\frac{d\theta_i}{dx}\,dx = \theta_i\,(2) - \int_1^2 (x+1)\frac{d\theta_0}{dx}\,\frac{d\theta_i}{dx}\,dx \quad \textbf{(2.16)}$$

for $i = 1, 2, \ldots, n$.

If we write

$$K_{ij} = \int_1^2 (x+1)\frac{d\theta_i}{dx}\,\frac{d\theta_j}{dx}\,dx \qquad i, j = 1, 2, \ldots, n \qquad \textbf{(2.17)}$$

and

$$F_i = \theta_i(2) - \int_1^2 (x+1)\frac{d\theta_0}{dx}\frac{d\theta_i}{dx}\,dx \qquad i = 1, 2, \ldots, n \qquad \mathbf{(2.18)}$$

then (2.16) has the form

$$\sum_{j=1}^{n} K_{ij}\, a_j = F_i \qquad i = 1, 2, \ldots, n \qquad \mathbf{(2.19)}$$

or

$$[K]\{a\} = \{F\}$$

which is clearly the matrix form of a system of n linear equations for the coefficients a_j. We see immediately from (2.17) that the coefficient matrix $[K]$ is **symmetric** i.e. $K_{ij} = K_{ji}$.

We now consider specific low order approximate solutions $\tilde{u}$, successively linear, quadratic and cubic, and obtain, for each case, the system (2.19).

- **Linear approximation**

We use a trial solution

$$\tilde{u} = \alpha_0 + \alpha_1 x$$

We can put this in the standard form (2.2) by requiring $\tilde{u}(1) = \alpha_0 + \alpha_1 = 1$ from the given essential boundary condition. Hence

$$\begin{aligned}\tilde{u} &= 1 - \alpha_1 + \alpha_1 x \\ &= 1 + a_1(1 - x)\end{aligned}$$

where we have re-labelled α_1 as $-a_1$. For this trial solution

$$\theta_0(x) = 1 \text{ and } \theta_1(x) = 1 - x$$

where clearly $\theta_0(1) = 1$ and $\theta_1(1) = 0$ as necessary. Note again that the suppressible (derivative) boundary condition $du/dx = 1/3$ at $x = 1$ is not satisfied by θ_0 nor is the corresponding homogeneous form $du/dx = 0$ at $x = 1$ satisfied by θ_1.

Since $n = 1$, the system of equations (2.19) will in this case reduce to the single equation

$$K_{11}\, a_1 = F_1$$

where, using (2.17),

$$K_{11} = \int_1^2 (x+1)\frac{d\theta_1}{dx}\frac{d\theta_1}{dx}dx$$

$$= \int_1^2 (x+1)dx = \frac{5}{2}$$

and by (2.18), since $\dfrac{d\theta_0}{dx} = 0$,

$$F_1 = \theta_1(2) = -1$$

Hence (2.19) becomes

$$\frac{5}{2}a_1 = -1 \quad \text{or} \quad a_1 = -\frac{2}{5}$$

and our one-parameter approximation is

$$\tilde{u} = 1 - \frac{2}{5}(1-x) \tag{2.20}$$

- **Quadratic approximation**

We try

$$\tilde{u} = \alpha_0 + \alpha_1\, x + \alpha_2\, x^2$$

where, to make $\tilde{u}(1) = 1$ as necessary,

$$\alpha_0 = 1 - \alpha_1 - \alpha_2$$

so that

$$\begin{aligned}\tilde{u} &= 1 - \alpha_1 - \alpha_2 + \alpha_1\, x + \alpha_2\, x \\ &= 1 + a_1(1-x) + a_2(1-x^2)\end{aligned}$$

where we have relabelled α_1 as $-a_1$ and α_2 as $-a_2$. This is now in the standard form (2.2) with

$$\theta_0(x) = 1 \quad \text{so} \quad \frac{d\theta_0}{dx} = 0$$

$$\theta_1(x) = 1 - x \quad \text{so} \quad \frac{d\theta_1}{dx} = -1 \text{ and } \theta_1(2) = -1$$

$$\theta_2(x) = 1 - x^2 \quad \text{so} \quad \frac{d\theta_2}{dx} = -2x \text{ and } \theta_2(2) = -3$$

The system (2.19) for determining the parameters a_1 and a_2 is then

$$\begin{aligned} K_{11}\,a_1 + K_{12}\,a_2 &= F_1 \\ K_{21}\,a_1 + K_{22}\,a_2 &= F_2 \end{aligned} \qquad \textbf{(2.21)}$$

where, from (2.17),

$$K_{11} = \frac{5}{2} \text{ (from the linear approximation)}$$

$$\begin{aligned} K_{12} = K_{21} &= \int_1^2 (x+1)(-1)(-2x)dx \\ &= 2\int_1^2 (x^2+x)dx = \frac{23}{3} \end{aligned}$$

$$\begin{aligned} K_{22} &= \int_1^2 (x+1)(-2x)^2 dx \\ &= 4\int_1^2 (x^3+x^2)dx = \frac{73}{3} \end{aligned}$$

From (2.18)

$$\begin{aligned} F_1 &= -1 \text{ (from the linear approximation)} \\ F_2 &= \theta_2(2) = -3 \end{aligned}$$

Hence (2.21) becomes

$$\frac{5}{2}\,a_1 + \frac{23}{3}\,a_2 = -1$$

$$\frac{23}{3}\,a_1 + \frac{73}{3}\,a_2 = -3$$

giving $a_1 = -0.6486$, $a_2 = 0.0811$ and hence a quadratic approximation

$$\tilde{u} = 1 - 0.6486(1-x) + 0.0811(1-x^2)$$

- **Cubic approximation**

Following the pattern of the linear and quadratic approximations we try

$$\tilde{u} = \alpha_0 + \alpha_1 x + \alpha_2 x^2 + \alpha_3 x^3$$

with

$$\tilde{u}(1) = \alpha_0 + \alpha_1 + \alpha_2 + \alpha_3 = 1$$

so α_0, say, can be expressed in terms of the other parameters. Doing this, and re-labelling the parameters we obtain

$$\tilde{u} = 1 + a_1(1-x) + a_2(1-x^2) + a_3(1-x^3)$$

i.e.

$$\theta_0(x) = 1 \quad \text{so} \quad \frac{d\theta_0}{dx} = 0$$

$$\theta_1(x) = 1 - x \quad \text{so} \quad \frac{d\theta_1}{dx} = -1 \text{ and } \theta_1(2) = -1$$

$$\theta_2(x) = 1 - x^2 \quad \text{so} \quad \frac{d\theta_2}{dx} = -2x \text{ and } \theta_2(2) = -3$$

$$\theta_3(x) = 1 - x^3 \quad \text{so} \quad \frac{d\theta_3}{dx} = -3x^2 \text{ and } \theta_3(2) = -7$$

Using these expressions in (2.17) and (2.18) we can readily obtain the 3×3 system to calculate the coefficients. The only matrix components additional to those calculated in the quadratic case are

$$K_{33} = \int_1^2 (x+1)(-3x^2)^2 dx = \frac{387}{10}$$

$$K_{13} = K_{31} = \int_1^2 (x+1)(-1)(-3x^2)dx = \frac{73}{4}$$

$$K_{23} = K_{32} = \int_1^2 (x+1)(-2x)(-3x^2)dx = \frac{597}{10}$$

$$F_3 = -7$$

Hence the system of equations (2.19) is

$$\frac{5}{2}a_1 + \frac{23}{3}a_2 + \frac{73}{4}a_3 = -1$$

$$\frac{23}{3}a_1 + \frac{73}{3}a_2 + \frac{597}{10}a_3 = -3$$

$$\frac{73}{4}a_1 + \frac{597}{10}a_2 + \frac{387}{10}a_3 = -7$$

giving $a_1 = -0.64848$, $a_2 = 0.080316$, $a_3 = 0.000024$ and hence a cubic approximation

$$\tilde{u} = 1 - 0.64848\,(1 - x) + 0.080316\,(1 - x^2) + 0.000024\,(1 - x^3)$$

Note that the value of $\tilde{u}$ at $x = 2$ viz. $\tilde{u}(2) = 1.4075$ compares very well with the value of the exact solution

$$u(x) = 1 + \ln(x + 1)/2$$

at this point viz. $u(2) = 1.4055$. However, as indicated by the small value of a_3, in this particular example the cubic term has not significantly increased the accuracy over that obtained in the two-parameter case (see Figure 2.4).

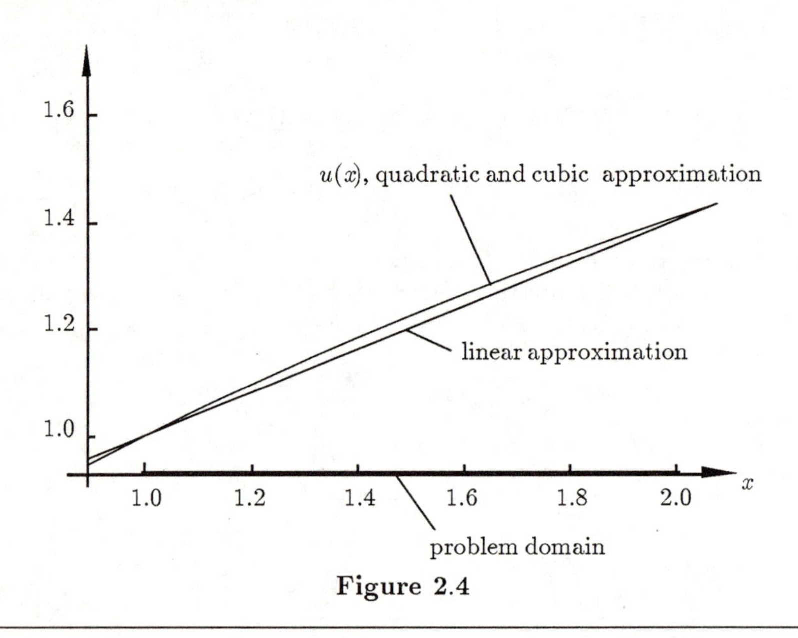

Figure 2.4

EXERCISES

2.1 Determine one- and two-parameter approximations to the boundary value problem:

$$\frac{d^2u}{dx^2} + u = -x \qquad 0 \le x \le 1$$

$$u(0) = u(1) = 0$$

using the standard Galerkin method. As trial solutions use (i) $\tilde{u}(x) = a_1 x(1 - x)$ and (ii) $\tilde{u}(x) = x(1 - x)(a_1 + a_2 x)$. Compare your answers with the exact solution.

2.2 The boundary value problem:

$$\frac{d^2u}{dx^2} + 1000x^2 = 0 \qquad 0 \le x \le 1$$

$$u(0) = 0 \qquad u(1) = 0$$

is used to model heat conduction in a bar of unit length with an initial heat source whose strength is proportional to x^2. The quantity u denotes the steady state temperature in the bar.
(i) Suggest why, on physical grounds, the trial solution $\tilde{u}(x) = a_1x(1-x^2)$ is not an unreasonable approximation to use.
(ii) Determine the parameter a_1 using the standard Galerkin method.
(iii) Repeat the problem using a two-parameter trial solution with

$$\theta_1(x) = x(1-x^2) \qquad \text{and} \qquad \theta_2(x) = x(1-x^4)$$

Compare your answers with the exact solution.

2.3 The boundary value problem:

$$\frac{d^2u}{dx^2} - 3(u-25) = 0 \qquad 0 \le x \le 2$$

$$u(0) = 150 \qquad u'(2) = 0$$

models the heat flow in a bar undergoing convection to the surroundings at a temperature of 25°C, the left hand end of the bar being kept at a constant temperature of 150°C and the right hand end being insulated.
(i) Suggest why the approximate trial solution $\tilde{u}(x) = 150 + a_1x(x-4)$ is physically appropriate.
(ii) Determine the parameter a_1 using the standard Galerkin method.

2.4 The deflection $u(x)$ of a simply supported beam with a uniformly distributed load p is governed by the boundary value problem:

$$\mathrm{EI}\frac{d^4u}{dx^4} - p = 0 \qquad 0 \le x \le \ell$$

$$u(0) = \frac{d^2u}{dx^2}(0) = 0 \qquad u(\ell) = \frac{d^2u}{dx^2}(\ell) = 0$$

Determine, using the standard Galerkin method, the unknown parameters in the approximate trial solution

$$\tilde{u}(x) = a_1 \sin(\frac{\pi x}{\ell}) + a_2 \sin(\frac{3\pi x}{\ell})$$

2.5 Explain why the one- and two-parameter trial solutions (i) $\tilde{u}(x) = a_1 x$ (ii) $\tilde{u}(x) = a_1 x + a_2 x^2$ cannot be used to determine a standard Galerkin approximation to the boundary value problem:

$$-\frac{d^2u}{dx^2} = \frac{1}{1+x} \qquad 0 \leq x \leq 1$$

$$u(0) = u'(1) = 0$$

However, show that these solutions **can** be used in a modified Galerkin approach and, in each case, determine the unknown parameters. Compare your approximations with the exact solution.

2.6 Using a two-parameter trial solution:

$$\tilde{u}(x) = a_1(1 - x^2) + a_2(1 - x^4)$$

determine an approximate solution to the boundary value problem:

$$\frac{d^2u}{dx^2} + (1 + x^2)u + 1 = 0 \qquad -1 \leq x \leq 1$$

$$u(-1) = 0 \qquad u(1) = 0$$

2.7 Consider the boundary value problem:

$$\frac{d^2u}{dx^2} = 2 \qquad 1 \leq x \leq 3$$

$$u(1) = 1 \qquad u'(3) = 6$$

(a) Determine modified Galerkin equations appropriate to this problem, distinguishing between essential and suppressible boundary conditions.

(b) Consider the trial solution: $\tilde{u}(x) = \theta_0(x) + a_1\theta_1(x) + a_2\theta_2(x)$ in which

$$\theta_0(x) = \begin{cases} 2 - x & 1 \leq x \leq 2 \\ 0 & 2 \leq x \leq 3 \end{cases}$$

$$\theta_1(x) = \begin{cases} x - 1 & 1 \leq x \leq 2 \\ 3 - x & 2 \leq x \leq 3 \end{cases}$$

$$\theta_2(x) = \begin{cases} 0 & 1 \leq x \leq 2 \\ x - 2 & 2 \leq x \leq 3 \end{cases}$$

In this case the given coordinate functions are piecewise linear functions.

(i) Sketch $\theta_0(x), \theta_1(x), \theta_2(x)$ and $\tilde{u}(x)$ and give interpretations to the parameters a_1 and a_2.

(ii) Using the modified Galerkin equations developed in (a) determine a_1, a_2. Compare your approximation with the exact solution.

2.8 The longitudinal vibrations of a bar can be modelled by the eigenvalue problem:

$$\frac{d^2u}{dx^2} = -\lambda^2 u \qquad 0 \le x \le \ell$$

in which $u(x)$ denotes the longitudinal displacement (i.e. along the bar). The boundary conditions to be imposed depend on the fixings:

(i) fixed at $x = 0$, $x = \ell$ then $u(0) = 0$, $u(\ell) = 0$

(ii) fixed at $x = 0$, free at $x = \ell$ then $u(0) = 0 \quad \frac{du}{dx}(\ell) = 0$

(a) Use one-parameter approximations to estimate λ for each of the cases (i) and (ii) and compare with the exact results.

(b) Repeat using a two-parameter approximation in case (i).

2.9 Using the standard Galerkin method find approximate values of the lowest magnitude eigenvalues of the eigenvalue problem:

$$x\frac{d^2u}{dx^2} + \lambda u = 0 \qquad 0 \le x \le 1$$

$$u(0) = u(1) = 0$$

Use the trial solution:

$$\tilde{u}(x) = a_1 x(1-x) + a_2 x^2(1-x)$$

(Strictly speaking only one parameter is required but better approximations to the lowest eigenvalue will be achieved if two or more parameters are used.)

2.10 Determine a two-parameter modified Galerkin approximation to the boundary value problem:

$$-\frac{d^2u}{dx^2} + u = 2\cos x \qquad \frac{\pi}{2} \le x \le \pi$$

$$u'(\frac{\pi}{2}) + 3u(\frac{\pi}{2}) = -1$$

$$u'(\pi) + 4u(\pi) = -4$$

Hint: $u(x)$ is an even function of x.

2.11 Using an n-parameter trial solution:

$$\tilde{u}(x) = \theta_0(x) + \sum_{i=1}^{n} a_i\theta_i(x)$$

formulate a general matrix approach, using the modified Galerkin technique, to determine the equations satisfied by the parameters $a_1, a_2, \ldots, a_n$ for the boundary value problem:

$$-\frac{d}{dx}(p(x)\frac{du}{dx}) + q(x)u = r(x) \qquad a \leq x \leq b$$

$$u(a) = \gamma_1 \qquad u(b) = \gamma_2$$

2.12 Apply the matrix procedure developed in Q2.11 to obtain an approximate solution to the boundary value problem:

$$\frac{d^2u}{dx^2} + u = e^x \qquad 0 \leq x \leq 1$$

$$u(0) = 0 \qquad u(1) = 0$$

Use a trial solution of the form:

$$\tilde{u}(x) = \alpha_0 + \alpha_1 x + \alpha_2 x^2 + \alpha_3 x^3$$

SUPPLEMENTS

2S.1 The Galerkin method – a justification

It is beyond the scope of this text to give a rigorous justification of the Galerkin method. However, it may be of some interest to discuss the technique informally. We do this by examining simple precursors involving, firstly, sums of products of numbers and, secondly, integrals of products of functions.

Galerkin's method is loosely based on the following simple result: 'If k and f are two numbers and it is demanded that the product kf should vanish for **all** values of k, that is,

$$kf = 0 \qquad \text{all } k$$

then we can conclude that $f \equiv 0$'.

All we need do to verify this statement is to choose (say) $k = 1$ and then $1.f = 0$ implies $f = 0$.

This rather trivial result may be generalized firstly to sums of pair-products of numbers and then to integrals of pair-products of functions.

If k_i and f_i, $i = 1, 2, \ldots, n$ are numbers and it is demanded that

$$\sum_{i=1}^{n} k_i f_i = 0 \qquad \text{all } k_i, \ i = 1, 2, \ldots, n$$

then it is easy to deduce that this statement can only be valid if $f_i \equiv 0$, $i = 1, 2, \ldots, n$. To show this we choose firstly

$$k_1 = 1, \quad k_j = 0 \qquad j = 2, 3, \ldots, n$$

to give

$$1.f_1 + 0.f_2 + 0.f_3 + \ldots + 0.f_n = 0$$

that is

$$f_1 = 0$$

Then by choosing $k_1 = 0, \quad k_2 = 1, \quad k_j = 0, \quad j = 3, 4, \ldots, n$ we obtain $f_2 = 0$ and so on.

The generalization of these basic results to integrals of pairs of functions requires considerably more care. As the result we obtain is fundamental not only to the Galerkin and other weighted residual methods but also to the so-called variational methods (to be covered in Chapter 8) we shall deduce the result, known as the **fundamental lemma of the variational calculus**, rigorously.

Let $f(x)$ be a continuous function in the interval $a \leq x \leq b$. Suppose $k(x)$ is a continuously differentiable function, i.e. a function with a continuous derivative.
If it is demanded that

$$\int_a^b f(x)k(x)\,dx = 0 \qquad \textbf{(2S.1a)}$$

for **every** such function $k(x)$ then it follows that

$$f(x) \equiv 0 \qquad a \leq x \leq b \qquad \textbf{(2S.1b)}$$

Proof

The proof is by contradiction. We suppose that (2S.1b) does not hold.

If $f(x)$ is not identically zero inside $a \leq x \leq b$ then there must be a point c, satisfying $a \leq c \leq b$, such that $f(c) \neq 0$. To be definite we suppose $f(c) > 0$. Now, since $f(x)$ is continuous there must be an interval (no matter how small) throughout which $f(x) > 0$. Let this interval be $a' < c < b'$

We now choose a particular function $k(x)$ satisfying the above conditions:

$$k(x) = \begin{cases} 0 & a \leq x \leq a' \\ (x - a')^2(x - b')^2 & a' \leq x \leq b' \\ 0 & b' \leq x \leq b \end{cases}$$

Here $k(x)$ is continuously differentiable throughout $a \leq x \leq b$.

Now for this particular $k(x)$, the integral (2S.1a) becomes

$$\int_{a'}^{b'} f(x)(x - a')^2(x - b')^2\, dx$$

However, since $f(x) > 0$ everywhere inside $a' \leq x \leq b'$ the integrand is positive everywhere inside this interval and so the integral must also be positive which contradicts (2S.1a). A similar contradiction can be obtained if it is assumed that $f(c) < 0$. The result (2S.1b) is therefore proved.

In the method of weighted residuals we appeal to a modified version of this result.

If it is demanded that

$$\int_a^b f(x)k_i(x)\, dx = 0 \qquad i = 1, 2, \ldots, n$$

for a sequence of n functions $k_i(x), i = 1, 2, \ldots, n$ then we can conclude intuitively from the fundamental lemma that $f(x)$ is 'small' throughout the range of the integral:

$$f(x) \approx 0 \qquad a \leq x \leq b$$

But this is precisely the assumption that we make in the Galerkin weighted residual procedure (see (2.3)) where we required that

$$\int_a^b \varepsilon(x)\theta_i(x)\, dx = 0 \qquad i = 1, 2, \ldots, n$$

for a sequence of n coordinate functions $\theta_i(x)$. We can therefore conclude that the residual function $\varepsilon(x)$ is such that

$$\varepsilon(x) \approx 0 \qquad a \leq x \leq b$$

Obviously the precise choice of the functions $\theta_i(x)$ and the number n of them used is crucially important when assessing how close $\varepsilon(x)$ is to zero and therefore in determining the **quality** of the approximation throughout the interval $a \leq x \leq b$.

2S.2 Modified Galerkin equations for general second-order boundary value problems

The most general second-order linear ordinary differential equation:

$$a_2(x)\frac{d^2u}{dx^2} + a_1(x)\frac{du}{dx} + a_0(x)u = g(x) \qquad a \le x \le b$$

can always be written in so-called **self-adjoint form**:

$$-\frac{d}{dx}\left(p(x)\frac{du}{dx}\right) + q(x)u = r(x) \qquad a \le x \le b \tag{2S.2a}$$

The proof that this rearrangement is always possible, in principle, is given in the notes following Theorem 8.1 in Chapter 8. (Note however that because the rearrangement is dependent on being able to find a certain integral the practical application of this result may be extremely complicated in certain cases.)

We are thus led to consider the following boundary value problem: the differential equation (2S.2a) together with boundary conditions:

$$\alpha_0 u(a) + \alpha_1 u'(a) = \gamma_1 \tag{2S.2b}$$

and

$$\beta_0 u(b) + \beta_1 u'(b) = \gamma_2 \tag{2S.2c}$$

in which $\alpha_0,\ \alpha_1,\ \beta_0,\ \beta_1,\ \gamma_1,\ \gamma_2$ are given constants.

To determine a modified Galerkin reformulation of (2S.2) we consider the approximation:

$$\tilde{u}(x) = \theta_0(x) + \sum_{i=1}^{n} a_i\theta_i(x) \tag{2S.3}$$

in which, at the outset, $\tilde{u}$ is chosen to satisfy both boundary conditions (2S.2b) and (2S.2c).

As usual we begin with the weighted residual equations:

$$\int_a^b \left\{-\frac{d}{dx}\left(p(x)\frac{d\tilde{u}}{dx}\right) + q(x)\tilde{u} - r(x)\right\}\theta_i(x)dx = 0 \quad i = 1, \ldots, n \tag{2S.4}$$

Integrating the first term by parts:

$$\int_a^b \left\{p(x)\frac{d\tilde{u}}{dx}\frac{d\theta_i}{dx} + (q(x)\tilde{u} - r(x))\theta_i(x)\right\} dx + \left[-p(x)\frac{d\tilde{u}}{dx}\theta_i(x)\right]_a^b = 0 \qquad i = 1, 2, \dots, n \quad \textbf{(2S.5)}$$

We now consider the following cases:

(a) $\alpha_1 \neq 0$, $\beta_1 \neq 0$ (both boundary conditions suppressible):
Here, from (2S.2b) and (2S.2c), $\tilde{u}$ satisfies

$$\tilde{u}'(a) = \frac{1}{\alpha_1}(\gamma_1 - \alpha_0\tilde{u}(a)) \qquad \tilde{u}'(b) = \frac{1}{\beta_1}(\gamma_2 - \beta_0\tilde{u}(b))$$

Hence (2S.5) becomes

$$\int_a^b \left\{p(x)\frac{d\tilde{u}}{dx}\frac{d\theta_i}{dx} + (q(x)\tilde{u} - r(x))\theta_i\right\} dx - \frac{p(b)\theta_i(b)}{\beta_1}(\gamma_2 - \beta_0\tilde{u}(b)) + \frac{p(a)\theta_i(a)}{\alpha_1}(\gamma_1 - \alpha_0\tilde{u}(a)) = 0$$

$$i = 1, 2, \dots, n \qquad \textbf{(2S.6)}$$

These are the modified Galerkin equations. We can now relax the constraints on $\tilde{u}$ and not demand that $\tilde{u}$ must satisfy either of the boundary conditions since these have been used directly in the derivation of (2S.6)

(b) $\alpha_1 \neq 0$, $\beta_1 = 0$ (suppressible boundary condition at $x = a$, essential condition at $x = b$):
Here, from (2S.2b) and (2S.2c),$\tilde{u}$ satisfies

$$\tilde{u}'(a) = \frac{1}{\alpha_1}(\gamma_1 - \alpha_0\tilde{u}(a)) \qquad \text{and} \quad \beta_0\tilde{u}(b) = \gamma_2$$

Since $\theta_0(x)$ is chosen to satisfy all the boundary conditions and $\theta_i(x)$ the corresponding homogeneous form of the boundary conditions it follows that

$$\theta_i(b) = 0 \qquad i = 1, 2, \dots, n$$

Hence (2S.5) becomes, in this case,

$$\int_a^b \left\{ p(x)\frac{d\tilde{u}}{dx}\frac{d\theta_i}{dx} + (q(x)\tilde{u} - r(x))\theta_i(x) \right\} dx$$

$$+\frac{p(a)\theta_i(a)}{\alpha_1}(\gamma_1 - \alpha_0\tilde{u}(a)) = 0 \qquad i = 1, 2, \ldots, n \quad \textbf{(2S.7)}$$

Here we can relax the constraint that the trial solution $\tilde{u}$ must satisfy the boundary condition at $x = a$ as this has been used in the formulation of (2S.7). However, we must still maintain that $\tilde{u}$ satisfies the essential boundary condition at $x = b$.

(c) $\alpha_1 = 0, \quad \beta_1 \neq 0$ (an essential boundary condition at $x = a$ and a suppressible condition at $x = b$):
This case is similar to (b). We obtain the modified Galerkin equations:

$$\int_a^b \left\{ p(x)\frac{d\tilde{u}}{dx}\frac{d\theta_i}{dx} + (q(x)\tilde{u} - r(x))\theta_i(x) \right\} dx$$

$$-\frac{p(b)\theta_i(b)}{\beta_1}(\gamma_2 - \beta_0\tilde{u}(b)) = 0 \qquad i = 1, 2, \ldots, n \quad \textbf{(2S.8)}$$

where it need only be demanded that the trial solution $\tilde{u}$ satisfies the essential boundary condition at $x = a$.

(d) $\alpha_1 = 0, \quad \beta_1 = 0$ (both boundary conditions essential)
Here the boundary terms in (2S.6) vanish because the boundary conditions imply

$$\theta_i(a) = 0, \qquad \theta_i(b) = 0 \qquad i = 1, 2, \ldots, n$$

Hence the modified Galerkin equations are

$$\int_a^b \left\{ p(x)\frac{d\tilde{u}}{dx}\frac{d\theta_i}{dx} + (q(x)\tilde{u} - r(x))\theta_i(x) \right\} dx \qquad i = 1, 2, \ldots, n \qquad \textbf{(2S.9)}$$

in which we demand that the trial solution $\tilde{u}$ satisfies both boundary conditions.

3

Shape Functions for One-dimensional Elements

PREVIEW

This chapter introduces one-dimensional finite elements and their associated shape functions. A local coordinate variable is introduced to allow shape functions to be written in a standard form. Simple applications of shape functions are considered, particularly the evaluation of ordinary integrals. Finally, the Gaussian quadrature technique is introduced and used in tandem with the element/shape function approach in the numerical evaluation of definite integrals.

3.1 Division of region into elements

In solving a one-dimensional boundary value problem using the method of finite elements we divide the domain of the problem say $a \le x \le b$ into shorter segments or subdomains (see Figure 3.1).

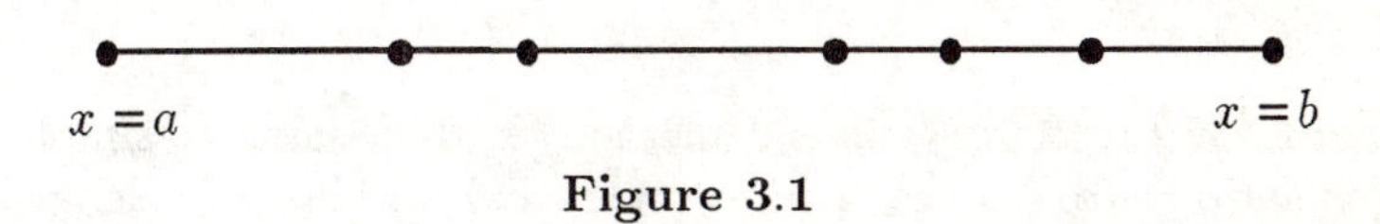

Figure 3.1

Points at which elements begin or end are called **nodes** and these are numbered as shown in Figure 3.2. The elements too are numbered, element numbers being enclosed in parentheses to distinguish them from node numbers.

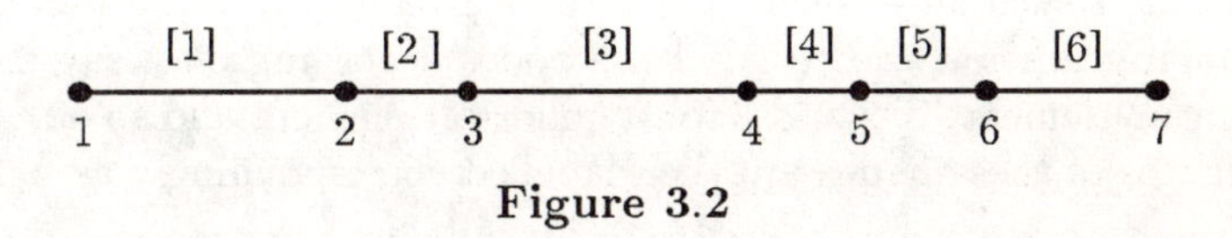

Figure 3.2

We see from Figure 3.2 that elements are not necessarily equal in length. The selection of node positions is dependent on the problem being solved. If, for example, it is suspected that the unknown function $u(x)$ is varying more rapidly in one part of the domain then nodes might be placed closer together in that part. Also, if one of the physical parameters of the problem changes abruptly at a particular point say $x = x_i$ then a node would be placed at this point.

Figure 3.2 which depicts nodes only at the boundaries of the elements is in fact only strictly correct for what we will call **linear elements**. Higher order elements are also used and these require more than two nodes per element. For example quadratic elements require one additional node placed somewhere, most commonly half-way, between the boundary nodes. Figure 3.3 depicts the domain $a \le x \le b$ divided into 3 quadratic elements.

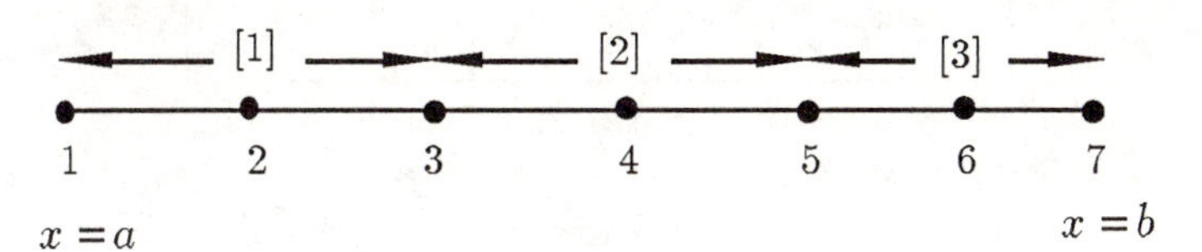

Figure 3.3

It is possible, particularly in two-dimensional problems, to use more than one type of element in the segmentation process if there are good reasons for so doing.

3.2 Local and global node numbers

The basis of the finite element method is to assume a simple trial approximation $\tilde{u}(x)$ to the true solution $u(x)$ within each element. In the case of a linear element, for example, we assume, as the name implies, a linear variation

$$\tilde{u} = a_0 + a_1 x \qquad x \text{ within element } [e]$$

Thus, initially at least, our attention will be focused on a single, typical element. For such an element it is often convenient to use a **local node numbering** scheme and to label the nodes accordingly as, say, 1 and 2 for a linear element, 1, 2 and 3 for a quadratic element and so on. The x-coordinates of these nodes are then labelled correspondingly as x_1, x_2, x_3 etc.

It is a simple matter to draw up tables connecting local node numbers and **global node numbers**, which is the name given to the overall numbering scheme for the entire domain.

For the six linear elements of Figure 3.2 the connectivity relations are given in Table 3.1.

Element [e]	Local Node 1	Local Node 2
[1]	1	2
[2]	2	3
[3]	3	4
[4]	4	5
[5]	5	6
[6]	6	7

Table 3.1

Here the table entries in columns 2 and 3 show the global node numbers. On the other hand, for the three quadratic elements of Figure 3.3 the connectivity relations are as in Table 3.2.

Element [e]	Local Node 1	Local Node 2	Local Node 3
[1]	1	2	3
[2]	3	4	5
[3]	5	6	7

Table 3.2

Note that some texts on the finite element method use letters to denote local nodes e.g. i and j for a two noded linear element. This is quite acceptable in a one-dimensional analysis but becomes impractical for two and three dimensions where some commonly used elements have many nodes e.g. 8, 10, 15 or more. Hence, in one dimension, we will use only numbers to label nodes as follows:

- Linear Element:
  Local nodes are numbered 1 and 2 at $x = x_1$ and $x = x_2$. The corresponding nodal values of the unknown labelled $\tilde{u}_1 = \tilde{u}(x_1)$ and $\tilde{u}_2 = \tilde{u}(x_2)$.
- Quadratic Element:
  Local nodes numbered 1,2,3 at $x = x_1, x = x_2, x = x_3$ respectively; nodal values labelled $\tilde{u}_1, \tilde{u}_2, \tilde{u}_3$.

3.3 The linear element

We have said that within an element we look for a simple trial approximation $\tilde{u}(x)$ to the problem at hand. The simplest such solution is a linear one, viz.

$$\tilde{u} = a_0 + a_1 x \tag{3.1}$$

within an element [e] of length $\ell = x_2 - x_1$.

For finite element analysis it is convenient to recast (3.1) in terms of the nodal values $\tilde{u}_1$ and $\tilde{u}_2$ (see Figure 3.4) and clearly we have

$$\tilde{u}_1 = a_0 + a_1\,x_1 \qquad \tilde{u}_2 = a_0 + a_1\,x_2$$

which are easily solved to give

$$a_0 = \frac{\tilde{u}_2 - \tilde{u}_1}{x_2 - x_1} \qquad a_1 = \frac{\tilde{u}_1\,x_2 - \tilde{u}_2\,x_1}{x_2 - x_1} \tag{3.2}$$

Substituting (3.2) into (3.1) and collecting terms in $\tilde{u}_1$ and $\tilde{u}_2$ gives

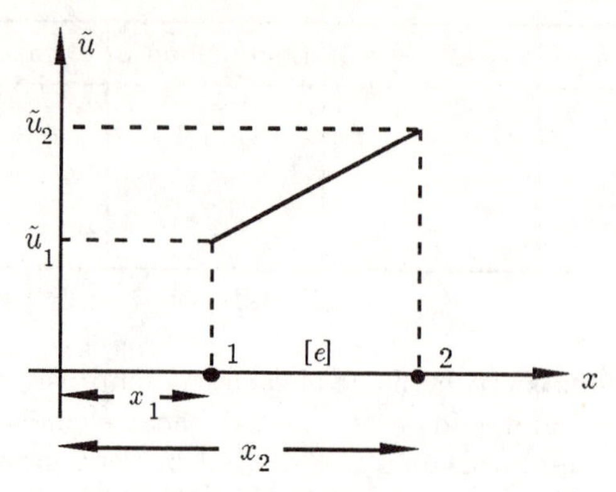

Figure 3.4

$$\tilde{u} = \left(\frac{x_2 - x}{x_2 - x_1}\right)\tilde{u}_1 + \left(\frac{x - x_1}{x_2 - x_1}\right)\tilde{u}_2$$

If we now define two linear polynomials

$$L_1(x) \equiv \frac{x_2 - x}{x_2 - x_1} = \frac{x_2 - x}{\ell} \tag{3.3}$$

$$L_2(x) \equiv \frac{x - x_1}{x_2 - x_1} = \frac{x - x_1}{\ell} \tag{3.4}$$

then we have the most useful form of the linear approximation for finite element analysis:

$$\tilde{u} = L_1(x)\tilde{u}_1 + L_2(x)\tilde{u}_2 \tag{3.5}$$

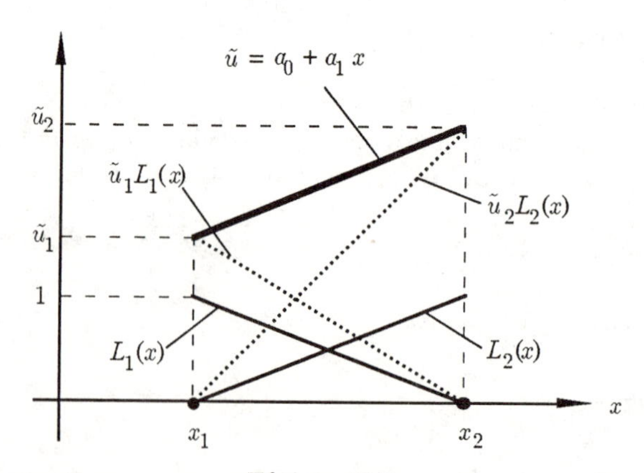

Figure 3.5

The relationship between $\tilde{u}, L_1(x)$ and $L_2(x)$ is shown in Figure 3.5. Clearly all that we are stating in (3.5) is that a straight line $\tilde{u} = a_0 + a_1 x$ can be thought of as a linear combination of the 'standard lines' $L_1(x)$ and $L_2(x)$ with the nodal values as weighting factors.

The polynomials $L_1(x)$ and $L_2(x)$ are called linear interpolation functions or, perhaps more commonly, **linear shape functions**. They, and their higher order one-dimensional and two-dimensional counterparts, play a major role in the remainder of this text.

3.4 Properties of linear shape functions

Clearly, from (3.3) and (3.4) we have

$$\begin{aligned} &\text{(i)}\ L_1(x_1) = 1 \quad \text{and} \quad L_1(x_2) = 0 \\ &\phantom{\text{(i)}\ } L_2(x_1) = 0 \quad \text{and} \quad L_2(x_2) = 1 \end{aligned} \tag{3.6}$$

(See Figure 3.5.)

Because of these properties we say that the shape function $L_1(x)$ is **associated** with node 1 and $L_2(x)$ with node 2.

(ii) $L_1(x) + L_2(x) = 1 \quad$ for all $x, \quad x_1 \leq x \leq x_2$

(iii) $L_1(x)x_1 + L_2(x)x_2 = x$

(iv) The derivatives dL_1/dx and dL_2/dx are constant and, further,
$dL_1/dx + dL_2/dx = -1/\ell + 1/\ell = 0 \quad$ for all $x, \ x_1 \leq x \leq x_2$

The shape functions $L_1(x)$ and $L_2(x)$ are polynomials of the same order as the original approximation (3.1). They are also different polynomials in different elements because of course x_1 and x_2 are different for each element. We should thus properly write $L_1^{[e]}(x)$ and $L_2^{[e]}(x)$ for an element $[e]$ and (3.5) should be written

$$\tilde{u}^{[e]} = L_1^{[e]}(x)\, \tilde{u}_1^{[e]} + L_2^{[e]}(x)\, \tilde{u}_2^{[e]}$$

3.5 Local coordinate systems

In using shape functions (whether linear or indeed of higher order) we will frequently be concerned with integrals taken over elements such as

$$\int_{[e]} L_1(x)\, L_2(x)\, dx, \qquad \int_{[e]} \frac{dL_1}{dx} \frac{dL_2}{dx}\, dx$$

For linear elements in one-dimension the integrands are simple polynomials and hence the integrals are relatively straightforward to evaluate. However, some simplification can be obtained by using appropriate coordinate systems local to an element. This simplification is more marked for higher order one-dimensional elements and in two-dimensional problems and consequently we will use **local coordinate systems** extensively in this text. A coordinate system local to an element is one where the origin is located at a point within that element.

A simple example of a one-dimensional local coordinate system is one where the origin is chosen at the left hand end of the element. This is equivalent to replacing the global coordinate x by $x' = x - x_1$ so that the linear shape functions (3.3) and (3.4) can be written as

$$L_1(x') = 1 - \frac{x'}{\ell} \qquad L_2(x') = \frac{x'}{\ell}$$

where $0 \le x' \le \ell$.

Perhaps it is more useful to use a local coordinate system whose origin is at the centre of the element and additionally to scale the coordinate so that it varies from -1 to $+1$ within the element. This latter feature is particularly useful when, as is often necessary, numerical integration over an element is used.

Thus we introduce a dimensionless or natural variable ξ defined by

$$\xi = \frac{2}{\ell}\left(x - (x_1 + \frac{\ell}{2})\right) \qquad \text{or} \qquad \xi = \frac{2}{\ell}(x' - \frac{\ell}{2}) \tag{3.7}$$

where clearly $\xi(x_1) = -1$ and $\xi(x_2) = +1$ (see Figure 3.6). We shall refer to the element depicted as the **standard linear element** in one dimension.

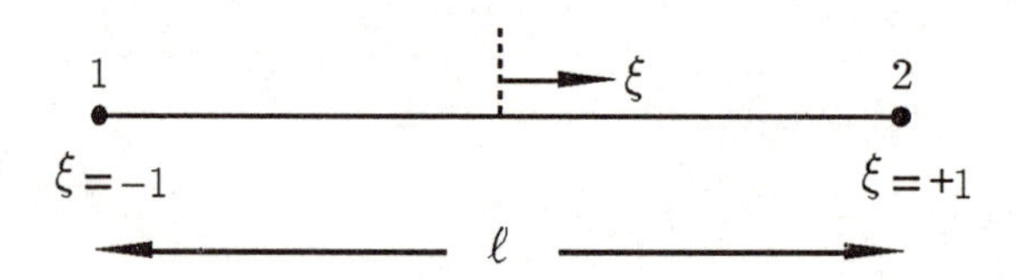

Figure 3.6

Using (3.7) enables us to express the linear shape functions as

$$L_1(\xi) = \frac{1}{2}(1 - \xi) \qquad L_2(\xi) = \frac{1}{2}(1 + \xi) \tag{3.8}$$

It is also useful to note that the transformation defined by (3.7) can be expressed in terms of these shape functions. For if we put

$$x = \frac{\ell}{2}\xi + x_1 + \frac{\ell}{2}$$

where $\ell = x_2 - x_1$, we find readily that

$$x = \frac{x_1}{2}(1-\xi) + \frac{x_2}{2}(1+\xi)$$

or

$$x = x_1 L_1(\xi) + x_2 L_2(\xi) \tag{3.9}$$

The relation in (3.9) is a simple way of representing the linear transformation from the x variable to the ξ variable i.e. of mapping an element whose endpoints have coordinates x_1, x_2 to one whose endpoints are -1 and $+1$.

Integrals over elements can be readily calculated using this ξ coordinate. For example

$$\begin{aligned}\int_{[e]} L_1^2(x)\,dx &= \int_{-1}^{1} L_1^2(\xi)\,\frac{dx}{d\xi}\,d\xi \\ &= \frac{\ell}{2}\int_{-1}^{1}\frac{1}{4}(1-\xi)^2\,d\xi \\ &= \frac{\ell}{24}\,\left[-(1-\xi)^3\right]_{-1}^{1} = \frac{\ell}{3}\end{aligned}$$

where ℓ is the length of the element. An identical result holds for the other linear shape function so we have

$$\int_{[e]} L_i^2(x)\,dx = \frac{\ell}{3} \qquad i = 1,2 \tag{3.10}$$

Products of linear shape functions are dealt with just as easily:

$$\begin{aligned}\int_{[e]} L_1(x)\,L_2(x)\,dx &= \frac{\ell}{2}\int_{-1}^{1} L_1(\xi)\,L_2(\xi)d\xi \\ &= \frac{\ell}{8}\int_{-1}^{1}(1-\xi^2)d\xi = \frac{\ell}{6}\end{aligned} \tag{3.11}$$

3.6 Quadratic elements

A quadratic element is a three-noded element where a quadratic (or second order) approximation is used for the unknown function $u(x)$ viz.

$$\tilde{u} = a_0 + a_1\,x + a_2\,x^2 \qquad x \text{ within } [e] \tag{3.12}$$

As in the case of the linear element we wish to rewrite (3.12) in terms of nodal values. We use a three-noded element with the middle node situated halfway between the two end-nodes (see Figure 3.7).

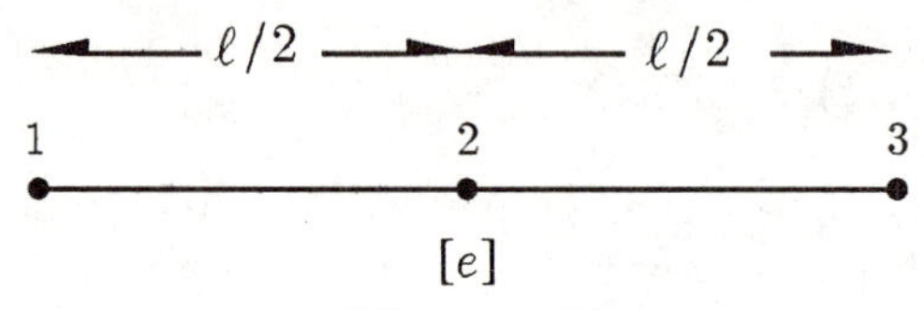

Figure 3.7

It is convenient to use a local coordinate system with a coordinate x' measured from node 1. We then have, instead of (3.12),

$$\tilde{u} = \alpha_1 + \alpha_2\,x' + \alpha_3(x')^2 \tag{3.13}$$

or, introducing the three nodal values,

$$\tilde{u}_1 = \alpha_1 \qquad \tilde{u}_2 = \alpha_1 + \alpha_2\,\frac{\ell}{2} + \alpha_3\left(\frac{\ell}{2}\right)^2 \qquad \tilde{u}_3 = \alpha_1 + \alpha_2\,\ell + \alpha_3\,\ell^2$$

We solve these equations for the α_i in terms of the nodal values $\tilde{u}_i$, substitute into (3.13) and collect terms. We obtain after straightforward algebra

$$\tilde{u} = Q_1(x')\,\tilde{u}_1 + Q_2(x')\,\tilde{u}_2 + Q_3(x')\,\tilde{u}_3 \tag{3.14}$$

where

$$Q_1(x') = \left(1 - \frac{2x'}{\ell}\right)\left(1 - \frac{x'}{\ell}\right) \qquad Q_2(x') = \frac{4x'}{\ell}\left(1 - \frac{x'}{\ell}\right)$$

$$Q_3(x') = \frac{x'}{\ell}\left(\frac{2x'}{\ell} - 1\right) \tag{3.15}$$

These quadratic polynomials are the second-order counterparts of the linear shape functions $L_i(x)$ discussed earlier. Clearly, in terms of the local coordinate x':

$$\text{(a)}\quad Q_1(x') = \begin{cases} 0 & \text{at } x' = \ell/2 \text{ and } x' = \ell \quad \text{(nodes 2 and 3)} \\ 1 & \text{at } x' = 0 \quad \text{(node 1)} \end{cases}$$

$$Q_2(x') = \begin{cases} 0 & \text{at nodes } 1,3 \\ 1 & \text{at node } 2 \end{cases}$$

$$Q_3(x') = \begin{cases} 0 & \text{at nodes } 1,2 \\ 1 & \text{at node } 3 \end{cases}$$

$$\text{(b)}\quad Q_1(x') + Q_2(x') + Q_3(x') = 1 \qquad 0 \le x' \le \ell$$

Again it is convenient to express these quadratic shape functions in terms of the natural coordinate ξ where $-1 \le \xi \le 1$ over the element. In this case $\xi = 0$ corresponds to the position of the central node. See Figure 3.8 for this **standard quadratic element**.

1 2 3

$\xi = -1$ $\xi = 0$ $\xi = +1$

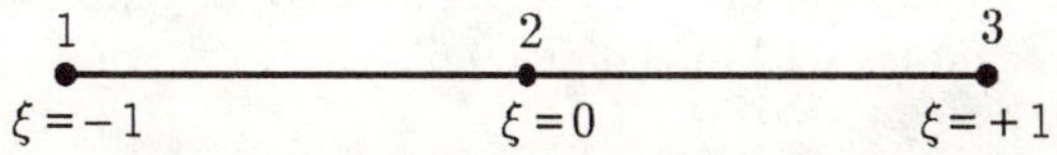

Figure 3.8

Using the transformation (3.7) in (3.15) we obtain expressions for the quadratic shape functions in natural coordinates:

$$Q_1(\xi) = \frac{\xi}{2}(\xi - 1) \qquad Q_2(\xi) = (1 - \xi)(1 + \xi)$$
$$Q_3(\xi) = \frac{\xi}{2}(\xi + 1) \tag{3.16}$$

These are sketched over the range $-1 \le \xi \le 1$ in Figure 3.9. All the graphs are of course parabolae.

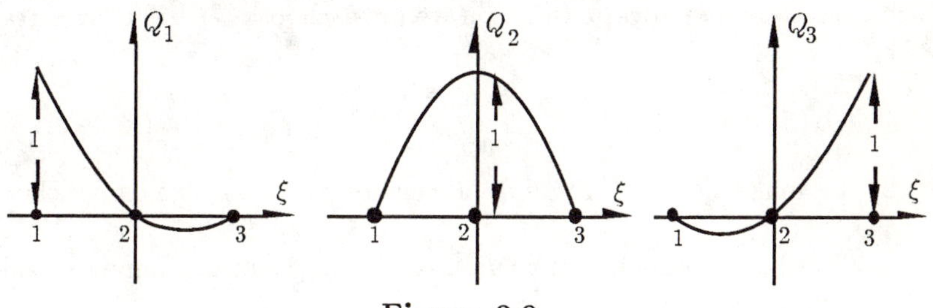

Figure 3.9

The trial solution (3.14) over the standard quadratic element can be written

$$\tilde{u} = Q_1(\xi)\,\tilde{u}_1 + Q_2(\xi)\,\tilde{u}_2 + Q_3(\xi)\,\tilde{u}_3 \qquad -1 \le \xi \le +1$$

or

$$\tilde{u} = \sum_{i=1}^{3} Q_i(\xi)\,\tilde{u}_i \qquad -1 \le \xi \le +1 \tag{3.17}$$

Alternative derivation of the shape functions

Clearly the procedure we have used for deriving linear and quadratic shape functions becomes more tedious to apply for still higher order elements. It is useful therefore to realize that simpler procedures exist based on the following properties of one-dimensional shape functions:

- Each element shape function has unit value at one node and is zero at the other nodes within the element.
- Each shape function is a polynomial of the same degree as the original trial solution.

In the quadratic case, for example, we seek three functions $Q_i(x)$, $i = 1, 2, 3$ and express each Q_i as a product of two functions:

$$Q_i = F_i\,G_i \quad i = 1, 2, 3$$

where F_i is a function which is zero at nodes other than node i and G_i is selected so that Q_i is indeed a quadratic polynomial. Hence, using the natural (ξ) coordinate system we select, for example,

$$Q_1(\xi) = \xi(\xi - 1)G_1$$

because this function is zero at nodes 2 and 3. Since, here, $F_1(\xi) = \xi^2 - \xi$ which is already a quadratic function we must choose G_1 to be a constant C_1 i.e.

$$Q_1(\xi) = C_1\xi(\xi - 1)$$

Finally to make $Q_1 = 1$ at node 1 (where $\xi = -1$) we must choose $C_1 = 1/2$.

Similar arguments can be used to obtain the other shape functions (3.16) very readily.

The derivation and analysis of shape functions in one- and two-dimensions is fully discussed in the Appendix, the early sections of which could be studied at this point. However this is not a necessary diversion in order to understand the following sections and chapters.

3.7 Some integrals involving quadratic shape functions

As already mentioned we often require in finite element analysis element integrals involving shape functions. For example, for quadratic elements

$$\int_{[e]} \frac{dQ_i}{dx}\frac{dQ_j}{dx}\,dx \qquad i,j = 1,2,3$$

Such integrals can be readily calculated by transforming the original element of length ℓ to the standard quadratic element shown in Figure 3.8. We then have

$$\begin{aligned}\int_{[e]} \frac{dQ_i}{dx}\frac{dQ_j}{dx}\,dx &= \int_{-1}^{1} \frac{dQ_i}{d\xi}\frac{d\xi}{dx}\frac{dQ_j}{d\xi}\frac{d\xi}{dx}\frac{dx}{d\xi}\,d\xi \\ &= \frac{2}{\ell}\int_{-1}^{1} \frac{dQ_i}{d\xi}\frac{dQ_j}{d\xi}\,d\xi\end{aligned}$$

since $\dfrac{d\xi}{dx} = \dfrac{2}{\ell}$. But from (3.16)

$$\frac{dQ_1}{d\xi} = \xi - \frac{1}{2} \qquad \frac{dQ_2}{d\xi} = -2\xi \qquad \frac{dQ_3}{d\xi} = \xi + \frac{1}{2}$$

Hence, for example,

$$\int_{[e]} \frac{dQ_1}{dx}\frac{dQ_1}{dx}\,dx = \frac{2}{\ell}\int_{-1}^{1}\left(\xi - \frac{1}{2}\right)^2 d\xi = \frac{7}{3\ell}$$

By similar calculations the remaining five integrals are

$$\begin{aligned}&\int_{[e]} \frac{dQ_2}{dx}\frac{dQ_2}{dx}\,dx = \frac{16}{3\ell} \qquad &&\int_{[e]} \frac{dQ_3}{dx}\frac{dQ_3}{dx}\,dx = \frac{7}{3\ell}\\ &\int_{[e]} \frac{dQ_1}{dx}\frac{dQ_2}{dx}\,dx = -\frac{8}{3\ell} \qquad &&\int_{[e]} \frac{dQ_1}{dx}\frac{dQ_3}{dx}\,dx = \frac{1}{3\ell}\\ &\int_{[e]} \frac{dQ_2}{dx}\frac{dQ_3}{dx}\,dx = -\frac{8}{3\ell} && \end{aligned} \qquad (3.18)$$

Even more straightforwardly we have for the areas under the quadratic shape function graphs in Figure 3.9:

$$\int_{[e]} Q_i(x)\,dx = \frac{\ell}{2}\int_{-1}^{1} Q_i(\xi)\,d\xi$$

from which using (3.16) we find

$$\int_{[e]} Q_1(x)\,dx = \int_{[e]} Q_3(x)\,dx = \frac{\ell}{6}$$

$$\int_{[e]} Q_2(x)\,dx = \frac{2\ell}{3} \tag{3.19}$$

Finally, for later use, we list all integrals of the type

$$\int_{[e]} Q_i(x)\,Q_j(x)\,dx$$

These are again readily obtained by transformation of the element $[e]$ of length ℓ to the standard quadratic element. The results are

$$\int_{[e]} Q_1\,Q_2\,dx = \frac{2\ell}{30} \qquad \int_{[e]} Q_1\,Q_3\,dx = -\frac{\ell}{30} \qquad \int_{[e]} Q_2\,Q_3 dx = \frac{2\ell}{30}$$

$$\int_{[e]} Q_1^2\,dx = \frac{4\ell}{30} \qquad \int_{[e]} Q_2^2\,dx = \frac{16\ell}{30} \qquad \int_{[e]} Q_3^2\,dx = \frac{4\ell}{30} \tag{3.20}$$

It should be emphasized that integrals of the type quoted in (3.18), (3.19) and (3.20), and the corresponding integrals involving linear shape functions quoted in (3.10) and (3.11), are mainly for use in hand calculations. Numerical integration techniques are more appropriate for computer implementations of finite elements.

3.8 Some applications of shape functions

To help us to become thoroughly familiar with linear and quadratic shape functions we shall work through a few simple examples involving approximations of **known** functions over just two elements.

Example 3.1

Construct a piecewise linear approximation to the function $\sin x$ using the two elements (0, 1) and (1,2).

Solution

We use the obvious element and global node numbering system shown in Figure 3.10.

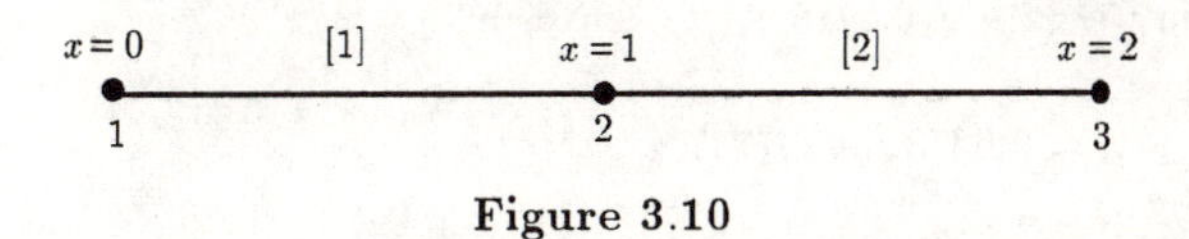

Figure 3.10

We use a natural coordinate ξ and, using (3.9), map each element in turn into the standard linear element of Figure 3.6.

• **Element [1]**:

In this case (3.9) becomes

$$x = 1.L_2(\xi) = (\xi + 1)/2$$

The nodal values are

$$u_1 = \sin 0 = 0 \qquad u_2 = \sin 1 = 0.8415$$

(Here we do not use the notation $\tilde{u}_1, \tilde{u}_2$ as we know the exact values at the nodes.) The shape functions (3.3) and (3.4) for this element are

$$L_1(x) = 1 - x, \quad L_2(x) = x$$

Hence (3.5) gives

$$\sin x \approx L_1(x)\, u_1 + L_2(x)\, u_2$$

$$= 0 + 0.8415\, x \qquad 0 \le x \le 1 \tag{3.21a}$$

or, in terms of ξ,

$$\sin x \approx 0.4208(\xi + 1) \qquad -1 \le \xi \le 1 \tag{3.21b}$$

- **Element [2]**:

The transformation $x = (\xi + 3)/2$ will map this element into the standard linear element. The nodal values are, using local node numbers,

$$u_1 = \sin 1 = 0.8415 \qquad u_2 = \sin 2 = 0.9093$$

In terms of x, the linear shape functions in element [2] are

$$L_1(x) = 2 - x, \quad L_2(x) = x - 1$$

so that, again using (3.5),

$$\sin x \approx (2 - x)0.8415 + (x - 1)(0.9093)$$

$$= 0.0678\, x + 0.7737 \qquad 1 \le x \le 2 \tag{3.22a}$$

or, in terms of ξ,

$$\sin x \approx 0.0339\, \xi + 0.8754 \qquad -1 \le \xi \le 1 \tag{3.22b}$$

Note that the nodal values were **known** here as we were simply trying to approximate a **given** function $\sin x$. In using finite elements to solve boundary value problems the nodal values are the basic **unknowns** to be determined.

Once the nodal values have been obtained, however, the chosen shape functions are used to obtain approximate values of u at non-nodal points. For example, using the above approximations, (3.21a) and (3.22a) respectively:

at $x = 0.5$ $\qquad \sin x \approx 0.8415\left(\frac{1}{2}\right) = 0.4208$

at $x = 1.5$ $\qquad \sin x \approx 0.0678(1.5) + 0.7737 = 0.8754$

These compare with the exact values

$$\sin(0.5) = 0.4794 \text{ and } \sin(1.5) = 0.9975.$$

The agreement between approximate and exact values is only fair, reflecting the unsuitability of linear approximations to a sinusoid.

Clearly other quantities can be estimated using linear approximations. We demonstrate by calculating $\int_0^2 \sin x\, dx$ and using natural coordinates:

- **Element [1]**:

$$\int_0^1 \sin x\, dx = \int_{-1}^{1} \sin x \frac{dx}{d\xi}\, d\xi$$

$$\approx \frac{1}{2}\int_{-1}^{1} 0.4208(\xi + 1)\, d\xi \qquad \text{using (3.21b)}$$

$$= 0.4208$$

- **Element [2]**:

$$\int_1^2 \sin x\, dx \approx \frac{1}{2}\int_{-1}^{1} (0.0339\,\xi + 0.8754)\, d\xi \qquad \text{using (3.22b)}$$

$$= 0.8754$$

Adding, we obtain

$$\int_0^2 \sin x\, dx \approx 1.2962$$

which compares with the exact value of 1.4161. Again, our result is only a moderate approximation to the true value of the integral. This is perhaps to be expected since only linear approximations are being used. We will show in the following example that a quadratic approximation gives much closer agreement with the true value.

Example 3.2

Repeat Example 3.1 using a two-element quadratic approximation to $\sin x$ and re-estimate the value of $\int_0^2 \sin x\, dx$.

Solution

Although we could choose elements of different lengths we will use two equal length three-noded elements, as shown in Figure 3.11, each with an additional central node.

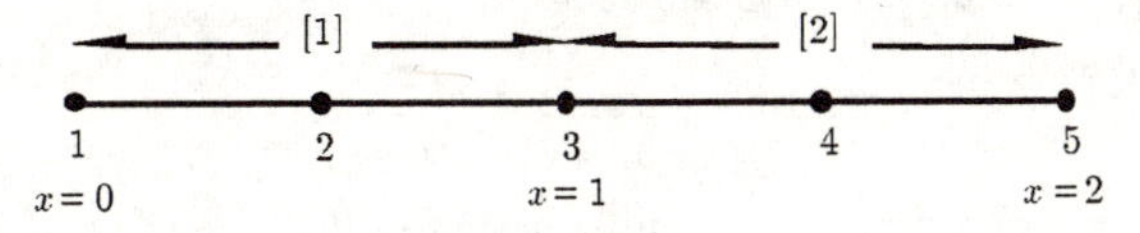

Figure 3.11

In element [1] we again map to the standard element using

$$x = \tfrac{1}{2}(\xi + 1)$$

and over this element we have, from (3.17),

$$\begin{aligned}\sin x &\approx Q_1(\xi)\,u_1 + Q_2(\xi)\,u_2 + Q_3(\xi)\,u_3 \\ &= 0 + Q_2(\xi)\,\sin(0.5) + Q_3(\xi)\,\sin 1 \\ &= (1-\xi^2)(0.4794) + \frac{1}{2}\xi(1+\xi)0.8415\end{aligned}$$

using the shape functions (3.16).

Hence, integrating over this element:

$$\begin{aligned}\int_0^1 \sin x\,dx &\approx \frac{1}{2}\int_{-1}^{1}\left\{0.4794(1-\xi^2) + \frac{0.8415}{2}(\xi+\xi^2)\right\}d\xi \\ &= 0.4595\end{aligned}$$

Similarly for element [2] we map to the standard element using, as in Example 3.1,

$$x = \tfrac{1}{2}\xi + \tfrac{3}{2}$$

Over this element

$$\begin{aligned}\sin x &\approx Q_1(\xi)\sin 1 + Q_2(\xi)\sin(1.5) + Q_3(\xi)\sin 2 \\ &= 0.8415\,\frac{\xi}{2}\,(\xi-1) + 0.9975(1-\xi^2) + 0.9093\,\frac{\xi}{2}\,(1+\xi)\end{aligned}$$

Then

$$\begin{aligned}\int_1^2 \sin x\;dx &= \frac{1}{2}\int_{-1}^{1} \sin x\;d\xi \\ &\approx 0.9568\end{aligned}$$

after integrating the approximating polynomial.

Adding, we obtain

$$\begin{aligned}\int_1^2 \sin x\;dx &\approx 0.4595 + 0.9568 \\ &= 1.4163\end{aligned}$$

which is now a very good approximation to the true value of 1.4161.

3.9 Numerical integration over elements

The integrals over elements in the previous section were all evaluated with little difficulty. However this is not always the case and, in both one-dimensional and two-dimensional applications of finite elements, numerical approximation of integrals has often to be used.

Numerical evaluation of an integral of the form $\int_a^b f(x)dx$ can proceed in one of two ways:

- using a formula based on evaluating the integrand at predetermined and usually equally spaced x-values. If there are n such values then we can pass a polynomial of order $(n-1)$ through the n points. Based on this idea we can obtain familiar numerical integration formulae such as the trapezium rule when n is 2 and Simpson's rule when n equals 3. The general approach here is known as Newton–Cotes quadrature.

- treating the values of x at which the integrand is evaluated as unknowns to be determined. This technique is known as **Gaussian** or **Gauss–Legendre quadrature** and it is the approach normally used in connection with the finite element method. This is because, for a given degree of accuracy, fewer function evaluations are required than when using Newton–Cotes formulae.

The details of the Gauss–Legendre approach are covered in standard texts on numerical methods. Basically for an nth order Gauss–Legendre integration we have the approximation (see Gerald and Wheatley, 1989)

$$\int_{-1}^{1} f(\xi)\,d\xi \approx \sum_{i=1}^{n} w_i\, f(\xi_i) \tag{3.23}$$

where the points $\xi_1, \xi_2, \ldots \xi_n$ are the (sample) points at which the function is evaluated and the coefficients $w_1, w_2, \ldots w_n$ are called weights. Note that the range of integration is from -1 to $+1$ and we have labelled the (dummy) integration variable as ξ to conform with normal finite element practice. The fact that standard tables of w_i and ξ_i are referred to an integration range $(-1, 1)$ is an additional reason for the use of standard elements over which the local coordinate ξ varies from -1 to $+1$.

Values of ξ_i and w_i for $n = 2, 3, .., 6$ are shown in Table 3.3. Note that an nth order formula contains $2n$ unknowns, viz. n values of ξ_i and n values of w_i, and consequently gives the exact result when applied to a polynomial of degree $2n - 1$. Note the symmetric sampling points ξ_i.

Number of points n	Values of ξ_i	Weighting factors w_i
2	$\pm$ 0.577350	1.0
3	$\pm$ 0.774597	0.555556
	0.0	0.888889
4	$\pm$ 0.861136	0.347855
	$\pm$ 0.339981	0.652145
5	$\pm$ 0.906180	0.236927
	0.0	0.568889
	$\pm$ 0.538469	0.478629
6	$\pm$ 0.932469	0.171324
	$\pm$ 0.661209	0.360762
	$\pm$ 0.238619	0.467914

Table 3.3

Example 3.3

(a) Determine $\int_0^1 \sin^3 x \, dx$ exactly by analytical techniques.
(b) Use Gaussian quadrature to determine an approximation to $\int_0^1 \sin^3 x \, dx$.
(c) Assuming that $0 \leq x \leq 1$ is a single element, express $\sin^3 x$ in terms of quadratic shape functions and calculate $\int_0^1 \sin^3 x \, dx$ numerically.

Solution

$$\text{(a) } I \equiv \int_0^1 \sin^3 x \, dx = \int_0^1 \sin x(1 - \cos^2 x) dx$$

Putting $u = \cos x$

$$I = \int_{\cos 1}^{0} (1 - u^2) du$$

$$= \left[u - \frac{u^3}{3} \right]_{\cos 1}^{0}$$

$$= 0.178941$$

(b) To use Gaussian quadrature we must convert the interval of integration from (0,1) to (−1,1). We can, using (3.9), accomplish this by putting

$$x = (\xi + 1)/2$$

Hence

$$\int_0^1 \sin^3 x \, dx = \frac{1}{2}\int_{-1}^1 \sin^3\left(\frac{1+\xi}{2}\right) d\xi$$

For the sake of example we shall use the 3-point Gaussian quadrature formula

$$\int_{-1}^1 f(\xi) d\xi = \sum_{i=1}^3 w_i f(\xi_i)$$

Using Table 3.3:

$$\xi_1 = -0.774597 \quad \text{so} \quad f(\xi_1) = \sin^3\left(\frac{1-0.774597}{2}\right)$$
$$= 0.001422$$

$$\xi_2 = 0 \quad \text{so} \quad f(\xi_2) = \sin^3\left(\frac{1}{2}\right) = 0.110195$$

$$\xi_3 = +0.774597 \quad \text{so} \quad f(\xi_3) = \sin^3\left(\frac{1+0.774597}{2}\right)$$
$$= 0.466149$$

Then, using the appropriate weighting factors,

$$\int_{-1}^1 f(\xi)\, d\xi = 0.555556\,(0.001422 + 0.466149)$$
$$+ (0.888889)(0.110195)$$
$$= 0.357713$$

leading to

$$\int_0^1 \sin^3 x \, dx = 0.178857$$

which agrees with the calculated analytical value to 4 decimal places.

(c) We again put $\xi = 2x - 1$ to convert to the standard quadratic element. The nodal values are

$$u_1 = \sin^3 0 = 0$$

$$u_2 = \sin^3(\frac{1}{2}) = 0.110195$$

$$u_3 = \sin^3 1 = 0.595823$$

Then using (3.17) and (3.16) we have the quadratic approximation

$$\sin^3\left(\frac{\xi+1}{2}\right) \approx Q_1(\xi)\,u_1 + Q_2(\xi)\,u_2 + Q_3(\xi)\,u_3$$
$$= 0 + (1-\xi^2)(0.110195) + \frac{1}{2}\xi(\xi+1)(0.595823) \quad \mathbf{(3.24)}$$

Although it is perfectly straightforward to integrate (3.24) analytically we will again demonstrate the use of the 3-point Gaussian quadrature formula which is in fact exact for this case.

Considering the first term and putting $f_1(\xi) = 1 - \xi^2$

$$I_1 \equiv \int_{-1}^{1} f_1(\xi) = 2 \times 0.555556\; f_1(0.774597)$$
$$+ 0.888889\; f_1(0)$$

(noting that the integrand is even). This gives

$$I_1 = 1.111112\,(1 - (0.774597)^2) + 0.888889$$
$$= 1.333333$$

Similarly with $f_2(\xi) = \xi^2 + \xi$ and again using the 3-point Gaussian formula

$$I_2 = \int_{-1}^{1} f_2(\xi)\,d\xi$$
$$= (0.555556)(0.774597^2 + (-0.774597))$$
$$+ (0.555556)(0.774597^2 + 0.774597)$$
$$= 0.666667$$

From (3.24) we then have

$$\int_{-1}^{1} \sin^3\left(\frac{\xi+1}{2}\right) d\xi \approx 0.110195\,I_1 + 0.5(0.595823)\,I_2$$
$$= 0.345534$$
$$\therefore \int_{0}^{1} \sin^3 x\,dx \approx 0.172767$$

While not as accurate as the calculation in (b) this is an acceptable approximation bearing in mind the one-element approximation.

EXERCISES

3.1 Develop the linear shape functions $L_1(\xi)$ and $L_2(\xi)$ directly by assuming an approximate solution

$$\tilde{u}(\xi) = \alpha_1 + \alpha_2 \xi$$

in the standard element $-1 \leq \xi \leq +1$ and solving for α_1 and α_2.

3.2 (a) Prove results (ii) and (iii) in Section 3.4 for the linear shape functions $L_1(x)$ and $L_2(x)$.

(b) Using the linear shape functions in natural coordinates show that

$$\text{(i)} \quad \sum_{i=1}^{2} L_i(\xi) = 1 \qquad \text{(ii)} \quad \sum_{i=1}^{2} L_i(\xi)\xi_i = \xi \qquad \text{(iii)} \quad \sum_{i=1}^{2} \frac{dL_i}{d\xi} = 0$$

(where $\xi_i = \xi(x_i)$)

3.3 Evaluate the following integrals for a linear element of length $x_2 - x_1 = \ell$:

$$\text{(i)} \quad \int_{x_1}^{x_2} L_i^2 L_j^2 \; dx \qquad \text{(ii)} \quad \int_{x_1}^{x_2} x L_i \; dx \qquad i, j = 1, 2$$

Express your answers to (i) as components of a 2×2 matrix and those to (ii) as a column matrix.

3.4 The nodal coordinates (cm) and the nodal values of a scalar function u for several linear elements are given below. Estimate u at the given value of x in each case.

x	x_1	x_2	$\tilde{u}_1$	$\tilde{u}_2$
2.0	1.5	2.5	6.7	12.8
2.8	1.0	3.5	11.4	1.9
3.1	1.0	5.0	32.6	12.4

3.5 Two adjacent linear elements [1] and [2] have global node numbers 1,2 and 2,3 respectively. Write down the approximate trial solutions $\tilde{u}^{[1]}(x)$ and $\tilde{u}^{[2]}(x)$.
Show that

(i) $\quad \tilde{u}^{[1]}(x_2) = \tilde{u}^{[2]}(x_2)$

(which is sometimes called C_0 compatibility)

(ii) $\dfrac{d\tilde{u}^{[1]}}{dx}(x_2) \neq \dfrac{d\tilde{u}^{[2]}}{dx}(x_2)$

3.6 Develop the quadratic shape functions $Q_i(\xi)$, $i = 1, 2, 3$, directly by considering a three noded standard element with nodes at $\xi = -1$, $\xi = 0$, $\xi = +1$, assuming a trial solution

$$\tilde{u}(\xi) = \alpha_0 + \alpha_1 \xi + \alpha_2 \xi^2$$

and solving for α_0, α_1 and α_2.

3.7 Verify the following results for the quadratic shape functions

(i) $\sum_{i=1}^{3} Q_i(\xi) = 1$ (ii) $\sum_{i=1}^{3} Q_i(\xi)\xi_i = \xi$

(iii) $\sum_{i=1}^{3} Q_i(\xi)\xi_i^2 = \xi^2$ (iv) $\sum_{i=1}^{3} \dfrac{dQ_i}{d\xi} = 0$

(where $\xi_i = \xi(x_i)$)

3.8 Verify the results in (3.18).

3.9 Verify the results in (3.19).

3.10 Verify the results in (3.20).

3.11 The nodal coordinates (cm) for various quadratic elements and the corresponding nodal values u_i are given below. Estimate u at the given value of x in each case.

x	x_1	x_2	x_3	$\tilde{u}_1$	$\tilde{u}_2$	$\tilde{u}_3$
0.25	0.0	1.0	2.0	17.6	18.3	22.9
3.10	0.5	2.5	4.5	5.1	3.8	6.7
1.25	1.0	1.5	2.0	2.1	0.1	−1.6

3.12 Using the procedure defined at the end of Section 3.6 develop the shape functions, $C_i(\xi)$ $i = 1, 2, 3, 4$, for the one dimensional standard cubic element shown in Figure Q3.12.

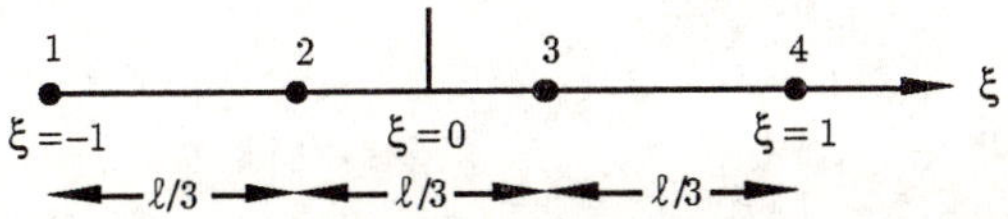

1 2 3 4 ξ

$\xi=-1$ $\xi=0$ $\xi=1$

$\ell/3$ $\ell/3$ $\ell/3$

Figure Q3.12

3.13 Using three-point Gaussian quadrature, obtain approximate values for the integrals

(i) $\displaystyle\int_1^2 x^5\,dx$ (ii) $\displaystyle\int_0^2 \frac{dx}{\sqrt{x}}$

Compare with the exact analytical values in each case.

3.14 Evaluate $\displaystyle\int_0^{\pi/2} \sin^2 x\,dx$

(i) by an exact analytical method

(ii) using four-point Gaussian quadrature

(iii) by dividing the interval $0 \leq x \leq \frac{\pi}{2}$ into two linear elements

(iv) by treating the interval $0 \leq x \leq \frac{\pi}{2}$ as one quadratic element.

Finite Element Solution of One-dimensional Boundary Value Problems

PREVIEW

In this chapter we look in detail at the application of the finite element method to the approximate solution of one-dimensional boundary value problems. We introduce the artifice of the roof function to show that the finite element method is a direct application of the Galerkin method and also develop a matrix formulation of the finite element equations. We apply the procedures to simple vibration problems and to a one-dimensional heat transfer problem. The first supplement contains a Fortran77 program which implements the finite element method for second-order boundary value problems.

4.1 Introduction

In this chapter we complete our study of linear one-dimensional boundary value problems of second order. We will segment or discretize the domain of the problem into line segments (elements) and use linear or quadratic trial solutions within each element. The modified Galerkin technique will then be used to determine the unknown parameters which in this case will be the values of the unknown function $u(x)$ at the nodes.

Firstly we investigate solutions of the **one-dimensional Poisson equation**

$$\frac{d^2u}{dx^2} = f(x)$$

where $f(x)$ is a known function. We make this choice because

(a) the problems can be readily solved by direct integration so we can check the finite element results.

(b) the equation arises as a model in a number of areas of science and engineering, for example heat transfer and electrostatics.

The specific boundary value problem involving Poisson's equation that we shall solve is

$$\frac{d^2u}{dx^2} = 2 \qquad 1 \le x \le 3 \tag{4.1a}$$

$$u(1) = 1 \qquad u'(3) = 6 \tag{4.1b}$$

Using the n-parameter approximation

$$\tilde{u}(x) = \theta_0(x) + \sum_{i=1}^{n} a_i\theta_i(x)$$

the modified Galerkin equations are obtained, as usual, by first considering the standard Galerkin equations:

$$\int_1^3 \{\frac{d^2\tilde{u}}{dx^2} - 2\}\theta_i \, dx = 0 \qquad i = 1, 2, \ldots, n$$

If we integrate the term involving the highest derivative and incorporate the derivative boundary condition we obtain, for the boundary value problem (4.1), the modified Galerkin equations

$$-\int_1^3 \{\frac{d\tilde{u}}{dx}\frac{d\theta_i}{dx} + 2\theta_i\} \, dx + 6\theta_i(3) = 0 \qquad i = 1, 2, \ldots, n \tag{4.2}$$

To perform a simple hand calculation we segment the domain into two linear elements of equal length (see Figure 4.1). (Note that in this and many other problems in this text we use segmentation of the domain into a very small number of elements in order to illustrate the basic approach. Standard computer packages for implementing the finite element method allow, of course, much larger numbers of elements to be used.)

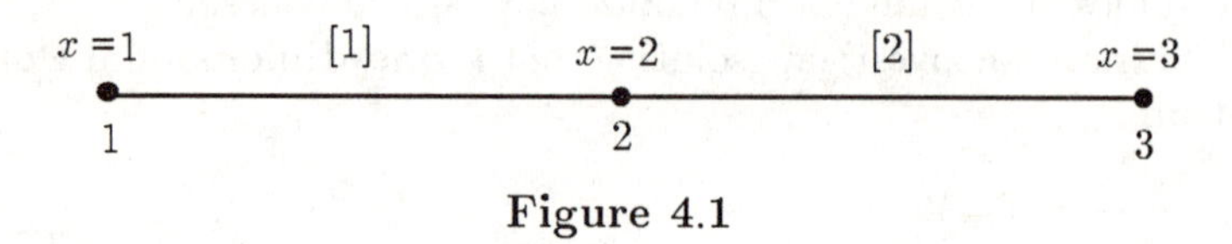

Figure 4.1

We demonstrate two similar but distinct methods for implementing the modified Galerkin technique and hence obtaining a pair of linear equations for the two unknown nodal values u_2 and u_3. In the first approach we focus our attention on the **complete domain** of the problem using what are known as **roof functions**. These are combinations of the shape functions already discussed in Chapter 3. With this approach the element subdivision only becomes important at the integration stage.

Our second approach, which is similar to the finite element approach to problems in structural mechanics and more closely akin to the techniques used in computer implementation of finite elements, will focus on each individual element. We shall derive for such an element a square matrix known, largely for historical reasons, as the **element stiffness matrix** and also a column matrix or column vector known as the **element force vector**.

The overall system of linear equations will then be expressible in matrix form as

$$[K]\{\tilde{u}\} = \{F\} + \{B\}$$

where $[K]$ is the sum of the element stiffness matrices, $\{F\}$ is the sum of the element force vectors and $\{B\}$ is a column vector which contains the information resulting from the imposition of suppressible boundary conditions. If both boundary conditions are essential then $\{B\}$ will be zero. $\{\tilde{u}\}$, of course, is the column matrix containing the unknown nodal values. $[K]$ and $\{F\}$ are often referred to as the **global stiffness matrix** and **global force vector** respectively. Note that we use square brackets to denote a square matrix and curly brackets to denote a column vector.

We will initially use linear elements and compare our answers with the exact solution to the boundary value problem (4.1) which is of course

$$u(x) = x^2 \qquad 1 \leq x \leq 3$$

We then show that using a single quadratic element we can, as would be expected, obtain this exact solution at all points in the domain.

4.2 Roof function approach for Poisson's equation

We know, from Chapter 3, that within an element with nodes at $x = x_1$ and $x = x_2$ we can use the linear trial approximation (3.5)

$$\tilde{u} = L_1(x)\,\tilde{u}_1 + L_2(x)\,\tilde{u}_2$$

where the shape functions are

$$L_1(x) = \frac{x_2 - x}{x_2 - x_1} \quad \text{and} \quad L_2(x) = \frac{x - x_1}{x_2 - x_1}$$

With the subdivision shown in Figure 4.1, viz. two elements of equal length, we have the following:

- Element [1]:

$$L_1(x) = 2 - x \qquad L_2(x) = x - 1$$

and

$$\tilde{u}^{[1]} = (2 - x)\,\tilde{u}_1 + (x - 1)\,\tilde{u}_2 \tag{4.3}$$

- Element [2]:

$$L_1(x) = 3 - x \qquad L_2(x) = x - 2$$

and

$$\tilde{u}^{[2]} = (3 - x)\,\tilde{u}_2 + (x - 2)\,\tilde{u}_3 \tag{4.4}$$

Over the complete domain this approximation is as shown in Figure 4.2.

We can conveniently re-write (4.3) and (4.4) using local node numbers and superscripts to denote elements:

$$\begin{aligned} \tilde{u}^{[1]} &= L_1^{[1]}(x)\,\tilde{u}_1 + L_2^{[1]}(x)\,\tilde{u}_2 \\ \tilde{u}^{[2]} &= L_1^{[2]}(x)\,\tilde{u}_2 + L_2^{[2]}(x)\,\tilde{u}_3 \end{aligned} \tag{4.5}$$

where

$$\left.\begin{aligned} L_1^{[1]}(x) &= 2 - x \\ L_2^{[1]}(x) &= x - 1 \end{aligned}\right\} \text{within element [1]}$$

$$\left.\begin{aligned} L_1^{[2]}(x) &= 3 - x \\ L_2^{[2]}(x) &= x - 2 \end{aligned}\right\} \text{within element [2]}$$

Note carefully that $L_2^{[1]}$ and $L_1^{[2]}$ are different functions even though both are shape functions associated with (i.e. equal to 1 at) global node 2.

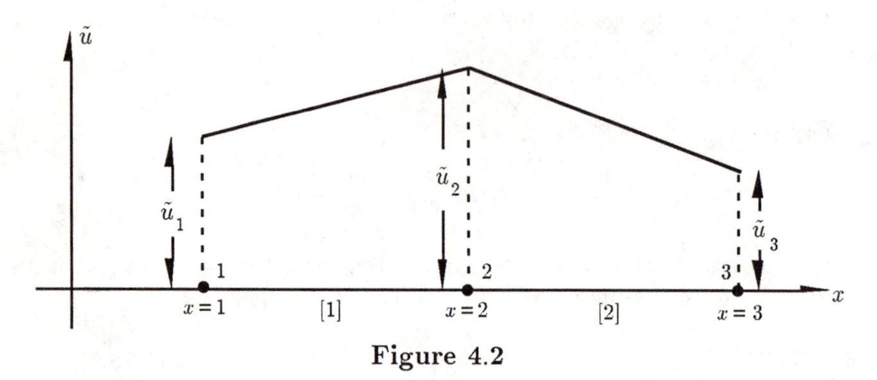

Figure 4.2

We now define, for each node i, a 'roof' or 'hat' function $R_i(x)$ which is a combination of the shape functions associated with that node. Thus

$$R_1(x) = \begin{cases} L_1^{[1]}(x) & \text{within element [1]} \\ 0 & \text{within element [2]} \end{cases} = \begin{cases} 2 - x & 1 < x < 2 \\ 0 & 2 < x < 3 \end{cases}$$

$$R_2(x) = \begin{cases} L_2^{[1]}(x) & \text{within element [1]} \\ L_1^{[2]}(x) & \text{within element [2]} \end{cases} = \begin{cases} x - 1 & 1 < x < 2 \\ 3 - x & 2 < x < 3 \end{cases}$$

$$R_3(x) = \begin{cases} 0 & \text{within element [1]} \\ L_2^{[2]}(x) & \text{within element [2]} \end{cases} = \begin{cases} 0 & 1 < x < 2 \\ x - 2 & 2 < x < 3 \end{cases}$$

See Figure 4.3 from which the reason for using the term 'roof function' becomes apparent.

The trial solutions $\tilde{u}^{[1]}$ and $\tilde{u}^{[2]}$ in (4.5) may now be combined into one expression valid over the whole domain of the problem using these roof functions:

$$\begin{aligned} \tilde{u} &= R_1(x)\,\tilde{u}_1 + R_2(x)\,\tilde{u}_2 + R_3(x)\,\tilde{u}_3 \\ &= \sum_{j=1}^{3} R_j(x)\,\tilde{u}_j \end{aligned} \tag{4.6}$$

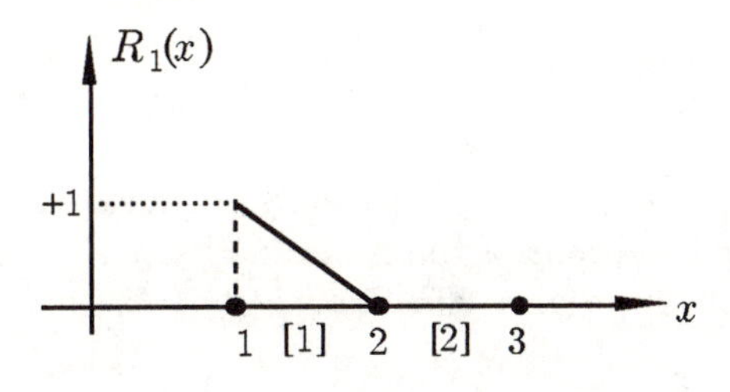

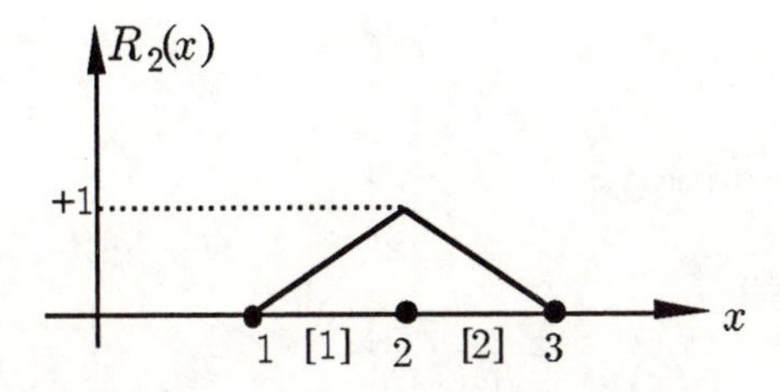

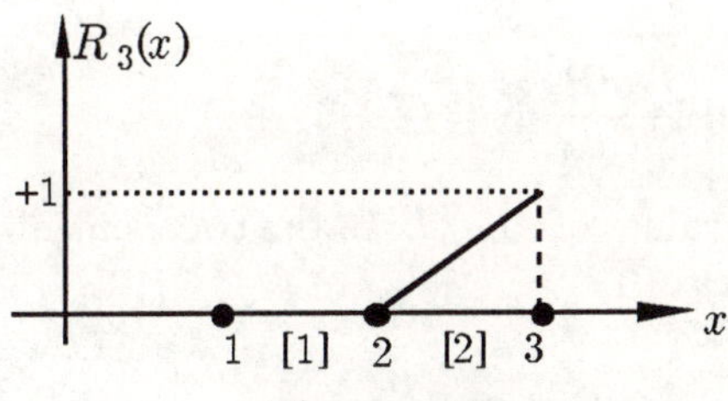

Figure 4.3

Essentially the use of roof functions is an artifice which allows functions defined in a piecewise manner, the trial solutions $\tilde{u}^{[1]}$ and $\tilde{u}^{[2]}$ in (4.5) in this case, to be considered as a single function, (4.6), defined over the complete domain of the boundary value problem. This allows us to establish a correspondence between (4.6) and the Galerkin approximation (2.2).

Since $R_1(x) = 1$ and $R_2(x) = R_3(x) = 0$ at $x = 1$ we see that $R_1(x)$ or, more generally, $R_1(x)\,u_1$ satisfies the given essential boundary condition at $x = 1$. In other words $R_1(x)$ is equivalent to the function $\theta_0(x)$ of equation (2.2). Similarly $R_2(x)$ and $R_3(x)$ are essentially the coordinate functions $\theta_1(x)$ and $\theta_2(x)$ since they satisfy the homogeneous form of this essential boundary condition. The nodal values $\tilde{u}_2$ and $\tilde{u}_3$ in (4.5) are clearly analogous to the parameters a_1 and a_2 in (2.2).

Thus from (4.2) the modified Galerkin equations are

$$-\int_1^3 \left\{ \frac{d\tilde{u}}{dx}\frac{dR_2}{dx} + 2R_2 \right\} dx + 6\,R_2\,(3) = 0 \tag{4.7}$$

$$-\int_1^3 \left\{ \frac{d\tilde{u}}{dx}\frac{dR_3}{dx} + 2R_3 \right\} dx + 6\, R_3\,(3) = 0 \qquad \textbf{(4.8)}$$

Both integrals must be split into separate integrals over elements [1] and [2] because of the definitions of the roof functions.

In element [1] we have effectively

$$\tilde{u} = \tilde{u}^{[1]} = R_1(x)\tilde{u}_1 + R_2(x)\tilde{u}_2$$

so

$$\frac{d\tilde{u}}{dx} = \frac{dR_1}{dx}\tilde{u}_1 + \frac{dR_2}{dx}\tilde{u}_2 = -1 + (1)\tilde{u}_2$$

Similarly in element [2]

$$\tilde{u} = \tilde{u}^{[2]} = R_2(x)\,\tilde{u}_2 + R_3(x)\,\tilde{u}_3$$

so

$$\frac{d\tilde{u}}{dx} = \frac{dR_2}{dx}\tilde{u}_2 + \frac{dR_3}{dx}\tilde{u}_3 = (-1)\tilde{u}_2 + (1)\tilde{u}_3$$

(Note the different values of dR_2/dx in the two elements. See Figure 4.3.) Hence (4.7) becomes

$$-\int_1^2 \{(\tilde{u}_2 - 1)(1) + 2(x-1)\}\, dx$$
$$- \int_2^3 \{(\tilde{u}_3 - \tilde{u}_2)(-1) + 2(3-x)\}\, dx + 0 = 0$$

or, by straightforward integration,

$$-\left[(\tilde{u}_2 - 1)x + (x-1)^2\right]_1^2 - \left[(\tilde{u}_2 - \tilde{u}_3)x - (3-x)^2\right]_2^3 = 0$$

i.e.

$$-2\tilde{u}_2 + \tilde{u}_3 = 1 \qquad \textbf{(4.9a)}$$

A similar calculation using (4.8) gives

$$\tilde{u}_2 - \tilde{u}_3 = -5 \qquad \textbf{(4.9b)}$$

Solving (4.9a) and (4.9b) we find

$$\tilde{u}_2 = 4 \qquad \text{and} \qquad \tilde{u}_3 = 9$$

which agree with the values obtained from the exact solution to (4.1) at the nodal points. We shall discuss this surprising result at the end of the next section after studying an alternative formulation of the above analysis.

4.3 A matrix approach for Poisson's equation

From the point of view of hand calculation the approach outlined in the previous section is adequate for a simple two element segmentation. However, a computer implementation of the finite element method will involve many elements (perhaps numbering in the thousands). It is appropriate therefore to rework the two element problem using a **matrix** formulation, a formulation which extends readily to the many-element situation.

We again use the trial solution (4.6) and simply **replace** the coordinate functions θ_i in (4.2) with R_i. This gives

$$-\int_1^3 \{\frac{d\tilde{u}}{dx}\frac{dR_i}{dx} + 2R_i\}\, dx + 6R_i(3) = 0 \qquad i = 1, 2, 3 \qquad \textbf{(4.10)}$$

where the roof functions $R_i(x)$ are as before, see Figure 4.3.

Equations (4.10) for $i = 2$ and $i = 3$ are of course (4.7) and (4.8) together with an identical equation for $i = 1$. In fact the equation for $i = 1$ is **invalid**. Its inclusion in (4.10) assumes (falsely) that R_1 is a coordinate function. However, including this invalid equation in (4.10) facilitates the matrix formulation of the problem and later in the solution procedure the invalidity will be removed.

We firstly re-arrange (4.10) as

$$\int_1^3 \frac{d\tilde{u}}{dx}\frac{dR_i}{dx} dx = -\int_1^3 2\, R_i\, dx + 6\, R_i\,(3) \qquad \textbf{(4.11)}$$

such that the term involving $\tilde{u}$, which contains the nodal values, $\tilde{u}_1, \tilde{u}_2, \tilde{u}_3$, is placed on the left hand side.

Using (4.6) we have

$$\frac{d\tilde{u}}{dx} = \sum_{j=1}^{3} \frac{dR_j}{dx}\, \tilde{u}_j$$

so (4.11) becomes

$$\sum_{j=1}^{3} \left\{ \int_1^3 \frac{dR_j}{dx}\,\frac{dR_i}{dx} dx \right\} \tilde{u}_j = -\int_1^3 2\, R_i\, dx + 6\, R_i\,(3) \quad i = 1, 2, 3 \quad \textbf{(4.12)}$$

If we define

$$K_{ij} = \int_1^3 \frac{dR_i}{dx}\frac{dR_j}{dx}dx \qquad i,j = 1,2,3 \tag{4.13}$$

$$F_i = -\int_1^3 2\,R_i\,dx \qquad i = 1,2,3 \tag{4.14}$$

then (4.12) has the simple form

$$\sum_{j=1}^{3} K_{ij}\,\tilde{u}_j = F_i + 6\,R_i\,(3) \qquad i = 1,2,3 \tag{4.15}$$

If the quantities K_{ij} are thought of as the component elements of a 3×3 global stiffness matrix $[K]$ and the quantities F_i are considered as the components of a 3×1 global force vector $\{F\}$ then (4.15) has the standard matrix form

$$[K]\{\tilde{u}\} = \{F\} + \{B\} \tag{4.16}$$

where $\{\tilde{u}\} = \{\tilde{u}_1, \tilde{u}_2, \tilde{u}_3\}^T$. The components, say B_i, of $\{B\}$ are $6R_i(3)$ and hence, from the definition of the roof functions ($R_1(x)$ and $R_2(x)$ both vanish at $x = 3$ and $R_3(x) = 1$ at $x = 3$) we have

$$\{B\} = \begin{Bmatrix} 0 \\ 0 \\ 6 \end{Bmatrix}$$

Note that $\{B\}$ contains the suppressible boundary condition at $x = 3$.

In calculating $[K]$ and $\{F\}$ using (4.13) and (4.14) we have to integrate separately over each element. Effectively therefore we are expressing $[K]$ as a sum of element stiffness matrices $[K^{[e]}]$ and $\{F\}$ as a sum of element force vectors $\{F^{[e]}\}$.

The individual terms of these element quantities are readily calculated as follows:

$$K_{ij}^{[1]} = \int_{[1]} \frac{dR_i}{dx}\frac{dR_j}{dx}\,dx \qquad i,j = 1,2,3$$

But within element [1] we have defined

$$R_1(x) = L_1(x), \qquad R_2(x) = L_2(x), \qquad R_3(x) = 0$$

Hence

$$K_{ij}^{[1]} = \begin{cases} \displaystyle\int_{[1]} \frac{dL_i}{dx}\frac{dL_j}{dx}\,dx & i,j = 1,2 \\ 0 & i \text{ or } j = 3 \end{cases}$$

from which, using the definitions (3.3) and (3.4) of the linear shape functions, it is a trivial calculation to show that

$$[K^{[1]}] = \begin{bmatrix} 1 & -1 & 0 \\ -1 & 1 & 0 \\ 0 & 0 & 0 \end{bmatrix}$$

is the element stiffness matrix for element [1].

Similarly, within element [2],

$$R_1(x) = 0, \qquad R_2(x) = L_1(x), \qquad R_3(x) = L_2(x)$$

(where we are using **local** node numbers for the shape functions).

Hence

$$K_{ij}^{[2]} = \begin{cases} 0 & i \text{ or } j = 1 \\ \displaystyle\int_{[2]} \frac{dL_{i-1}}{dx}\frac{dL_{j-1}}{dx}\,dx & i,j = 2,3 \end{cases}$$

whence

$$[K^{[2]}] = \begin{bmatrix} 0 & 0 & 0 \\ 0 & 1 & -1 \\ 0 & -1 & 1 \end{bmatrix}$$

By simple matrix addition we obtain the global stiffness matrix

$$[K] = [K^{[1]}] + [K^{[2]}] = \begin{bmatrix} 1 & -1 & 0 \\ -1 & 2 & -1 \\ 0 & -1 & 1 \end{bmatrix} \tag{4.17}$$

Proceeding in a similar manner for the components of $\{F\}$ we have

$$F_1 = F_1^{[1]} + F_1^{[2]}$$

where, from (4.14),

$$\begin{aligned} F_1 &= -\int_{[1]} 2\,R_1\,dx - \int_{[2]} 2\,R_1\,dx \\ &= -2\int_{[1]} L_1\,dx + 0 \\ &= -2\left(\frac{1}{2}\right) = -1 \end{aligned}$$

Similarly

$$\begin{aligned} F_2 &= -\int_{[1]} 2\,R_2\,dx - \int_{[2]} 2\,R_2\,dx \\ &= -1 - 1 \\ &= -2 \end{aligned}$$

and

$$F_3 = -1$$

Hence

$$\{F\} = \begin{Bmatrix} -1 \\ -2 \\ -1 \end{Bmatrix} \tag{4.18}$$

and the system (4.16) becomes

$$\begin{bmatrix} 1 & -1 & 0 \\ -1 & 2 & -1 \\ 0 & -1 & 1 \end{bmatrix} \begin{Bmatrix} \tilde{u}_1 \\ \tilde{u}_2 \\ \tilde{u}_3 \end{Bmatrix} = \begin{Bmatrix} -1 \\ -2 \\ -1 \end{Bmatrix} + \begin{Bmatrix} 0 \\ 0 \\ 6 \end{Bmatrix} \tag{4.19}$$

At this stage we insert the given essential boundary condition $\tilde{u}_1 = u(x = 1) = 1$ so that we may ignore the first row of (4.19) (remember, it is invalid), substitute for $\tilde{u}_1$ in the second row and obtain the reduced system

$$\begin{bmatrix} 2 & -1 \\ -1 & 1 \end{bmatrix} \begin{Bmatrix} \tilde{u}_2 \\ \tilde{u}_3 \end{Bmatrix} = \begin{Bmatrix} -1 \\ 5 \end{Bmatrix}$$

which of course agrees with the system (4.9) obtained via the roof function approach.

Note that while the solution to this system gives the correct nodal values

$$\begin{aligned} \tilde{u}_2 &= u(x = 2) = 4 \\ \tilde{u}_3 &= u(x = 3) = 9 \end{aligned}$$

the exact solution $u(x) = x^2$ is not obtained at non-nodal points. This is immediately apparent from Figure 4.4 which displays exact and approximate solutions.

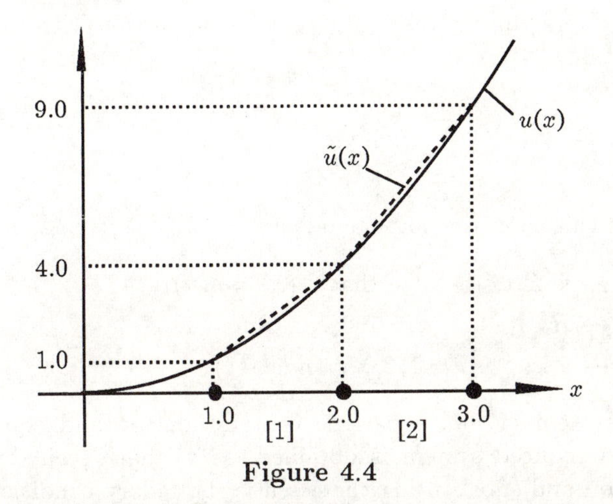

Figure 4.4

Clearly we would not expect to obtain a correct quadratic solution using linear approximations. It is in fact a characteristic of finite element approximations that the best accuracy is achieved at the nodal points.

Note finally that the same final system of linear equations as that obtained here by Galerkin techniques can be obtained by a quite different approach, via variational methods. This approach will be studied in Chapter 8.

Coordinate functions and roof functions

As we have seen, the basis of the matrix formulation of a one-dimensional finite element solution is the replacement of the functions $\theta_0(x), \theta_i(x)$, $i = 1, 2, \ldots$ which arise in a modified Galerkin formulation (such as (4.2)) by roof functions; this process is followed in turn, by the replacement of the roof functions by shape functions in each element so as to evaluate the integrals which define each matrix component.

Now, as detailed in the second supplement to Chapter 2, there are four possible sets of boundary conditions that can arise in solving a second-order boundary value problem over a domain $a \leq x \leq b$. We assume a finite element mesh with *nde* nodes so that $\tilde{u}_1 = \tilde{u}(a)$ and $\tilde{u}_{nde} = \tilde{u}(b)$.

- essential boundary condition at $x = a$, suppressible boundary condition at $x = b$ (as in the above example).

In this case the trial solution

$$\tilde{u}(x) = \sum_{i=1}^{nde} \tilde{u}_i R_i(x)$$

involves $(nde - 1)$ unknowns $\tilde{u}_2, \ldots, \tilde{u}_{nde}$. Comparison with the Galerkin approximation (2.2)

$$\tilde{u}(x) = \theta_0(x) + \sum_{i=1}^{n} a_i \theta_i(x)$$

implies, for this case, the correspondence

$$\begin{aligned} \theta_0(x) &\leftrightarrow \tilde{u}_1 R_1(x) && (\text{so that } \theta_0(x) = \tilde{u}(a)) \\ \theta_i(x) &\leftrightarrow R_{i+1}(x) && i = 1, 2, \ldots, (nde - 1) \\ a_i &\leftrightarrow \tilde{u}_{i+1} && i = 1, 2, \ldots, (nde - 1) \end{aligned}$$

The final system of nde equations contains one invalid equation (the first) and a reduced system is obtained, as we have seen, by ignoring this equation and substituting the essential boundary condition into the remainder.

• essential boundary conditions at $x = a$ and $x = b$.

In this case there are $(nde - 2)$ unknowns $\tilde{u}_2, \ldots, \tilde{u}_{nde-1}$. The correspondence is

$$\begin{aligned} \theta_0(x) &\leftrightarrow \tilde{u}(a) R_1(x) + \tilde{u}(b) R_{nde}(x) \\ \theta_i(x) &\leftrightarrow R_{i+1}(x) && i = 1, 2, \ldots, (nde - 2) \\ a_i &\leftrightarrow \tilde{u}_{i+1} && i = 1, 2, \ldots, (nde - 2) \end{aligned}$$

The nde equations now contain two invalid equations – the first and last – and are reduced to $(nde - 2)$ equations using the essential boundary conditions

• suppressible boundary conditions at $x = a$ and $x = b$.

In this case all nodal values are unknown. A function θ_0 is not required (since the suppressible conditions have already been dealt with). The correspondence is

$$\begin{aligned} \theta_i(x) &\leftrightarrow R_i(x) && i = 1, 2, \ldots, nde \\ a_i(x) &\leftrightarrow \tilde{u}_i(x) && i = 1, 2, \ldots, nde \end{aligned}$$

The final nde equations are all valid.

• suppressible boundary condition at $x = a$, essential boundary condition at $x = b$.

In this case there are $(nde - 1)$ unknowns $\tilde{u}_1, \ldots, \tilde{u}_{nde-1}$. The final equation of the system is invalid.

Assembling $[\mathrm{K}]$ and $\{\mathrm{F}\}$

In the example above we have obtained the global matrix $[K]$ by literally adding the individual element matrices $[K^{[e]}]$ and have similarly found $\{F\}$ from the individual vectors $\{F^{[e]}\}$. The size of $[K]$ was 3×3 or, more generally, $nde \times nde$ and $\{F\}$ is of dimension $(nde \times 1)$.

However each element matrix $[K^{[e]}]$ contains, for linear elements, only four non-zero terms because of the properties of the roof functions used in the definition of the components $K_{ij}^{[e]}$. The construction of $[K]$ from the individual $[K^{[e]}]$ can thus alternatively be carried out by the following process: we write out only the non-zero components of each $[K^{[e]}]$ but label each component with the **global** node numbers of the element to ensure the correct positioning of that component in the global matrix $[K]$.

Thus, in the above two-element three-node example, we would write

$$[K^{[1]}] = \begin{bmatrix} 1 & -1 \\ -1 & 1 \end{bmatrix} \begin{matrix} \scriptstyle 1 \\ \scriptstyle 2 \end{matrix} \qquad [K^{[2]}] = \begin{bmatrix} 1 & -1 \\ -1 & 1 \end{bmatrix} \begin{matrix} \scriptstyle 2 \\ \scriptstyle 3 \end{matrix}$$

and copy these successively into a 3×3 matrix which is initialized at zero:

$$[K] = \begin{bmatrix} 0 & 0 & 0 \\ 0 & 0 & 0 \\ 0 & 0 & 0 \end{bmatrix} \begin{matrix} \scriptstyle 1 \\ \scriptstyle 2 \\ \scriptstyle 3 \end{matrix}$$

Copying in $[K^{[1]}]$ produces

$$[K] = \begin{bmatrix} 1 & -1 & 0 \\ -1 & 1 & 0 \\ 0 & 0 & 0 \end{bmatrix} \begin{matrix} \scriptstyle 1 \\ \scriptstyle 2 \\ \end{matrix}$$

then copying in $[K^{[2]}]$ produces

$$[K] = \begin{bmatrix} 1 & -1 & 0 \\ -1 & 1+1 & 0 \\ 0 & -1 & 1 \end{bmatrix} \begin{matrix} \\ \scriptstyle 2 \\ \scriptstyle 3 \end{matrix}$$

which is (4.17). This process can be readily extended to more elements.

A similar approach can be used to obtain the vector $\{F\}$. The reader is invited to carry out the details to obtain (4.18).

4.4 Quadratic solution for Poisson's equation

We have noted that the boundary value problem that we are studying:

$$\frac{d^2u}{dx^2} = 2 \qquad 1 \le x \le 3$$

$$u(1) = 1 \qquad \text{and} \qquad u'(3) = 6$$

has the exact solution $u = x^2$. Hence we may expect to obtain the exact solution at **all** points using a quadratic 'approximation'.

We will simply take the complete domain to be one element and use quadratic shape functions i.e. assume the trial solution:

$$\tilde{u}(x) = \tilde{u}_1\, Q_1(x) + \tilde{u}_2\, Q_2(x) + \tilde{u}_3\, Q_3(x) \tag{4.20}$$

and the model shown in Figure 4.5.

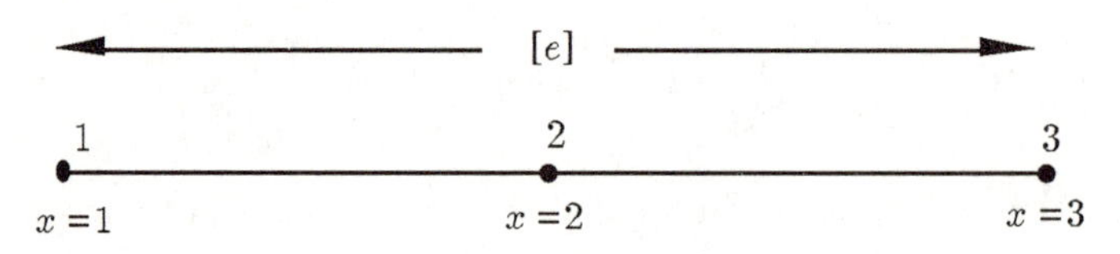

Figure 4.5

Comparison with (4.6) tells us immediately that for this case the link between roof functions and shape functions is simply

$$R_i(x) = Q_i(x), \; i = 1, 2, 3$$

Hence, using (4.13), the components of the stiffness matrix are

$$K_{ij} = \int_{[e]} \frac{dQ_i}{dx}\frac{dQ_j}{dx}\, dx \qquad i, j = 1, 2, 3 \tag{4.21}$$

and, from (4.14), the components of the force vector are

$$F_i = -\int_{[e]} 2\, Q_i\, dx \qquad i = 1, 2, 3 \tag{4.22}$$

All the integrals needed for evaluating (4.21) and (4.22) have already been quoted in Section 3.7. Using (3.18) with $\ell = 2$ here we find

$$[K] = \frac{1}{6}\begin{bmatrix} 7 & -8 & 1 \\ -8 & 16 & -8 \\ 1 & -8 & 7 \end{bmatrix}$$

and, by (3.19),

$$[F] = \{-2/3, -8/3, -2/3\}^T$$

Hence the governing system (4.16) becomes

$$\begin{bmatrix} 7 & -8 & 1 \\ -8 & 16 & -8 \\ 1 & -8 & 7 \end{bmatrix}\begin{Bmatrix} \tilde{u}_1 \\ \tilde{u}_2 \\ \tilde{u}_3 \end{Bmatrix} = \begin{Bmatrix} -4 \\ -16 \\ +32 \end{Bmatrix}$$

Inserting the essential condition $\tilde{u}_1 = 1$ (and again ignoring the first, invalid, equation) we have the 2×2 system

$$\begin{bmatrix} 16 & -8 \\ -8 & 7 \end{bmatrix}\begin{Bmatrix} \tilde{u}_2 \\ \tilde{u}_3 \end{Bmatrix} = \begin{Bmatrix} -8 \\ 31 \end{Bmatrix}$$

with the solutions $\tilde{u}_2 = 4$ and $\tilde{u}_3 = 9$, these of course being the exact solution at the nodes.

Also, using these calculated nodal values in the trial solution (4.20) and using the normalized variable $\xi = x - 2$ we obtain

$$\begin{aligned} \tilde{u}(\xi) &= Q_1(\xi) + 4\,Q_2(\xi) + 9\,Q_3(\xi) \\ &= \frac{\xi}{2}(\xi - 1) + 4\,(1 - \xi^2) + 9\,\frac{\xi}{2}(\xi + 1) \\ &= \xi^2 + 4\,\xi + 4 \\ &= (\xi + 2)^2 \end{aligned}$$

so that $\tilde{u}(x) = x^2$. In other words, the quadratic approximation gives us, as expected, the exact solution at **all** points in the domain.

4.5 A general one-dimensional equilibrium problem

The differential equation

$$D\,\frac{d^2u}{dx^2} + Gu + Q = 0 \qquad \textbf{(4.23a)}$$

arises in a variety of situations in science and engineering involving time-independent phenomena. The quantities D, G and Q must be constants or functions only of x if (4.23a) is to be a linear equation. In our examples, they will all be assumed constant. The special case $G \equiv 0$ is Poisson's equation:

$$D\frac{d^2u}{dx^2} = -Q$$

and we have already discussed finite element formulations for this problem.

Another special case viz. $Q \equiv 0$ gives rise to **Helmholtz's equation**:

$$D\frac{d^2u}{dx^2} + Gu = 0$$

which arises in vibration theory.

An application of (4.23a) where all three terms are present is in heat transfer where, with appropriate interpretations of D, G and Q, the equation is used as a model for calculating the temperature of a one-dimensional fin which is losing heat by both conduction and convection to its surroundings.

These applications will be discussed later but we will first study the general finite element formulation of boundary value problems involving (4.23a).

As a specific problem we will use the boundary conditions

$$u(0) = 0 \qquad \text{(essential)} \qquad \textbf{(4.23b)}$$

$$\frac{du}{dx}(1) = 2 \quad \text{(suppressible)} \qquad \textbf{(4.23c)}$$

and hence seek solutions over the domain $0 \leq x \leq 1$. The modified Galerkin equations appropriate to the boundary value problem (4.23) are easily obtained (either directly, as for boundary value problem (4.1), or by using the results derived in the second supplement to Chapter 2). We find

$$\int_0^1 \left\{-D\frac{d\tilde{u}}{dx}\frac{d\theta_i}{dx} + G\tilde{u}\theta_i + Q\theta_i\right\} dx + 2D\theta_i(1) = 0 \qquad i = 1, \ldots, n \quad \textbf{(4.24)}$$

We will firstly segment the domain into three linear elements of equal length i.e. $\ell = \ell^{[1]} = \ell^{[2]} = \ell^{[3]} = 1/3$ (see Figure 4.6).

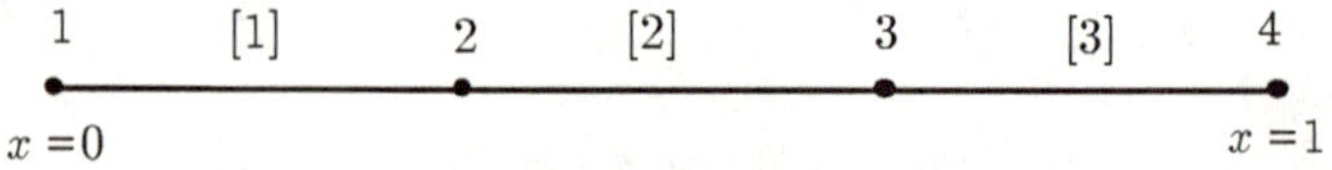

1 [1] 2 [2] 3 [3] 4

$x = 0$ $x = 1$

Figure 4.6

As there are four nodes we seek a trial solution of the form:

$$\tilde{u}(x) = \sum_{j=1}^{4} R_j(x)\,\tilde{u}_j \qquad \textbf{(4.25)}$$

where $R_j(x)$ is the roof function associated with node j (see Figure 4.7).

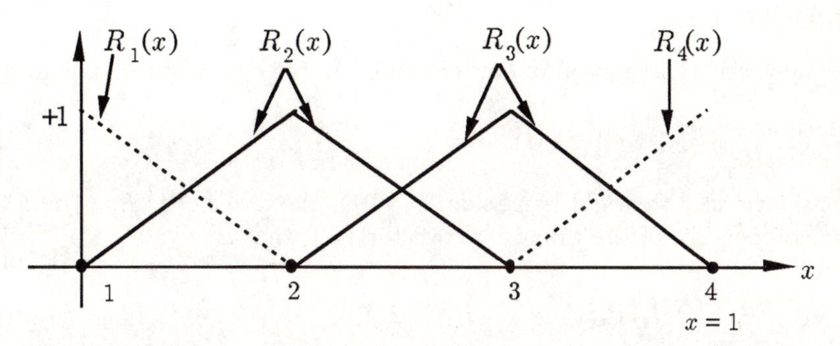

Figure 4.7

Note again that (4.25) is analogous to the Galerkin form (2.2) with

$$\theta_0(x) \longleftrightarrow R_1(x)\,\tilde{u}_1$$

so that

$$\theta_0(0) = (1)(0) = 0$$

that is, $\theta_0(x)$ satisfies, as it must, the essential boundary condition (4.23b) at $x = 0$.

Also by comparison of (4.25) with (2.2)

$$\theta_i(x) \longleftrightarrow R_{i+1}(x) \qquad i = 1, 2, 3$$
$$a_i \longleftrightarrow \tilde{u}_{i+1} \qquad i = 1, 2, 3$$

and clearly

$$\theta_i(0) = 0 \qquad i = 1, 2, 3$$

from the properties of the roof functions.

We now use the trial solution (4.25) and **replace** the coordinate functions θ_i in (4.24) with R_i. This gives (multiplying through by -1)

$$\int_0^1 \{D\frac{d\tilde{u}}{dx}\frac{dR_i}{dx} - G\tilde{u}R_i - QR_i\}dx - 2DR_i(1) = 0 \qquad i = 1, 2, 3, 4 \quad \textbf{(4.26)}$$

or, rearranging

$$D\int_0^1 \frac{d\tilde{u}}{dx}\frac{dR_i}{dx}dx - G\int_0^1 \tilde{u}\,R_i\,dx$$

$$= Q\int_0^1 R_i(x)\,dx + 2DR_i(1) \qquad i = 1,2,3,4 \qquad \textbf{(4.27)}$$

We now write this system of four equations in the equivalent matrix form

$$[K]\{\tilde{u}\} = \{F\} + \{B\} \qquad \textbf{(4.28)}$$

Consider firstly the right hand side of (4.27): the terms $Q\int_0^1 R_i(x)\,dx$ are the components of the global force vector $\{F\}$, that is

$$F_i = Q\int_0^1 R_i(x)\,dx \qquad i = 1,2,3,4 \qquad \textbf{(4.29)}$$

These integrals, and those needed to obtain $[K]$, are readily found using the following connections between roof functions and shape functions within the three elements (see Figure 4.7).

Element	$R_1(x)$	$R_2(x)$	$R_3(x)$	$R_4(x)$
[1]	$L_1(x)$	$L_2(x)$	0	0
[2]	0	$L_1(x)$	$L_2(x)$	0
[3]	0	0	$L_1(x)$	$L_2(x)$

Note that we have used **local** node numbers for the shape functions.

Hence, from (4.29),

$$F_1 = Q\int_0^1 R_1(x)\,dx$$

$$= Q\int_{[1]} L_1(x)\,dx = \frac{Q\ell}{2}$$

$$F_2 = Q\left\{\int_{[1]} L_2(x)\,dx + \int_{[2]} L_1(x)\,dx\right\} = Q\ell$$

Similarly $F_3 = Q\ell$ and $F_4 = \dfrac{Q\ell}{2}$

that is,

$$\{F\} = \frac{Q\ell}{2}\begin{Bmatrix}1\\2\\2\\1\end{Bmatrix} \tag{4.30}$$

is the global force vector.

The quantity $\{B\}$ in (4.28) has, by (4.27), components

$$B_i = 2DR_i(1) \qquad i = 1, 2, 3, 4$$

Using the properties of the roof functions we obtain

$$\{B\} = \{0, 0, 0, 2D\}^T \tag{4.31}$$

Calculation of the global stiffness matrix [K]

This matrix arises from the left hand side of the system (4.27) which, when we use (4.25) and its derivative

$$\frac{d\tilde{u}}{dx} = \sum_{j=1}^{4} \frac{dR_j}{dx}\,\tilde{u}_j$$

has the form

$$\sum_{j=1}^{4}\left\{D\int_0^1 \frac{dR_j}{dx}\frac{dR_i}{dx}dx\right\}\tilde{u}_j - \sum_{j=1}^{4}\left\{G\int_0^1 R_j\,R_i\,dx\right\}\tilde{u}_j \tag{4.32}$$

Now, putting

$$K_{ij}^D = D\int_0^1 \frac{dR_j}{dx}\frac{dR_i}{dx}\,dx \tag{4.33}$$

and

$$K_{ij}^G = G\int_0^1 R_j\,R_i\,dx \tag{4.34}$$

we see that (4.32) has the form

$$\sum_{j=1}^{4} (K_{ij}^{D} - K_{ij}^{G})\, \tilde{u}_j \equiv \sum_{j=1}^{4} K_{ij}\, \tilde{u}_j \quad \text{say}$$

In other words the global stiffness matrix $[K]$ has two components

$$[K] = [K^D] - [K^G]$$

The calculation of the matrix $[K^D]$ from (4.33) is essentially the same as the calculation of the matrix $[K]$ arising from Poisson's equation in Section 4.3. In the current problem with three elements of length $\ell = 1/3$ we have

$$[K^D] = [K^{D[1]}] + [K^{D[2]}] + [K^{D[3]}]$$

where

$$[K^{D[1]}] = \frac{D}{\ell}\begin{bmatrix} 1 & -1 & 0 & 0 \\ -1 & 1 & 0 & 0 \\ 0 & 0 & 0 & 0 \\ 0 & 0 & 0 & 0 \end{bmatrix} \tag{4.35}$$

$$[K^{D[2]}] = \frac{D}{\ell}\begin{bmatrix} 0 & 0 & 0 & 0 \\ 0 & 1 & -1 & 0 \\ 0 & -1 & 1 & 0 \\ 0 & 0 & 0 & 0 \end{bmatrix} \tag{4.36}$$

$$[K^{D[3]}] = \frac{D}{\ell}\begin{bmatrix} 0 & 0 & 0 & 0 \\ 0 & 0 & 0 & 0 \\ 0 & 0 & 1 & -1 \\ 0 & 0 & -1 & 1 \end{bmatrix} \tag{4.37}$$

so that, on adding these three element matrices,

$$[K^D] = \frac{D}{\ell}\begin{bmatrix} 1 & -1 & 0 & 0 \\ -1 & 2 & -1 & 0 \\ 0 & -1 & 2 & -1 \\ 0 & 0 & -1 & 1 \end{bmatrix} \tag{4.38}$$

The elements of the matrix $[K^G]$ are calculated similarly from (4.34) using the link between roof functions and shape functions already discussed.

Clearly

$$[K^G] = \sum_{e=1}^{3} [K^{G[e]}]$$

where, firstly,

$$K_{ij}^{G[1]} = \begin{cases} G\displaystyle\int_{[1]} R_i R_j \, dx & i,j = 1,2 \\ 0 & \text{otherwise} \end{cases}$$

(because only roof functions $R_1(x)$ and $R_2(x)$ are non-zero in element [1]). Hence

$$K_{ij}^{G[1]} = \begin{cases} G\displaystyle\int_{[1]} L_i L_j \, dx & i,j = 1,2 \\ 0 & \text{otherwise} \end{cases}$$

Using (3.10) and (3.11) to evaluate the integrals gives

$$[K^{G[1]}] = \frac{G\ell}{6} \begin{bmatrix} 2 & 1 & 0 & 0 \\ 1 & 2 & 0 & 0 \\ 0 & 0 & 0 & 0 \\ 0 & 0 & 0 & 0 \end{bmatrix} \qquad \textbf{(4.39)}$$

Also

$$K_{ij}^{G[2]} = \begin{cases} G\displaystyle\int_{[2]} R_i R_j \, dx & i,j = 2,3 \\ 0 & \text{otherwise} \end{cases}$$

(because only roof functions $R_2(x)$ and $R_3(x)$ are non-zero in element [2]).

The four individual integrals here are the same as those that arise in the matrix $K^{G[1]}$ because, in element [2], the roof functions are such that $R_2(x) \equiv L_1(x)$ and $R_3(x) \equiv L_2(x)$.

Hence

$$[K^{G[2]}] = \frac{G\ell}{6} \begin{bmatrix} 0 & 0 & 0 & 0 \\ 0 & 2 & 1 & 0 \\ 0 & 1 & 2 & 0 \\ 0 & 0 & 0 & 0 \end{bmatrix} \qquad \textbf{(4.40)}$$

By similar reasoning

$$[K^{G[3]}] = \frac{G\ell}{6} \begin{bmatrix} 0 & 0 & 0 & 0 \\ 0 & 0 & 0 & 0 \\ 0 & 0 & 2 & 1 \\ 0 & 0 & 1 & 2 \end{bmatrix} \qquad \textbf{(4.41)}$$

so, after addition, we have

$$[K^G] = \frac{G\ell}{6}\begin{bmatrix} 2 & 1 & 0 & 0 \\ 1 & 4 & 1 & 0 \\ 0 & 1 & 4 & 1 \\ 0 & 0 & 1 & 2 \end{bmatrix} \tag{4.42}$$

Subtracting $[K^G]$ in (4.42) from $[K^D]$ in (4.38) gives us the stiffness matrix $[K]$ and hence, together with $\{F\}$ from (4.30) and $\{B\}$ from (4.31), we now have the full 4×4 system:

$$[K]\{\tilde{u}\} = \{F\} + \{B\} \tag{4.43}$$

Finally, the insertion of the essential boundary condition $\tilde{u}_1 = 0$ enables us to reduce the system (4.43) to

$$\frac{D}{\ell}\begin{bmatrix} 2 & -1 & 0 \\ -1 & 2 & -1 \\ 0 & -1 & 1 \end{bmatrix}\begin{Bmatrix} \tilde{u}_2 \\ \tilde{u}_3 \\ \tilde{u}_4 \end{Bmatrix} - \frac{G\ell}{6}\begin{bmatrix} 4 & 1 & 0 \\ 1 & 4 & 1 \\ 0 & 1 & 2 \end{bmatrix}\begin{Bmatrix} \tilde{u}_2 \\ \tilde{u}_3 \\ \tilde{u}_4 \end{Bmatrix}$$

$$= \frac{Q\ell}{2}\begin{Bmatrix} 2 \\ 2 \\ 1 \end{Bmatrix} + \begin{Bmatrix} 0 \\ 0 \\ 2D \end{Bmatrix}$$

which, after insertion of numerical values for the parameters and putting $\ell = 1/3$, would be soluble by standard numerical techniques such as Gaussian elimination.

Note that we have ignored the invalid first equation in (4.43).

4.6 Summary of the finite element approach

It is perhaps useful at this point to review the main stages in using finite elements to solve second-order one-dimensional linear boundary value problems.

- 1 An equivalent, integral form, of the problem is obtained using Galerkin's weighted residual method and integration by parts. (Modified Galerkin method.)

- 2 Derivative boundary conditions, if any, are absorbed. This gives rise to the column vector $\{B\}$.

- 3 The domain of the problem is subdivided (or 'meshed' or 'discretized') into smaller segments ('elements').

- 4 Nodes and elements are numbered.

- 5 The unknown function $u(x)$ is approximated by a simple polynomial (usually linear or quadratic) within each element. This involves the introduction of shape functions.

- 6 An approximation valid over the whole domain is obtained using roof functions which are combinations of shape functions.

- 7 The integrals are evaluated as a sum of integrals over each element.

- 8 Using a matrix formulation the process leads to the construction of a symmetric global matrix $[K]$ and global column vector $\{F\}$, calculated in terms of individual element quantities.

- 9 Essential boundary conditions are imposed to produce a reduced system of linear equations.

- 10 The system of equations is solved for the nodal values by standard numerical techniques.

- 11 Values of u at non-nodal points are estimated, if needed, using the calculated nodal values and the assumed element approximation of u.

4.7 Application of finite elements to a vibration problem

One dimensional oscillations – for example the transverse oscillations of a string, the torsional oscillations of a shaft, the longitudinal vibrations of a bar – are governed by the **one-dimensional wave equation**

$$\frac{\partial^2 U}{\partial x^2} = \frac{1}{v^2}\frac{\partial^2 U}{\partial t^2} \tag{4.44}$$

where U is the displacement (an angular displacement in the case of torsional oscillations) and v is a constant. An outline of the derivation of (4.44) and its two-dimensional equivalent is given in the third supplement to this chapter.

In (4.44) U is a function of distance x and time t. However, a particularly common situation is one in which the time dependence is sinusoidal with frequency ω and period $T = 2\pi/\omega$. In this case using complex exponentials we can put

$$U(x,t) = u(x)e^{i\omega t}$$

so that (4.44) becomes the ordinary differential equation

$$\frac{d^2u}{dx^2} = -\frac{\omega^2}{v^2}u$$

or

$$\frac{d^2u}{dx^2} + \beta^2 u = 0 \tag{4.45}$$

where $\beta^2 = \omega^2/v^2$.

We recognize (4.45) as one case ($Q = 0$) of the general one-dimensional equation (4.23a) whose finite element formulation with linear elements has been fully discussed in the previous section. The parameters D and G in (4.23) are, of course, 1 and β^2 respectively here. Equation (4.45) is called Helmholtz's equation.

Let us consider the specific boundary value problem

$$\frac{d^2u}{dx^2} + \beta^2 u = 0 \qquad 0 < x < 1 \tag{4.46a}$$

$$u(0) = u(1) = 0 \tag{4.46b}$$

Physically this corresponds to oscillations with fixed end-points and is an example of a homogeneous boundary value problem as introduced in Chapter 1.

The analytical solution to this problem is straightforward since the general solution to (4.46a) is readily shown to be

$$u = a\cos\beta x + b\sin\beta x$$

The boundary conditions (4.46b) tell us that $a = 0$ and $\beta = n\pi$; in other words the only possible **modes** of oscillation are

$$u = b\sin n\pi x \qquad n = 1, 2, 3....$$

In more mathematical language the permitted values $\beta = n\pi$ are known as **eigenvalues** and correspond physically to the fact that only specific frequencies of oscillation are permitted viz. $\omega = \omega_n$ where

$$\omega_n^2 = n^2\pi^2 v^2 \qquad n = 1, 2, 3, ...$$

The solutions

$$u = u_n = b_n \sin n\pi x \qquad n = 1, 2, 3, \ldots$$

are known as **eigenvectors**.

Finite element solution

Let us consider a finite element model for the boundary value problem (4.46) involving segmentation of the domain $0 \leq x \leq 1$ into four linear elements of equal length $\ell = 1/4$ (see Figure 4.8).

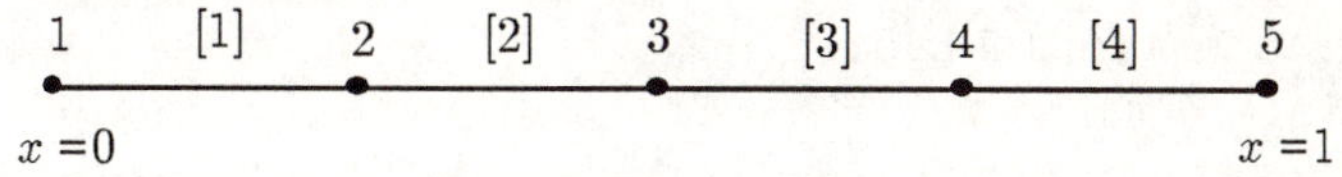

1 [1] 2 [2] 3 [3] 4 [4] 5

$x = 0$ $x = 1$

Figure 4.8

The development of a system of linear equations for the nodal values $\tilde{u}_1, \tilde{u}_2, \ldots, \tilde{u}_5$ is similar to that carried out in Section 4.5 for the general equilibrium problem. The only differences to note are:

1. Since the right hand side of (4.46a) is zero the force vector $\{F\}$ is identically zero.
2. Since both boundary conditions (4.46b) are essential the column vector $\{B\}$ is zero.
3. The individual element matrices $[K^{D[e]}]$ and $[K^{G[e]}]$ will be 5×5 matrices since there are five nodes and the global matrices $[K^D]$ and $[K^G]$ will each be the sum of four such element matrices.
   Thus,

$$[K^D] = \sum_{e=1}^{4} [K^{D[e]}]$$

where, for example,

$$[K^{D[1]}] = 4 \begin{bmatrix} 1 & -1 & 0 & 0 & 0 \\ -1 & 1 & 0 & 0 & 0 \\ 0 & 0 & 0 & 0 & 0 \\ 0 & 0 & 0 & 0 & 0 \\ 0 & 0 & 0 & 0 & 0 \end{bmatrix}$$

(compare (4.35)) and

$$[K^{D[4]}] = 4\begin{bmatrix} 0 & 0 & 0 & 0 & 0 \\ 0 & 0 & 0 & 0 & 0 \\ 0 & 0 & 0 & 0 & 0 \\ 0 & 0 & 0 & 1 & -1 \\ 0 & 0 & 0 & -1 & 1 \end{bmatrix}$$

Adding the four element matrices gives

$$[K^D] = 4\begin{bmatrix} 1 & -1 & 0 & 0 & 0 \\ -1 & 2 & -1 & 0 & 0 \\ 0 & -1 & 2 & -1 & 0 \\ 0 & 0 & -1 & 2 & -1 \\ 0 & 0 & 0 & -1 & 1 \end{bmatrix}$$

Similarly

$$[K^G] = \sum_{e=1}^{4} [K^{G[e]}]$$

where, for example,

$$[K^{G[1]}] = \frac{\beta^2}{24}\begin{bmatrix} 2 & 1 & 0 & 0 & 0 \\ 1 & 2 & 0 & 0 & 0 \\ 0 & 0 & 0 & 0 & 0 \\ 0 & 0 & 0 & 0 & 0 \\ 0 & 0 & 0 & 0 & 0 \end{bmatrix}$$

(compare (4.39)).
Adding the four element matrices gives

$$[K^G] = \frac{\beta^2}{24}\begin{bmatrix} 2 & 1 & 0 & 0 & 0 \\ 1 & 4 & 1 & 0 & 0 \\ 0 & 1 & 4 & 1 & 0 \\ 0 & 0 & 1 & 4 & 1 \\ 0 & 0 & 0 & 1 & 2 \end{bmatrix}$$

Finally as noted earlier $\{B\} = 0$ and so the final system of equations has the form

$$[K]\{\tilde{u}\} = \{0\} \qquad \textbf{(4.47a)}$$

where $[K] = [K^D] - [K^G]$. The insertion of the essential boundary conditions

$$\tilde{u}_1 = u(x = 0) = 0 \qquad \text{and} \qquad \tilde{u}_5 = u(x = 1) = 0$$

gives the final reduced system

$$4\begin{bmatrix} 2 & -1 & 0 \\ -1 & 2 & -1 \\ 0 & -1 & 2 \end{bmatrix}\begin{Bmatrix} \tilde{u}_2 \\ \tilde{u}_3 \\ \tilde{u}_4 \end{Bmatrix} = \frac{\beta^2}{24}\begin{bmatrix} 4 & 1 & 0 \\ 1 & 4 & 1 \\ 0 & 1 & 4 \end{bmatrix}\begin{Bmatrix} \tilde{u}_2 \\ \tilde{u}_3 \\ \tilde{u}_4 \end{Bmatrix} \qquad \textbf{(4.47b)}$$

where we have ignored both the first and fifth equations in the system (4.47a) since both are invalid. The homogeneous system of linear equations (4.47b) when written in the form $[A]\{\tilde{u}\} = 0$ will possess a non-trivial solution, i.e. not all the $\tilde{u}_i$'s zero, only if the determinant of $[A]$ is zero. That is, for non-trivial solutions we demand:

$$\begin{vmatrix} 8 - \frac{1}{6}\beta^2 & -4 - \frac{1}{24}\beta^2 & 0 \\ -4 - \frac{1}{24}\beta^2 & 8 - \frac{1}{6}\beta^2 & -4 - \frac{1}{24}\beta^2 \\ 0 & -4 - \frac{1}{24}\beta^2 & 8 - \frac{1}{6}\beta^2 \end{vmatrix} = 0$$

or

$$(8 - \frac{1}{6}\beta^2)\frac{1}{288}(7\beta^4 - 960\,\beta^2 + 9216) = 0$$

with solutions, in order of increasing frequencies,

$$\beta_1^2 = 10.39, \qquad \beta_2^2 = 48, \qquad \beta_3^2 = 126.76$$

These must be compared with the analytical values obtained earlier viz.

$$\beta_1^2 = \pi^2 = 9.87 \qquad \beta_2^2 = 4\pi^2 = 39.5 \qquad \beta_3^2 = 9\pi^2 = 88.8$$

The corresponding solutions obtained from (4.47b) are, respectively,

$$\begin{Bmatrix} \tilde{u}_2 \\ \tilde{u}_3 \\ \tilde{u}_4 \end{Bmatrix} = \begin{Bmatrix} \frac{1}{\sqrt{2}} \\ 1 \\ \frac{1}{\sqrt{2}} \end{Bmatrix}, \quad \begin{Bmatrix} 1 \\ 0 \\ -1 \end{Bmatrix}, \quad \begin{Bmatrix} \frac{1}{\sqrt{2}} \\ -1 \\ \frac{1}{\sqrt{2}} \end{Bmatrix}$$

Graphs of the lowest three vibration modes taken from the exact analysis (solid lines) are compared to those obtained from our finite element analysis (dashed lines) in Figure 4.9. Note that the general modal shapes are modelled quite well by the finite element solutions even though the accuracy obtained for the modal frequencies (apart from the lowest) is quite poor. If it is desired to improve the accuracy of the eigenvalue estimation then more elements must be employed.

For an analysis which used N linear elements, and hence $N + 1$ nodes, the lowest $N + 1$ modal frequencies would be approximated, the greatest accuracy being achieved for the lowest frequency with a corresponding decrease in accuracy as the frequency increases.

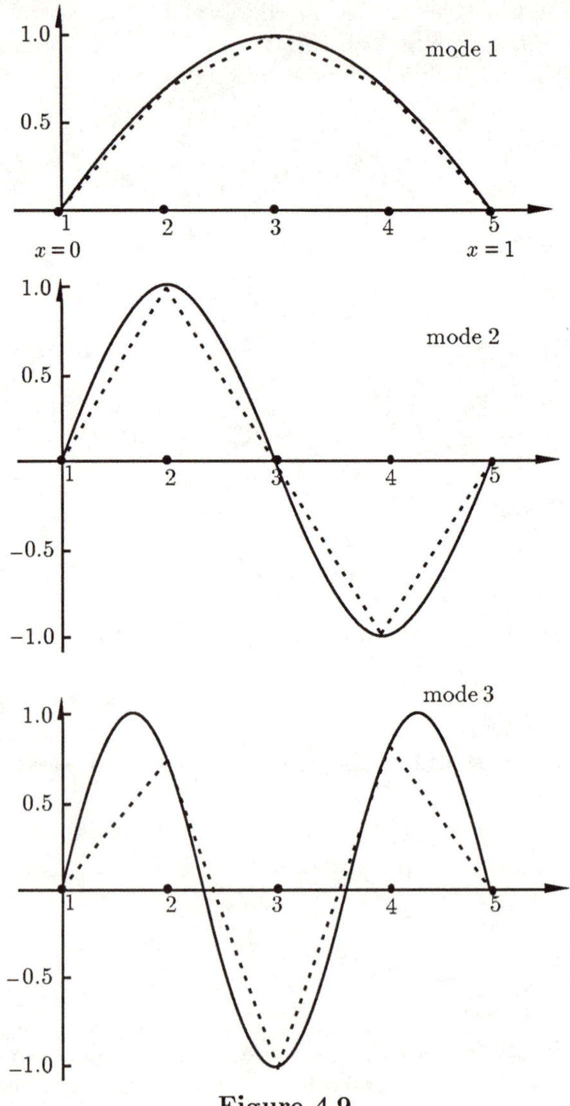

Figure 4.9

4.8 Finite elements in one-dimensional heat transfer

The governing differential equation for heat transfer by conduction and convection in one dimension is

$$k_x\, A \frac{d^2u}{dx^2} + h\, P\, (u_s - u) = 0 \tag{4.48}$$

The equation is often used, for example, to deduce the temperature u in a 'fin' (see Figure 4.10) whose material has a **thermal conductivity** k_x and also a **convection coefficient** h for transfer of heat by convection from the surface to the surroundings which are assumed to be at a constant temperature u_s. The fin is assumed to have a constant cross-sectional area A and perimeter P but it is perfectly possible to model a tapering fin using finite elements. For an outline discussion of heat transfer and derivation of (4.48) see the second supplement to this chapter.

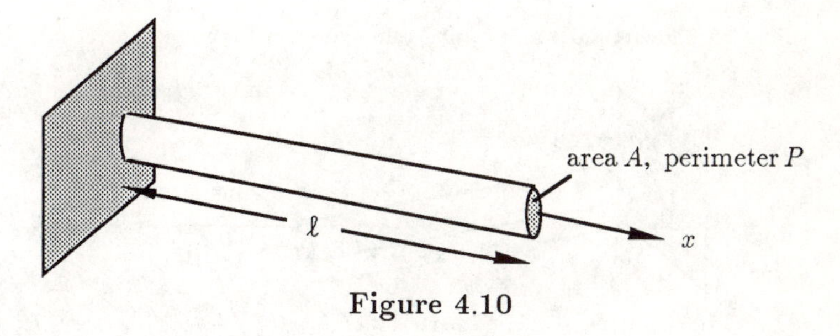

Figure 4.10

One end of the fin say $x = 0$ is connected as shown to a heat source at a known temperature. As well as heat being lost to the surroundings by convection from the surface of the fin it may also be lost by convection from the free end $x = \ell$. The mathematical description of the latter is by a boundary condition

$$-k_x\, A \frac{du}{dx} = h\, A\, (u_b - u_s) \qquad \text{at } x = \ell$$

where u_b denotes the temperature at the free end $x = \ell$. A simpler situation arises when the end $x = \ell$ is thermally insulated in which case we have

$$\frac{du}{dx} = 0 \qquad \text{at } x = \ell$$

In using (4.48) we are assuming that a steady state situation has been reached and that the temperature is independent of time.

Clearly (4.48) is a specific example of the general equation (4.23a) with

$$D = k_x A \qquad G = -hP \qquad Q = hPu_s$$

We will, for consistency, continue to write $[K^D]$ and $[K^G]$ for the overall stiffness matrices obtainable from the corresponding terms in (4.48).

For segmentation into linear elements, the matrices $[K^D]$ and $[K^G]$ have essentially been obtained in Section 4.5 as has the force vector $\{F\}$ which here comes from the term hPu_s in (4.48). We will now, however, demonstrate the use of both linear and quadratic elements by means of the following example.

Example 4.1

Consider the circular fin of radius 1cm shown in Figure 4.11.

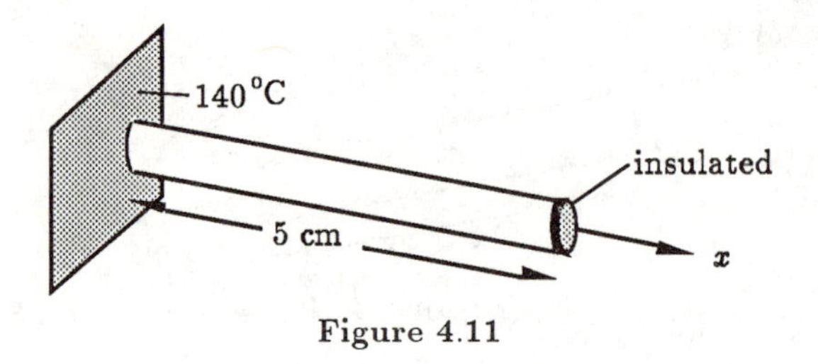

Figure 4.11

Assume parameter values for the material of the fin: thermal conductivity $k_x = 70\,\text{Wcm}^{-1}\,\text{K}^{-1}$, convection coefficient $h = 5\,\text{Wcm}^{-2}\,\text{K}^{-1}$. Assume the surrounding temperature is 40°C.
Using

(a) a finite element model consisting of two linear elements

(b) a finite element model of one quadratic element

(c) a finite element model involving two quadratic elements

obtain estimates of the temperature u at $x = 2.5\,\text{cm}$ and at the right hand end of the fin, $x = 5\,\text{cm}$.

Solution

In terms of a boundary value problem we are solving (4.48) with the given parameters and subject to the boundary conditions

$$u(0) = 140 \qquad u'(5) = 0$$

(a) Two linear elements model.

We have the model shown in Figure 4.12 where $\ell^{[e]} = 2.5\text{cm}$.

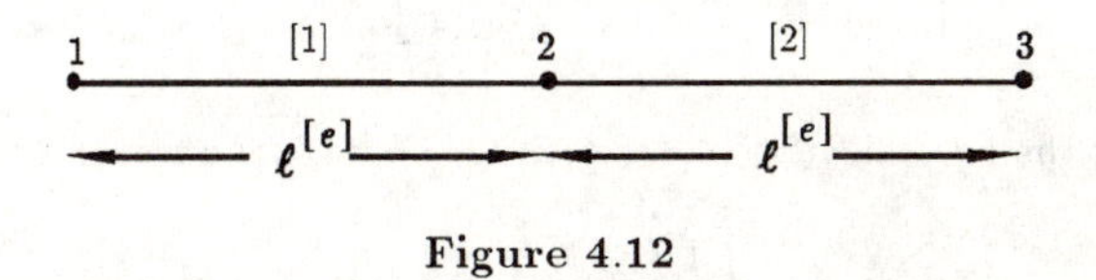

Figure 4.12

The global stiffness matrix is

$$[K] = [K^D] - [K^G]$$

Proceeding in exactly the same way as in Section 4.5 but with $D = k_x A$ and just two elements (three nodes):

$$[K^D] = \frac{k_x A}{\ell^{[e]}} \begin{bmatrix} 1 & -1 & 0 \\ -1 & 1 & 0 \\ 0 & 0 & 0 \end{bmatrix} + \frac{k_x A}{\ell^{[e]}} \begin{bmatrix} 0 & 0 & 0 \\ 0 & 1 & -1 \\ 0 & -1 & 1 \end{bmatrix}$$

(cf. (4.35) and (4.36))
Inserting given numerical values, $A = \pi(1^2)\,\text{cm}^2$; $k_x A/\ell^{[e]} = 87.96$ and so

$$[K^D] = 87.96 \begin{bmatrix} 1 & -1 & 0 \\ -1 & 2 & -1 \\ 0 & -1 & 1 \end{bmatrix}$$

Similarly, with $G = -hP$

$$[K^G] = -\frac{hP\ell^{[e]}}{6} \begin{bmatrix} 2 & 1 & 0 \\ 1 & 2 & 0 \\ 0 & 0 & 0 \end{bmatrix} - \frac{hP\ell^{[e]}}{6} \begin{bmatrix} 0 & 0 & 0 \\ 0 & 2 & 1 \\ 0 & 1 & 2 \end{bmatrix}$$

(cf. (4.39) and (4.40)).
Again inserting numerical values

$$\frac{hP\ell^{[e]}}{6} = 13.08$$

so

$$[K^{(G)}] = -13.08 \begin{bmatrix} 2 & 1 & 0 \\ 1 & 4 & 1 \\ 0 & 1 & 2 \end{bmatrix}$$

Hence

$$[K] = 87.96 \begin{bmatrix} 1 & -1 & 0 \\ -1 & 2 & -1 \\ 0 & -1 & 1 \end{bmatrix} + 13.08 \begin{bmatrix} 2 & 1 & 0 \\ 1 & 4 & 1 \\ 0 & 1 & 2 \end{bmatrix}$$

$$= \begin{bmatrix} 114.14 & -74.88 & 0 \\ -74.88 & 228.28 & -74.88 \\ 0 & -74.88 & 114.14 \end{bmatrix} \tag{4.49}$$

Using the equivalent of (4.30) with $Q = hPu_s$ we have

$$\{F\} = \frac{hPu_s}{2}\,\ell^{[e]} \begin{Bmatrix} 1 \\ 2 \\ 1 \end{Bmatrix}$$

$$= 1570.8 \begin{Bmatrix} 1 \\ 2 \\ 1 \end{Bmatrix} \tag{4.50}$$

Also, using the equivalent of (4.31) for this problem,

$$\{B\} = \begin{Bmatrix} 0 \\ 0 \\ +k_x\, A \dfrac{du}{dx}\Big|_{x=5} \end{Bmatrix} \tag{4.51}$$

We now have, from (4.49), (4.50) and (4.51) all the ingredients of the system

$$[K]\{\tilde{u}\} = \{F\} + \{B\}$$

where

$$\{\tilde{u}\} = \begin{Bmatrix} \tilde{u}_1 \\ \tilde{u}_2 \\ \tilde{u}_3 \end{Bmatrix}$$

Inserting both the suppressible boundary condition $[du/dx]_{x=5} = 0$ and the essential boundary condition $\tilde{u}_1 = 140$ and ignoring the invalid first equation we obtain the reduced system

$$\begin{bmatrix} 228.28 & -74.88 \\ -74.88 & 114.14 \end{bmatrix} \begin{Bmatrix} \tilde{u}_2 \\ \tilde{u}_3 \end{Bmatrix} = \begin{Bmatrix} 13624.8 \\ 1570.8 \end{Bmatrix}$$

which has solution

$$\tilde{u}_2 = u(x = 2.5) = 81.8°\text{C} \tag{4.52}$$

$$\tilde{u}_3 = u(x = 5.0) = 67.4°\text{C} \tag{4.53}$$

(b) One quadratic element model.

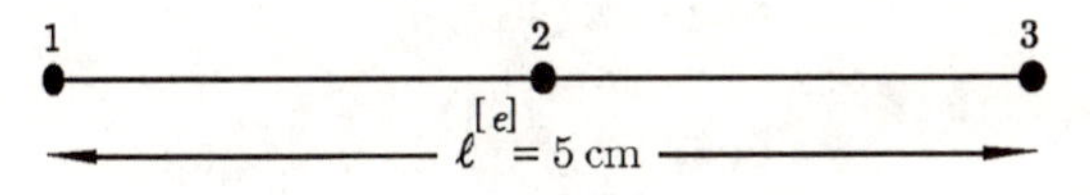

Figure 4.13

As already noted in Section 4.4, for a one-element model the shape functions are identical with the roof functions i.e.

$$R_i(x) = Q_i(x) \qquad i = 1, 2, 3$$

where the Q_i are the quadratic shape functions.

Hence the first of the two matrices making up the stiffness matrix has, using the equivalent of (4.33), components

$$K_{ij}^D = k_x\, A \int_0^5 \frac{dR_i}{dx}\frac{dR_j}{dx}\, dx \qquad i, j = 1, 2, 3 \tag{4.54}$$

$$= k_x\, A \int_{[e]} \frac{dQ_i}{dx}\frac{dQ_j}{dx}\, dx \tag{4.55}$$

These integrals have all been quoted in (3.18) (and already used in Section 4.4). The matrix obtained is

$$[K^{(D)}] = \frac{k_x\, A}{3\ell^{[e]}} \begin{bmatrix} 7 & -8 & 1 \\ -8 & 16 & -8 \\ 1 & -8 & 7 \end{bmatrix}$$

$$= 14.66 \begin{bmatrix} 7 & -8 & 1 \\ -8 & 16 & -8 \\ 1 & -8 & 7 \end{bmatrix} \tag{4.56}$$

Similarly from the equivalent of (4.34)

$$K_{ij}^G = -hP \int_0^5 R_i\, R_j\, dx \tag{4.57}$$

$$= -hP \int_{[e]} Q_i\, Q_j\, dx \qquad i, j = 1, 2, 3$$

which, using (3.20), gives rise to a matrix

$$[K^G] = \frac{-hP\ell^{[e]}}{30}\begin{bmatrix} 4 & 2 & -1 \\ 2 & 16 & 2 \\ -1 & 2 & 4 \end{bmatrix}$$

$$= -5.232\begin{bmatrix} 4 & 2 & -1 \\ 2 & 16 & 2 \\ -1 & 2 & 4 \end{bmatrix} \tag{4.58}$$

Hence

$$[K] = [K^D] - [K^G] = \begin{bmatrix} 123.5 & -106.8 & 9.43 \\ -106.8 & 318.3 & -106.8 \\ 9.43 & -106.8 & 123.5 \end{bmatrix} \tag{4.59}$$

Also, using the equivalent of (4.29), the components of the force vector are

$$\begin{aligned} F_i &= hPu_s \int_0^5 R_i(x)\,dx \\ &= hPu_s \int_{[e]} Q_i(x)\,dx \qquad i = 1, 2, 3 \end{aligned} \tag{4.60}$$

and using (3.19) for the integrals

$$\begin{aligned} \{F\} &= hPu_s \frac{\ell}{6}\begin{Bmatrix} 1 \\ 4 \\ 1 \end{Bmatrix} \\ &= 1047.2\begin{Bmatrix} 1 \\ 4 \\ 1 \end{Bmatrix} \end{aligned} \tag{4.61}$$

Finally, from the equivalent of (4.31),

$$\{B\} = \begin{Bmatrix} 0 \\ 0 \\ \left[k_x A \dfrac{du}{dx}\right]_5 \end{Bmatrix} = \begin{Bmatrix} 0 \\ 0 \\ 0 \end{Bmatrix} \tag{4.62}$$

after imposing the suppressible boundary condition at $x = 5\,\text{cm}$. Equations (4.59), (4.61) (and noting (4.62)) define a 3×3 system

$$[K]\{\tilde{u}\} = \{F\} + \{B\}$$

However, inserting the essential condition $\tilde{u}_1 = u(x = 0) = 140$ and ignoring the invalid first equation reduces the system to

$$\begin{bmatrix} 318.3 & -106.8 \\ -106.8 & 123.5 \end{bmatrix} \begin{Bmatrix} \tilde{u}_2 \\ \tilde{u}_3 \end{Bmatrix} = \begin{Bmatrix} 19140.8 \\ -273.0 \end{Bmatrix}$$

with solutions $\tilde{u}_2 = 83.7$ and $\tilde{u}_3 = 70.3$ which are reasonably consistent with the linear element solutions (4.52) and (4.53).

(c) Two quadratic elements model

We divide the domain into two quadratic elements of equal length. In each of the elements the central node is chosen to be at the mid-point (see Figure 4.14).

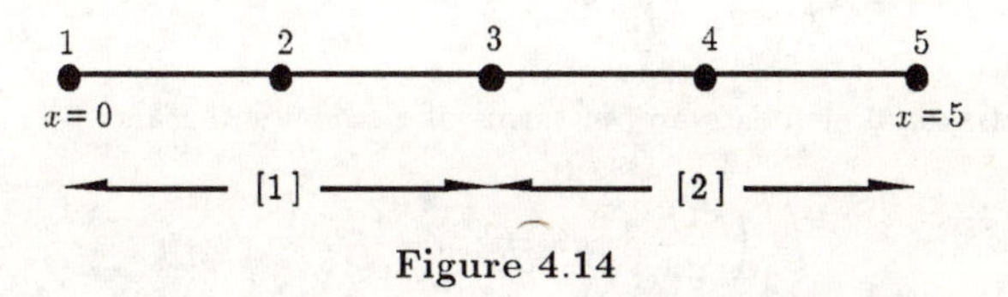

Figure 4.14

We first obtain as usual the relation between the roof functions $R_i(x), i = 1, 2, .., 5$ and the shape functions in order to evaluate the integrals that give us the stiffness matrix and force vector for the system.

In terms of roof functions, the trial solution over the whole domain can be written

$$\tilde{u} = \sum_{i=1}^{5} R_i(x)\tilde{u}_i \tag{4.63}$$

On the other hand, in terms of the quadratic shape functions we have within the individual elements

$$\tilde{u}^{[1]} = Q_1^{[1]}(x)\,\tilde{u}_1 + Q_2^{[1]}(x)\,\tilde{u}_2 + Q_3^{[1]}(x)\,\tilde{u}_3$$

$$\tilde{u}^{[2]} = Q_1^{[2]}(x)\,\tilde{u}_3 + Q_2^{[2]}(x)\,\tilde{u}_4 + Q_3^{[2]}(x)\tilde{u}_5 \tag{4.64}$$

Comparison of (4.63) and (4.64) then gives us the required correspondence which is as follows:

Roof Function	Corresponding Shape Function	
$R_1(x)$	$Q_1^{[1]}(x)$	
$R_2(x)$	$Q_2^{[1]}(x)$	in element [1]
$R_3(x)$	$Q_3^{[1]}(x)$	
$R_3(x)$	$Q_1^{[2]}(x)$	
$R_4(x)$	$Q_2^{[2]}(x)$	in element [2]
$R_5(x)$	$Q_3^{[2]}(x)$	

The components of the matrices $[K^D]$ and $[K^G]$ are calculated from the equivalents of (4.54) and (4.57) respectively and using the above table.

We have firstly

$$K_{ij}^D = k_x A \int_0^5 \frac{dR_i}{dx}\frac{dR_j}{dx}\,dx \qquad i, j = 1, 2, ..., 5$$

The diagonal elements are, in terms of element integrals

$$K_{11}^D = k_x A \int_{[1]} \left(\frac{dQ_1}{dx}\right)^2 dx = k_x A \frac{7}{3\ell^{[1]}}$$

$$K_{22}^D = k_x A \int_{[1]} \left(\frac{dQ_2}{dx}\right)^2 dx = k_x A \frac{16}{3\ell^{[1]}}$$

$$K_{33}^D = k_x A \left\{ \int_{[1]} \left(\frac{dQ_3}{dx}\right)^2 dx + \int_{[2]} \left(\frac{dQ_1}{dx}\right)^2 dx \right\}$$

$$= k_x A \left\{ \frac{7}{3\ell^{[1]}} + \frac{7}{3\ell^{[2]}} \right\}$$

$$K_{44}^D = k_x A \int_{[2]} \left(\frac{dQ_2}{dx}\right)^2 dx = k_x A \frac{16}{3\ell^{[2]}}$$

$$K_{55}^D = k_x A \int_{[2]} \left(\frac{dQ_3}{dx}\right)^2 dx = k_x A \frac{7}{3\ell^{[2]}}$$

(Note that we have supressed element superscripts for notational convenience.) We have again used (3.18) to write down the values of the integrals. Similar calculations give us the remaining matrix terms. Since $\ell^{[1]} = \ell^{[2]} = \ell^{[e]}$ say, we can write

$$[K^D] = \frac{k_x A}{3\ell^{[e]}} \begin{bmatrix} 7 & -8 & 1 & 0 & 0 \\ -8 & 16 & -8 & 0 & 0 \\ 1 & -8 & 14 & -8 & 1 \\ 0 & 0 & -8 & 16 & -8 \\ 0 & 0 & 1 & -8 & 7 \end{bmatrix} \qquad \textbf{(4.65)}$$

Comparison of (4.65) with the corresponding matrix (4.56) for the one-element quadratic model is interesting and suggests strongly the form of the matrix $[K^D]$ for four-, eight- and sixteen-element models. Note of course that the value of $\ell^{[e]}$ in (4.65) is 2.5 cm whereas $\ell^{[e]} = 5$ cm in (4.56).

A similar situation arises when we calculate the matrix $[K^G]$ for the two-element model. We have

$$K_{ij}^G = -hP \int_0^5 R_i\, R_j\, dx \qquad i,j = 1,2,\ldots,5$$

using the equivalent of (4.57). The diagonal elements are

$$K_{11}^G = -hP \int_{[1]} Q_1^2\, dx = -hP\,\frac{4\ell^{[1]}}{30}$$

$$K_{22}^G = -hP \int_{[1]} Q_2^2\, dx = -hP\,\frac{16\ell^{[1]}}{30}$$

$$K_{33}^G = -hP \left\{ \int_{[1]} Q_3^2\, dx + \int_{[2]} Q_1^2\, dx \right\} = -hP \left\{ \frac{4\ell^{[1]}}{30} + \frac{4\ell^{[2]}}{30} \right\}$$

$$K_{44}^G = -hP \int_{[2]} Q_2^2\, dx = -hP\,\frac{16\ell^{[2]}}{30}$$

$$K_{55}^G = -hP \int_{[2]} Q_3^2\, dx = -hP\,\frac{4\ell^{[2]}}{30}$$

We have used (3.20) to write down the integrals. Similar calculations give us the remainder of the matrix $[K^G]$ which is, since $\ell^{[1]} = \ell^{[2]} = \ell^{[e]}$

$$[K^G] = -\frac{hP\ell^{[e]}}{30}\begin{bmatrix} 4 & 2 & -1 & 0 & 0 \\ 2 & 16 & 2 & 0 & 0 \\ -1 & 2 & 8 & 2 & -1 \\ 0 & 0 & 2 & 16 & 2 \\ 0 & 0 & -1 & 2 & 4 \end{bmatrix} \qquad \textbf{(4.66)}$$

Again comparison of (4.66) with (4.58) for the one quadratic element model is instructive.

Inserting the given numerical values (remembering that $\ell^{[e]} = 2.5$ cm) we find from (4.65) and (4.66) that

$$[K] = [K^D] - [K^G]$$

$$= \begin{bmatrix} 215.722 & -229.336 & 26.7035 & 0 & 0 \\ -229.336 & 511.0318 & -229.336 & 0 & 0 \\ 26.7035 & -229.336 & 431.445 & -229.336 & 26.7035 \\ 0 & 0 & -229.336 & 511.0318 & -229.336 \\ 0 & 0 & 26.7035 & -229.336 & 215.722 \end{bmatrix} \qquad \textbf{(4.67)}$$

The force vector $\{F\}$ for this model is calculated from (4.60):

$$F_i = hP\,u_s \int_0^5 R_i(x)\,dx \qquad i = 1, 2, \ldots, 5$$

Hence

$$F_1 = hP\,u_s \int_{[1]} Q_1(x)\,dx = hP\,u_s \frac{\ell^{[e]}}{6}$$

$$F_2 = hP\,u_s \int_{[1]} Q_2(x)\,dx = hP\,u_s \frac{2\ell^{[e]}}{3}$$

$$F_3 = hP\,u_s \left\{ \int_{[1]} Q_3(x)\,dx + \int_{[2]} Q_1(x)\,dx \right\} = hP\,u_s \frac{\ell^{[e]}}{3}$$

$$F_4 = hP\,u_s \int_{[2]} Q_2(x)\,dx = hP\,u_s \frac{2\ell^{[e]}}{3}$$

$$F_5 = hP\,u_s \int_{[2]} Q_3(x)\,dx = hP\,u_s \frac{\ell^{[e]}}{6}$$

that is,

$$\{F\} = hP\,u_s \frac{\ell^{[e]}}{6} \begin{Bmatrix} 1 \\ 4 \\ 2 \\ 4 \\ 1 \end{Bmatrix} = 523.6 \begin{Bmatrix} 1 \\ 4 \\ 2 \\ 4 \\ 1 \end{Bmatrix} \tag{4.68}$$

(Compare (4.61) for the one quadratic-element model.) Finally we have a boundary term matrix

$$\{B\} = \begin{Bmatrix} 0 \\ 0 \\ 0 \\ 0 \\ k_x\,A \left.\dfrac{d\tilde{u}}{dx}\right|_5 \end{Bmatrix} = \begin{Bmatrix} 0 \\ 0 \\ 0 \\ 0 \\ 0 \end{Bmatrix} \tag{4.69}$$

where we have inserted the suppressible boundary condition at $x = 5$.

Equations (4.67), (4.68) and (4.69) define fully a 5×5 system:

$$[K]\{\tilde{u}\} = \{F\} + \{B\}$$

Inserting the essential boundary condition $\tilde{u}_1 = 140$ and ignoring the invalid first equation reduces this system to

$$\begin{bmatrix} 511.0318 & -229.336 & 0 & 0 \\ -229.336 & 431.445 & -229.336 & 26.7035 \\ 0 & -229.336 & 511.0318 & -229.336 \\ 0 & 26.7035 & -229.336 & 215.722 \end{bmatrix} \begin{Bmatrix} \tilde{u}_2 \\ \tilde{u}_3 \\ \tilde{u}_4 \\ \tilde{u}_5 \end{Bmatrix} = \begin{Bmatrix} 34201.44 \\ -2691.29 \\ 2094.4 \\ 523.6 \end{Bmatrix}$$

Solving gives

$$\begin{aligned} \tilde{u}_2 &= 104.5216 \\ \tilde{u}_3 &= 83.77418 \\ \tilde{u}_4 &= 72.91744 \\ \tilde{u}_5 &= 69.57611 \end{aligned}$$

These results compare very well with the values of the exact solution:

$$u = 100(\cosh\frac{\sqrt{7}x}{7} - \tanh\frac{5\sqrt{7}}{7}\sinh\frac{\sqrt{7}x}{7}) + 40$$

evaluated at the nodes:

$$\begin{aligned} u(1.25) &= 104.535 \\ u(2.5) &= 83.7460 \\ u(3.75) &= 72.9044 \\ u(5.0) &= 69.5452 \end{aligned}$$

Figure 4.15 displays the exact solution in comparison with the various finite element approximations obtained in this example.

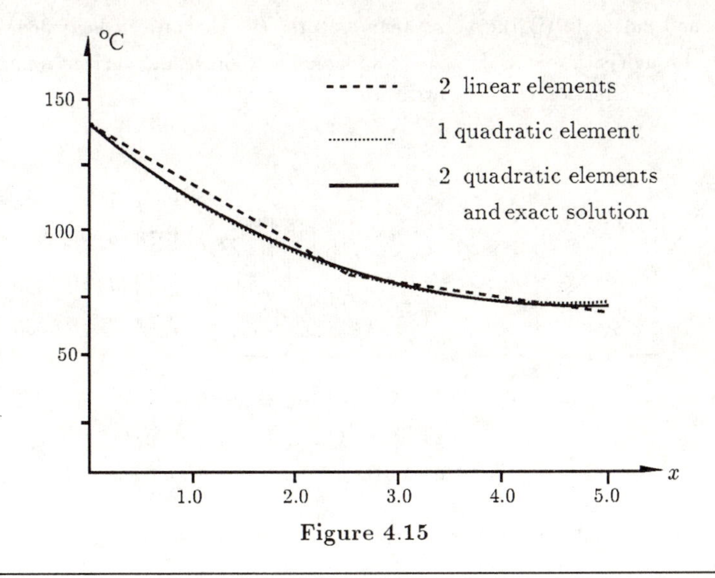

Figure 4.15

EXERCISES

4.1 Consider the boundary value problem:

$$\frac{d^2u}{dx^2} = 3 \qquad 0 \le x \le 2$$

$$u(0) = 2 \qquad u(2) = 0.5$$

(i) By using two linear elements of equal length calculate $\tilde{u}(1)$ and show that it agrees with the exact solution. Calculate $\tilde{u}(0.5)$ and $\tilde{u}(1.5)$ for this element model and show that the values do not agree with the analytic solution.
(ii) Repeat the problem using a four linear element model and again show that agreement with the analytic solution is obtained at the nodal points.
(iii) Modelling the domain as one quadratic element show that the analytic solution is obtained at all points.

4.2 In modelling steady state heat transfer by conduction through a wall composed of different materials the differential equation

$$k\frac{d^2u}{dx^2} = 0$$

is applicable to each section, u denoting the temperature and k the thermal conductivity of the material. For the wall shown in Figure Q4.2 and the boundary temperatures shown, model each section as a single linear element and calculate (i) the nodal temperatures (ii) the rate of heat flow $-k\frac{du}{dx}$ through each material. (The wall is assumed to have unit surface area normal to the direction of heat flow.)

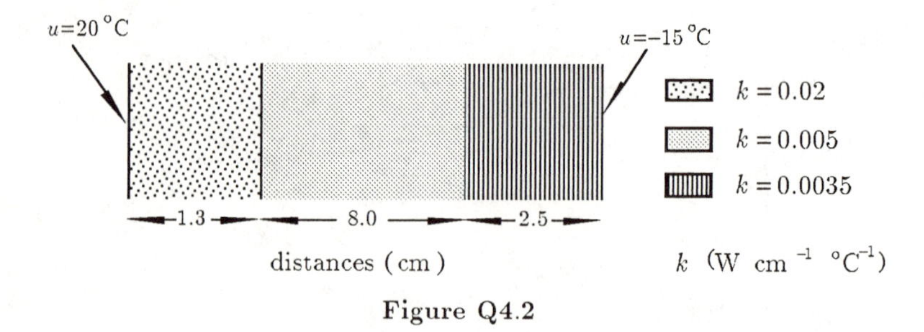

Figure Q4.2

4.3 The deflection u of a beam is governed by the differential equation

$$EI\frac{d^2u}{dx^2} - M(x) = 0 \qquad 0 \leq x \leq \ell$$

where EI is the bending stiffness term and $M(x)$ is the internal bending moment. If the beam is fixed at its ends (simply supported) then

$$u(0) = 0 \quad u(\ell) = 0$$

A particular beam of length 8 m is reinforced over its centre portion $200 \leq x \leq 600$ cm which has stiffness value 4×10^{10} N cm^2 whereas the outer sections have a stiffness 2.4×10^{10} N cm^2. The bending moment $M(x)$ is fixed at 10^6 N cm. Using a finite element model of four linear elements, all of length 200 cm, calculate the deflection in cm at $x = 200, 400$ and 600 cm.

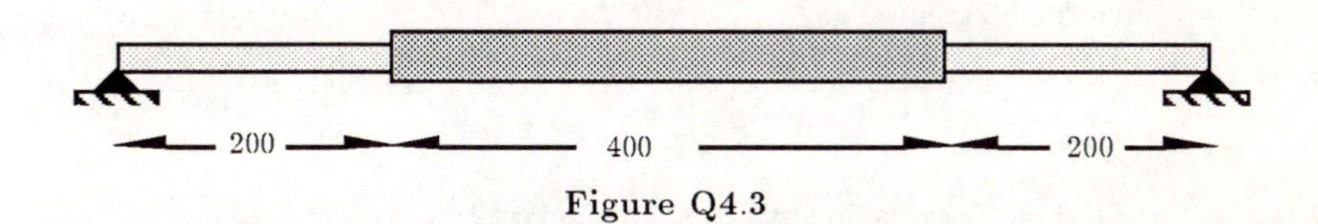

Figure Q4.3

4.4 Repeat Q4.3 for a beam also of length 800cm but with $EI = 2 \times 10^{10}$ N cm^2 for its whole length and $M(x)$ as shown in Figure Q4.4.

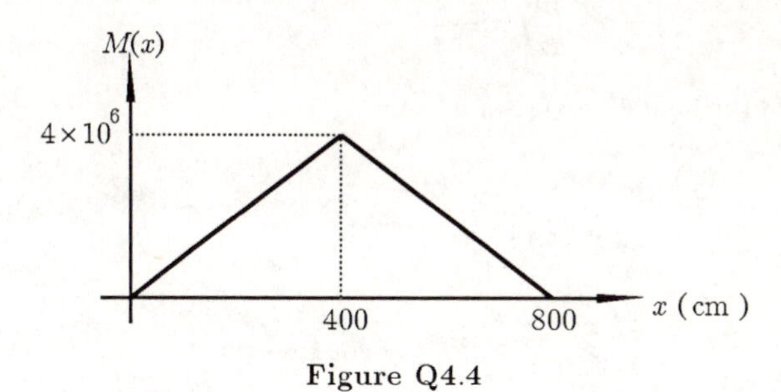

Figure Q4.4

4.5 (a) Solve the boundary value problem

$$\frac{d^2u}{dx^2} + 1000x^2 = 0 \qquad 0 \leq x \leq 1$$

$$u(0) = 0 \quad u(1) = 0$$

using (i) two linear elements (ii) one quadratic element
(b) Repeat but using boundary conditions

$$u(0) = 0 \qquad \frac{du}{dx}(1) = 1$$

Compare your answers with the analytic solutions in all cases.

4.6 Solve the boundary value problem

$$\frac{d^2u}{dx^2} + u + x = 0 \qquad 0 \le x \le 1$$

$$u(0) = 0 \quad u(1) = 0$$

using (i) two and then three linear elements (ii) one and then two quadratic elements.

4.7 Solve the differential equation

$$-\frac{d^2u}{dx^2} - u + x^2 = 0 \qquad 0 \le x \le 1$$

using three linear elements. Use as boundary conditions

(i) $u(0) = 0 \quad u(1) = 0$

(ii) $u(0) = 0 \quad \dfrac{du}{dx}(1) = 1$

(iii) $\dfrac{du}{dx}(0) = 1 \quad \dfrac{du}{dx}(1) = \dfrac{4}{3}$

4.8 Solve the boundary value problem:

$$\frac{d}{dx}\left(x\frac{du}{dx}\right) = \frac{2}{x^2} \qquad 1 \le x \le 2$$

$$u(1) = 2 \qquad \frac{du}{dx}(2) = -\frac{1}{4}$$

using
(i) two equal length linear elements
(ii) four equal length linear elements.

4.9 In modelling heat flow by conduction in an insulated rod with heat being lost by convection at the right hand end the resultant boundary value problem is:

$$kA\frac{d^2u}{dx^2} = 0 \qquad 0 \le x \le \ell$$

$$u(0) = u_0 \qquad kA\frac{du}{dx} = -hA(u - u_s) \quad \text{at } x = \ell$$

In a particular problem the parameter values are:
thermal conductivity $k = 0.5$ W cm^{-1} °C^{-1}
area of cross section $A = 2$ cm^2
length of rod $\ell = 10$ cm
convection coefficient $h = 12.5$ W cm^{-2} °C^{-1}
temperature of surroundings $u_s = 5$°C
temperature $u_0 = 50$°C
Using two linear elements determine $\tilde{u}$ at $x = \ell/2$ and $x = \ell$.

4.10 Repeat Q4.9 but using boundary conditions:

$$kA\frac{du}{dx} = hA(u - u_s) \quad \text{at } x = 0$$

(i.e. left hand end of the rod is losing heat by convection)

$$u(\ell) = u_\ell$$

The parameter values are $A = 1$ cm^2, $h = 0.1$ W cm^{-2} °C^{-1}, $u_s = -5$°C, $u_\ell = 20$°C. The rod is made up of two materials as shown in Figure Q4.10.

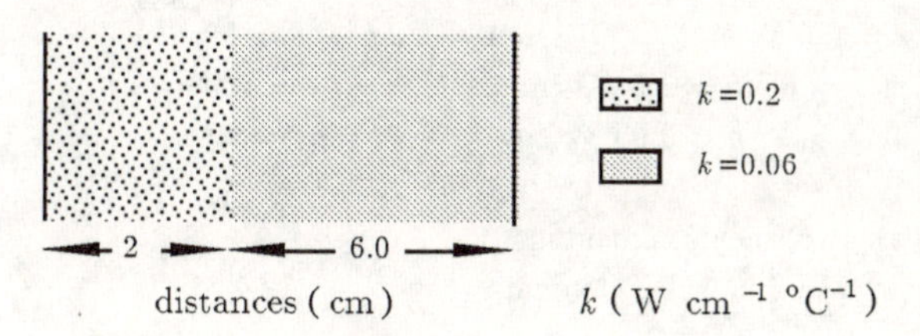

Figure Q4.10

Determine $\tilde{u}(0)$ and $\tilde{u}(2)$ using two linear elements.

4.11 In modelling steady state heat transfer in a fin which is losing heat to its surroundings by lateral convection and from its right hand end the governing differential equation is (see Section 4.8, Equation (4.48))

$$kA\frac{d^2u}{dx^2} + hP(u_s - u) = 0 \qquad 0 \leq x \leq \ell$$

with boundary conditions

$$u(0) = u_0 \qquad kA\frac{du}{dx} = -hA(u - u_s) \quad \text{at } x = \ell.$$

In a particular problem the parameter values are
thermal conductivity $k = 3$ W cm^{-1} °C^{-1}

area of cross section of fin $A = 4$ cm^2
perimeter of fin $P = 10$ cm
length of fin $\ell = 8$ cm
convection coefficient $h = 0.1$ W cm^{-2} °C^{-1}
temperature of surroundings $u_s = 20$°C
temperature $u_0 = 80$°C

Using a model of four linear elements calculate $\tilde{u}$ at x=2, 4, 6 and 8 cm.

4.12 Rework Q4.11 but using four linear elements of unequal length to determine $\tilde{u}$ at $x = 1, 3, 5$ and 8 cm.

4.13 Rework Q4.11 for a circular fin of radius 1cm which is to be modelled as one quadratic element. Use the following parameter values:
$k = 70$ W cm^{-1} °C^{-1}, $\ell = 5$ cm, $h = 5$ W cm^{-2} °C^{-1}, $u_s = 40$°C, $u_0 = 140$°C.

4.14 Calculate the temperatures at each end of a rod which is losing heat by convection from both ends but which is insulated to prevent lateral convection. The parameters are:

$k = 0.5$ W cm^{-1} °C^{-1} $\quad A = 1$ cm^2 $\quad \ell = 4$ cm
at $x = 0$ $\quad h = 1.5$ W cm^{-2} °C^{-1} $\quad u_s = 5$°C
at $x = \ell$ $\quad h = 0.05$ W cm^{-2} °C^{-1} $\quad u_s = 20$°C

Use (i) two linear elements
(ii) one quadratic element.

SUPPLEMENTS

4S.1 Computer program

As we have mentioned previously, the practical implementation of the finite element method is dependent upon the availability of powerful computing facilities. In this regard we have reproduced in this supplement a computer program, written in the Fortran77 language, which is applicable to a general linear second-order one-dimensional boundary value problem. The program uses three-noded quadratic elements but the reader should have little difficulty amending the program to employ elements of his/her choice. The program incorporates a simple 'mesh generator' which ensures that only the minimum of information need be input as initial data. (Here, the reader may find the first section of Chapter 9 helpful.) The program is accompanied by brief comments.

Fortran77 program to solve the boundary value problem:

$$-\frac{d}{dx}(p(x)\frac{du}{dx}) + q(x)u = r(x) \qquad a \le x \le b$$

$$\alpha_0 u(a) + \alpha_1 u'(a) = \gamma_1$$

$$\beta_0 u(b) + \beta_1 u'(b) = \gamma_2$$

```
      implicit double precision(a-h,o-z)
      dimension displ(60),F(60)
      dimension A(60,60),aband(60,3)
      dimension nelseg(10),xpoint(20)
      integer hbw
      common/geom/xnode(60),AIE(3,3),FIE(60)
      common/intg/xi(3),w(3)
      common/geom1/nenn(60,3)
      character*9 dat,out
```

User needs to define the form of $p(x)$ here. In this specimen program we have treated the boundary value problem of Q4.5 and have chosen to define $p(x)$ using a statement function which requires the parameter x to appear in the definition even if the function is constant. This explains the appearance of the 'redundant' expression $x - x$ in the definition.

```
      p(x)=-1.+x-x
```

Input data is contained in file '*dat*' (specified by user). The domain $a \le x \le b$ is divided into '*nseg*' segments. Within each segment the user chooses a number of elements $nelseg(i)$, $i = 1, nseg$. The x-coordinate of segment end-points is recorded in $xpoint(i)$, $i = 1, nseg+1$. The coefficients $\alpha_0, \alpha_1, \beta_0, \beta_1, \gamma_1, \gamma_2$ are contained in $a0, a1, b0, b1, g1, g2$ respectively.

```
      print*,'type the data file mame (must already exist)'
      print*,'and occupy the first 9 spaces (right justified)'
      print*,'then type output file name (file must not exist)'
      print*,'and occupy the next 9 spaces (right justified)'
      read(*,1000)dat,out
1000  format(a9,a9)
      open(5,file=dat,status='old')
```

```
      open(6,file=out,status='new',form='formatted')
      read(5,*,err=1001)nseg,(nelseg(i),i=1,nseg),
     1(xpoint(i),i=1,nseg+1)
      read(5,*)a0,a1,b0,b1,g1,g2
      goto 1003
1001  print*,'error detected in inputting information'
      stop
1003  continue
```

Here we use a simple 'mesh-generator'. This provides nel, nde, global node numbers for each element e ($nenn(e,i)$ $i = 1,2,3$) and the x-coordinate of each node ($xnode(i)$, $i = 1, nde$).

```
      k=1
      m=1
      do 97 i=1,nseg
      ll=nelseg(i)+m-1
      do 96 j=m,ll
      nenn(j,1)=k
      nenn(j,2)=k+1
      nenn(j,3)=k+2
96    k=k+2
      m=j
97    continue
      k=0
      do 95 i=1,nseg
      h=(xpoint(i+1)-xpoint(i))/(2.*nelseg(i))
      nde=2*nelseg(i)+1
      do 94 j=1,nde
94    xnode(j+k)=h*(j-1)+xpoint(i)
      k=nde+k-1
95    continue
      nde=0
      do 93 i=1,nseg
93    nde=nde+nelseg(i)
      nel=nde
      nde=2*nel+1
```

---

na, nb are node numbers of first and last nodes.

```
      na=nenn(1,1)
      nb=nenn(nel,3)
```

Here, by examining $a1, b1$ the boundary conditions are classed as either suppressible ($nsupa = 1, nsupb = 1$) or essential ($nsupa = 0, nsupb = 0$) or some combination.

```
        nsupb=1
        nsupa=1
        if(abs(a1).lt..1d-10)nsupa=0
        if(abs(b1).lt..1d-10)nsupb=0
```

Sample points and weights for third-order Gaussian quadrature (see Table 3.3).

```
        xi(1)=-0.774596669241483
        xi(2)=0.d 00
        xi(3)=-xi(1)
        w(1)=0.5555555555555555
        w(2)=0.8888888888888888
        w(3)=w(1)
```

Initialize global matrices A and F.

```
        do 1 i=1,nde
        do 1 j=1,nde
        A(i,j)=0.0
        F(j)=0.0
1       continue
```

Output mesh information.

```
        write(6,8)
        write(6,9)(i,xnode(i),i=1,nde)
        write(6,10)
        write(6,11)(i,(nenn(i,j),j=1,3),i=1,nel)
8       format(' node x-coord')
9       format(1h ,i3,d15.5)
10      format(1h ,//,1h ' element node numbers',/,' elmt 1 2 3')
11      format(4i5)
```

Half-bandwidth is 3 (for 'obvious' node-numbering) for 1-D quadratic elements.

```
      hbw=3
```

Superposition of local element matrices into global counterparts.

```
      do 7 i=1,nel
      call elmt2(i)
      do 6 j=1,3
      jj=nenn(i,j)
      F(jj)=F(jj)+FIE(j)
      do 6 k=1,3
      kk=nenn(i,k)
      A(jj,kk)=A(jj,kk)+AIE(j,k)
6     continue
7     continue
```

Depending on the type of boundary conditions particular terms in the overall stiffness matrix A and force vector F may need amending. If there is at least one essential boundary condition then the matrix A will be singular. In this case we proceed differently from when working by hand. Assume an essential boundary condition at $x = a$, so $u(a) = g1/a0$. We **replace** the first equation by the identity $u(a) = g1/a0$ by arranging

$$A_{11} = 1, \quad A_{1k} = 0, \; k = 1, \ldots, nde - 1 \quad \text{and} \quad F_1 = g1/a0$$

We then incorporate $u(a) = g1/a0$ in each of the remaining equations by arranging

$$A_{k1} = 0 \quad \text{and} \quad F(k) = F(k) - A_{k1} g1/a0 \quad k = 2, \ldots, nde$$

```
      if(nsupa.eq.0)goto 22
      if(nsupb.eq.0)goto 18
      A(na,na)=A(na,na)-p(xnode(na))*a0/a1
      F(na)=F(na)-p(xnode(na))*g1/a1
      A(nb,nb)=A(nb,nb)+p(xnode(nb))*b0/b1
      F(nb)=F(nb)+p(xnode(nb))*g2/b1
      goto 31
18    A(na,na)=A(na,na)-p(xnode(na))*a0/a1
      F(na)=F(na)-p(xnode(na))*g1/a1
      do 20 k=1,nde
```

```
20      F(k)=F(k)-A(k,nb)*g2/b0
        do 21 k=1,nde
        A(k,nb)=0.0
        A(nb,k)=0.0
21      continue
        A(nb,nb)=1.0
        F(nb)=g2/b0
        goto 31
22      continue
        if(nsupb.eq.0)goto 26
        A(nb,nb)=A(nb,nb)+p(xnode(nb))*b0/b1
        F(nb)=F(nb)+p(xnode(nb))*g2/b1
        do 24 k=1,nde
24      F(k)=F(k)-A(k,na)*g1/a0
        do 25 k=1,nde
        A(k,na)=0.0
        A(na,k)=0.0
25      continue
        A(na,na)=1.0
        F(na)=g1/a0
        goto 31
26      continue
        do 28 k=1,nde
        F(k)=F(k)-A(k,nb)*g2/b0
28      continue
        do 29 k=1,nde
        F(k)=F(k)-A(k,na)*g1/a0
29      continue
        do 30 k=1,nde
        A(k,nb)=0.0
        A(nb,k)=0.0
        A(k,na)=0.0
        A(na,k)=0.0
30      continue
        A(na,na)=1.0
        A(nb,nb)=1.0
        F(na)=g1/a0
        F(nb)=g2/b0
31      continue
```

The following procedure forms a banded matrix *aband* of size nde*hbw from the matrix A.

```
      do 34 iii=1,nde
      do 33 k=1,hbw
      aband(iii,k)=A(iii,k+iii-1)
33    continue
34    continue
      do 35 i=0,hbw-2
      do 36 k=i+1,hbw-1
      aband(nde-i,k+1)=0.
36    continue
35    continue
```

We now solve the system of equations using subroutine *band.*

```
      call band(aband,f,nde,hbw,displ)
      write(6,14)(displ(i),i=1,nde)
14    format(4d16.8)
      call exit
      end
```

Here we evaluate (for each x and corresponding natural coordinate xi) the expression $p(x)\frac{dQ_i}{dx}\frac{dQ_j}{dx} + q(x)Q_iQ_j$. Here the user must supply the functions $p(x)$, $q(x)$ (held as px, qx).

```
      function fun2a(x,xi,i,j,h)
      implicit double precision(a-h,o-z)
      px=-1.
      qx=0.
      call shape(i,Q,xi)
      call dshape(i,dQ,xi)
      Qi=Q
      dQi=2.*dQ/h
      call shape(j,Q,xi)
      call dshape(j,dQ,xi)
      Qj=Q
      dQj=2.*dQ/h
      fun2a=px*dQi*dQj+qx*Qi*Qj
      return
      end
```

Here we evaluate (for each x and corresponding natural coordinate xi)

the expression $r(x)Q_i$. Here the user must supply the function $r(x)$ (held as rx).

```
function fun2f(x,xi,i,h)
implicit double precision(a-h,o-z)
call shape(i,Q,xi)
rx=-1000.*x*x
Qi=Q
fun2f=rx*Qi
return
end
```

Quadratic shape functions (see (3.16)).

```
subroutine shape(n,Q,xi)
implicit double precision(a-h,o-z)
dimension Qn(3)
Qn(1)=0.5*(xi-1.)*xi
Qn(2)=-(xi-1.)*(xi+1.)
Qn(3)=0.5*(xi+1.)*xi
Q=Qn(n)
return
end
```

Shape function derivatives.

```
subroutine dshape(n,dQ,xi)
implicit double precision(a-h,o-z)
dimension dQn(3)
dQn(1)=xi-0.5
dQn(2)=-2.*xi
dQn(3)=xi+0.5
dQ=dQn(n)
return
end
```

This subroutine evaluates (using three-point Gaussian quadrature) the integrals:

$$\int_{[ne]} \left\{ p(x)\frac{dQ_i}{dx}\frac{dQ_j}{dx} + q(x)Q_iQ_j \right\} dx \quad \text{and} \quad \int_{[ne]} \{r(x)Q_i\}\, dx$$

```
      subroutine elmt2(ne)
      implicit double precision(a-h,o-z)
      dimension xn(3)
      common/geom/xnode(60),AIE(3,3),FIE(60)
      common/intg/xi(3),w(3)
      common/geom1/nenn(60,3)
      do 1 l=1,3
      i=nenn(ne,l)
      xn(l)=xnode(i)
1     continue
      h=abs(xn(3)-xn(1))
      do 3 k=1,3
      do 2 m=1,3
      AIE(k,m)=0.0
      FIE(k)=0.0
2     continue
3     continue
      do 6 k=1,3
      do 5 m=k,3
      do 4 l=1,3
      x=0.5*(xn(3)-xn(1))*xi(l)+0.5*(xn(1)+xn(3))
      AIE(k,m)=AIE(k,m)+(h/2.)*w(l)*fun2a(x,xi(l),k,m,h)
4     continue
      AIE(m,k)=AIE(k,m)
5     continue
6     continue
      do 8 k=1,3
      do 9 l=1,3
      x=0.5*(xn(3)-xn(1))*xi(l)+0.5*(xn(1)+xn(3))
      FIE(k)=FIE(k)+(h/2.)*w(l)*fun2f(x,xi(l),k,h)
9     continue
8     continue
      return
      end
```

This subroutine solves a system of banded equations storing result in *displ*. The routine uses an amended form of the Gaussian elimination procedure.

```
      subroutine band(sk,ff,nde,hbw,displ)
      implicit double precision(a-h,o-z)
      dimension aband(60,3),ff(60),displ(60)
      integer hbw
```

```
      m1=nde
      m2=m1-1
      do 3 i=1,m2
      i1=i+1
      i2=i+hbw-1
      if(i2.gt.m1)i2=m1
      do 2 k=i1,i2
      jj=k-i+1
      fact=aband(i,jj)/aband(i,1)
      nsf1=hbw-k+i
      do 1 j=1,nsf1
      jj1=j+k-i
      aband(k,j)=aband(k,j)-fact*aband(i,jj1)
1     continue
      ff(k)=ff(k)-fact*ff(i)
2     continue
3     continue
      do 5 i=1,m1
      ii=m1-i+1
      pivot=aband(ii,1)
      aband(ii,1)=0.0
      do 4 j=1,hbw
      k=ii+j-1
      if(k.gt.m1)goto 4
      ff(ii)=ff(ii)-aband(ii,j)*displ(k)
4     continue
      displ(ii)=ff(ii)/pivot
5     continue
      return
      end
```

4S.2 Basic equations of heat transfer

One-dimensional models

In the body of this chapter we have solved boundary value problems involving heat transfer in a fin (see Figure 4.10). Here, as background material, we discuss briefly the derivation of the governing equation (4.48) and the physical significance of the boundary conditions.

The heat transfer mechanisms modelled are

• **Conduction**

This is governed by Fourier's Law which states that the rate of heat flow q_1 by conduction in a material is proportional to

(i) the area A through which the heat flows A being measured perpendicular to the direction of flow

(ii) the temperature gradient, du/dx, in the direction x of heat flow.

Mathematically,

$$q_1 = -k_x A \frac{du}{dx} \tag{4S.1}$$

where the proportionality constant k_x depends on the material and is known as the thermal conductivity. The negative sign is because the direction of heat flow is along the direction in which du/dx decreases.

It follows from (4S.1) that in the SI system the units of thermal conductivity are $\mathrm{W\,m^{-1}\,K^{-1}}$.

• **Convection**

This is governed by Newton's law which states that the rate of heat transfer q_2 by convection between a surface of area A at a temperature u and a fluid at a temperature u_s is proportional both to A and to the temperature difference $(u - u_s)$.

Mathematically,

$$q_2 = hA(u - u_s) \tag{4S.2}$$

where the proportionality constant h is called the convective heat transfer coefficient.

It follows from (4S.2) that the SI units of h are $\mathrm{W\,m^{-2}\,K^{-1}}$.

Heat transfer in a fin of uniform cross section

Consider a fin attached to a wall whose surface is at a constant temperature u_0. See Figure 4.16. We assume that the fin is cooled along its surface by a surrounding fluid whose temperature u_s is constant.

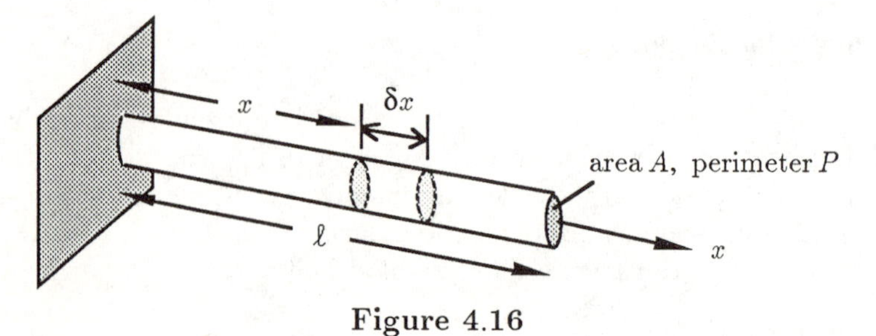

Figure 4.16

We assume that the transverse temperature gradients are small enough for the temperature at any cross section to be considered uniform i.e. we take $u = u(x)$. To derive (4.48) we simply write a heat balance equation for a small element δx of fin. We assume that heat flows by conduction into the left face of the element and out by conduction through the right face and by convection from the surface of the element.

Then, assuming that steady state conditions have been attained, we have, using (4S.1) and (4S.2),

$$\underset{\text{conduction in}}{\underset{\uparrow}{-k_x A\frac{du}{dx}\bigg|_x}} = \underset{\text{conduction out}}{\underset{\uparrow}{-k_x A\frac{du}{dx}\bigg|_{x+\delta x}}} + \underset{\text{convection out}}{\underset{\uparrow}{hP\delta x(u - u_s)}} \tag{4S.3}$$

Here P is the perimeter of the fin so that $P\delta x$ is the surface area between the sections x and $x + \delta x$.

For small enough δx we can write

$$-k_x A\frac{du}{dx}\bigg|_{x+\delta x} \approx -k_x A\frac{du}{dx}\bigg|_x + \frac{d}{dx}\left(-k_x A\frac{du}{dx}\right)\delta x \tag{4S.4}$$

so that, if k_x is constant, (4S.3) becomes

$$-k_x A\frac{d^2u}{dx^2}\delta x + hP\delta x(u - u_s) = 0$$

from which (4.48) follows immediately.

Boundary conditions for one-dimensional heat transfer problems

Suppose the end $x = \ell$ of the fin is perfectly insulated so that no heat passes through. It follows from (4S.1) that

$$\frac{du}{dx} = 0 \qquad \text{at } x = \ell$$

is the appropriate boundary condition. (Compare Example 4.1).

On the other hand, if the end $x = \ell$ is losing heat by convection it follows from (4S.1) and (4S.2) that

$$-k_x A\frac{du}{dx}\bigg|_{x=\ell} = hA(u(\ell) - u_s)$$

where we have balanced the conduction heat flow at this end with the convection heat loss.

Two and three-dimensional models of steady-state heat conduction

Consider a small rectangular parallelepiped with edges δx, δy and δz parallel respectively to the x, y and z axes (see Figure 4.17). The element is assumed to be part of a solid body through which heat is being conducted.

To obtain the governing equation for the temperature $u(x, y, z)$ in this case we again use an energy balance equation. We will allow now for the possibility of (a constant) internal generation of heat within the element at a rate $\dot{q}$ per unit volume.

Let q_x, q_y, q_z denote the rates of flow of heat across unit areas perpendicular to the x, y and z axes respectively so that

$$q_x = -k_x \frac{\partial u}{\partial x} \qquad q_y = -k_y \frac{\partial u}{\partial y} \qquad q_z = -k_z \frac{\partial u}{\partial z}$$

where we allow for the possibility of different thermal conductivities along the three coordinate directions.

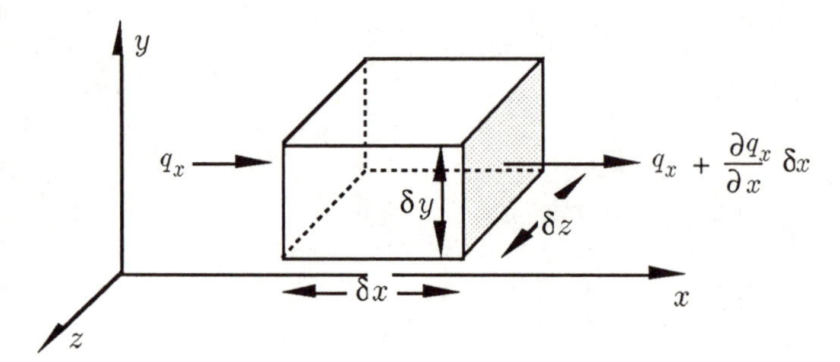

Figure 4.17

The rate at which heat flows by conduction into the element through its left hand face is

$$q_x \delta y\, \delta z = -k_x \frac{\partial u}{\partial x} \delta y\, \delta z$$

Similarly the rate of loss of heat by conduction through the right hand face is

$$q_{x+\delta x} \delta y\, \delta z \approx \left[-k_x \frac{\partial u}{\partial x} + \frac{\partial}{\partial x}\left(-k_x \frac{\partial u}{\partial x} \right) \delta x \right] \delta y\, \delta z$$

using (4S.4).

Hence the net rate at which heat is being gained or lost by the element through these two faces is approximately

$$\frac{\partial}{\partial x}\left(k_x\frac{\partial u}{\partial x}\right)\delta x\,\delta y\,\delta z$$

Performing similar calculations for the other four faces and using energy balance we have, assuming that the steady state has been reached so that the net rate of gain of heat by the element is zero,

$$\frac{\partial}{\partial x}\left(k_x\frac{\partial u}{\partial x}\right)+\frac{\partial}{\partial y}\left(k_y\frac{\partial u}{\partial y}\right)+\frac{\partial}{\partial z}\left(k_z\frac{\partial u}{\partial z}\right)+\dot{q}=0$$

where we have divided through by $\delta x\,\delta y\,\delta z$.

If $k_x = k_y = k_z = k$, a constant, we obtain

$$\nabla^2 u \equiv \frac{\partial^2 u}{\partial x^2}+\frac{\partial^2 u}{\partial y^2}+\frac{\partial^2 u}{\partial z^2}=-\frac{\dot{q}}{k} \qquad \textbf{(4S.5)}$$

which is Poisson's Equation in three dimensions.

If the system contains no heat sources, the steady state temperature satisfies

$$\nabla^2 u = 0$$

which is Laplace's Equation.

We will discuss the finite element solution of two-dimensional boundary value problems involving these equations in Chapter 7. Note that in one dimension (4S.5) is

$$\frac{d^2u}{dx^2}+\frac{\dot{q}}{k}=0$$

and we have of course in this chapter discussed the solution of this equation using finite elements.

4S.3 Governing equations for vibrations of continuous materials

One-dimensional models

The situation of interest is a vibrating system where the displacement U depends on only **one** space coordinate, say x, as well as on the time t. Typical applications for such a model might be the torsional oscillations of a shaft, the longitudinal vibrations of a bar and the transverse oscillations of a string. We focus on the last of these because it is perhaps the most familiar and the easiest to visualize although the governing equation obtained is valid for the other cases mentioned.

Consider then a stretched string of original length ℓ fixed at its ends, $x = 0$ and $x = \ell$, and executing oscillations. See Figure 4.18(a) for a possible 'snapshot' of the oscillation at a particular time t_0.

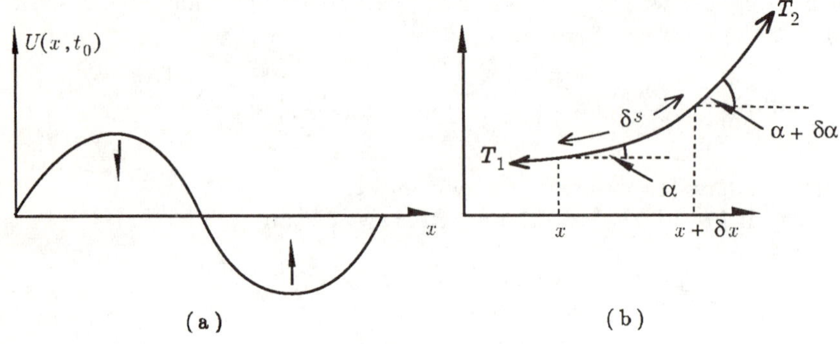

Figure 4.18

We will assume

(i) that the displacement U is entirely in the Ux plane and that U is always small compared with ℓ.
(ii) that the longitudinal displacement of a point on the string can be neglected i.e. we are modelling transverse oscillations only.
(iii) that no damping forces are present.

If ρ is the mass density (mass per unit length) of the string then a small segment δs (see Figure 4.18(b)) has mass $\rho\delta s$ and we can obtain the acceleration $\dfrac{\partial^2 U}{\partial t^2}$ of a particle on this element from the net transverse force on this element. This is

$$T_2 \sin(\alpha + \delta\alpha) - T_1 \sin\alpha$$

since the tension in the string is assumed to be the only force present. Consequently, by Newton's second law of motion

$$\frac{\partial^2 U}{\partial t^2} = \frac{T_2 \sin(\alpha + \delta\alpha) - T_1 \sin\alpha}{\rho\delta s}$$

However since we are modelling small transverse oscillations we note several simplifications:

- the net longitudinal force on the segment is zero:

$$T_1 \cos\alpha = T_2 \cos(\alpha + \delta\alpha) = T$$

- the length of the segment δs is approximately δx.

Consequently

$$\frac{\partial^2 U}{\partial t^2} = \frac{T\tan(\alpha+\delta\alpha) - T\tan\alpha}{\rho\delta s}$$

Further,

$$\frac{\rho}{T}\frac{\partial^2 U}{\partial t^2} \approx \frac{1}{\delta x}\left[\left.\frac{\partial U}{\partial x}\right|_{x+\delta x} - \left.\frac{\partial U}{\partial x}\right|_{x}\right]$$

or, in the limit as $\delta x \to 0$,

$$\frac{\rho}{T}\frac{\partial^2 U}{\partial t^2} = \frac{\partial^2 U}{\partial x^2}$$

The quantity T/ρ has the dimension of $(\textit{velocity})^2$ so we may write it as v^2 obtaining, finally, the one-dimensional wave equation

$$\frac{\partial^2 U}{\partial x^2} = \frac{1}{v^2}\frac{\partial^2 U}{\partial t^2} \tag{4S.6}$$

The only context in which we are solving this equation in this text is when the following conditions are satisfied:
(i) $U = 0$ at $x = 0$ and $x = \ell$, which corresponds to the string (or other vibrating system) being fixed at its end points;
(ii) when the time dependence is sinusoidal.

The latter condition is described mathematically by putting $U(x,t) = u(x)e^{i\omega t}$ so that (4S.6) becomes

$$\frac{d^2 u}{dx^2} = -\frac{\omega^2}{v^2}u$$

which is the one-dimensional Helmholtz equation whose finite element formulation has been described in Section 4.7

Two-dimensional models

In two dimensions, the archetypal vibration problem is that of a vibrating membrane (such as a drum-head) which is tightly stretched and held at its edges by a rigid frame. If this membrane is set into transverse vibration then, by a similar proof to that used in the case of a stretched string i.e.

assuming 'small deflections', the displacement $U(x,y,t)$ is found to satisfy the two-dimensional wave equation

$$\frac{\partial^2 U}{\partial x^2} + \frac{\partial^2 U}{\partial y^2} = \frac{1}{v^2}\frac{\partial^2 U}{\partial t^2} \tag{4S.7}$$

Again, if we assume sinusoidal time variation this reduces to the two-dimensional Helmholtz equation

$$\frac{\partial^2 u}{\partial x^2} + \frac{\partial^2 u}{\partial y^2} = -\frac{\omega^2}{v^2}u \tag{4S.8}$$

The finite element formulation of (4S.8) is discussed in the exercises to Chapter 7.

5

Finite Elements and Linear Elasticity

PREVIEW

In this chapter, which is largely self contained, we examine the application of finite elements in linear elasticity. We introduce the concepts of strain energy and potential energy and deduce the governing equations for two-dimensional elasticity (plane strain). Finite elements are then introduced in a natural way by developing the theory of pin-jointed structures in both two and three dimensions. The remainder of the chapter introduces the application of the finite element technique to plane strain. In the supplement we describe the computer implementation of finite element techniques to pin-jointed assemblages.

5.1 Introduction to linear elasticity

The mathematical analysis of structures provides a natural approach to the finite element method since a major step that must be taken when applying finite elements, viz. discretization, is automatically present in a structure.

Almost any structure encountered in practice is naturally segmented into elements and nodes, with the elements ranging in complication from simple springs, through struts and beams to complex two-dimensional components such as plates and shells. It is not surprising therefore that finite element procedures were first developed for structural engineering analysis. Indeed finite elements is now firmly established as an essential tool of civil, aeronautical and mechanical engineers.

It is appropriate, therefore, in an introductory text like this to give an introduction to finite elements as applied in structural mechanics as well as studying the method applied to the solution of boundary value problems. The latter application is, in some senses, the more general if perhaps the less obvious at first sight.

Exact solutions to problems in engineering are rarely possible and in seeking approximate solutions a complex engineering problem should, where possible, be replaced by a simpler one. Unfortunately there is often a price to be paid when this naive approach is used – that of poor accuracy. If accurate solutions are required then it is necessary to replace the complex problem by a **large number** of simpler problems. In essence of course, this is the underlying tenet of the finite element method.

As we have seen, albeit briefly, in Chapter 1, analysis of the humble spring provides a natural introduction to finite elements with many of the basic characteristics of the method emerging. In this chapter we extend that **direct approach** to finite elements by considering a more sophisticated structural element – the pin-jointed strut – both in one and two dimensions. We also consider the application of finite elements to two-dimensional continua by examining problems in 'plane strain'.

In order to appreciate this approach to finite elements it is first necessary to outline the foundations of solid mechanics. These, essentially, are concerned with the interplay of three physical quantities:

- **particle displacement** which is represented by a column vector $\{d\}$:

$$\{d\} = \{u, v, w\}^T$$

- material **stress** which is represented by a square symmetric matrix $[\sigma]$:

$$[\sigma] = \begin{bmatrix} \sigma_{11} & \sigma_{12} & \sigma_{13} \\ \sigma_{21} & \sigma_{22} & \sigma_{23} \\ \sigma_{31} & \sigma_{32} & \sigma_{33} \end{bmatrix}$$

- material **strain** also represented by a square symmetric matrix $[\varepsilon]$:

$$[\varepsilon] = \begin{bmatrix} \varepsilon_{11} & \varepsilon_{12} & \varepsilon_{13} \\ \varepsilon_{21} & \varepsilon_{22} & \varepsilon_{23} \\ \varepsilon_{31} & \varepsilon_{32} & \varepsilon_{33} \end{bmatrix}$$

(Note that in common with the notation used throughout this text a column vector is denoted by curly brackets. Square brackets are used in the description of other matrices.)

A thorough investigation of solid mechanics would necessitate a Cartesian tensor approach. In our treatment we shall, by a careful choice of subject matter, need to employ only the more amenable matrix notation. To a large extent, this chapter is self-contained. Those finite elements that are introduced emerge in a natural way and, in the interests of clarity, no sustained effort will be made to relate the material to the finite element treatment of boundary value problems.

We begin our study with the treatment of stress.

5.2 Introduction to stress

The analysis of stress is fundamental to the study of the mechanical behaviour of all materials whether they behave plastically, viscoelastically or, as in this chapter, in a **linearly elastic** manner.

The occurrence of stress is commonplace. For example suppose we apply equal and opposite forces to either end of a straight bar of uniform (but arbitrary) cross section. It is a matter of common experience that after a very short time during which it suffers a slight elongation the bar will be in equilibrium (see Figure 5.1). If the bar is made of a 'stiff' material such as metal or wood the elongation is not likely to be visible but could be detected by sensitive equipment. More obvious results are obtained with a highly elastic material such as rubber. (An elastic band extended between each hand provides an excellent example.)

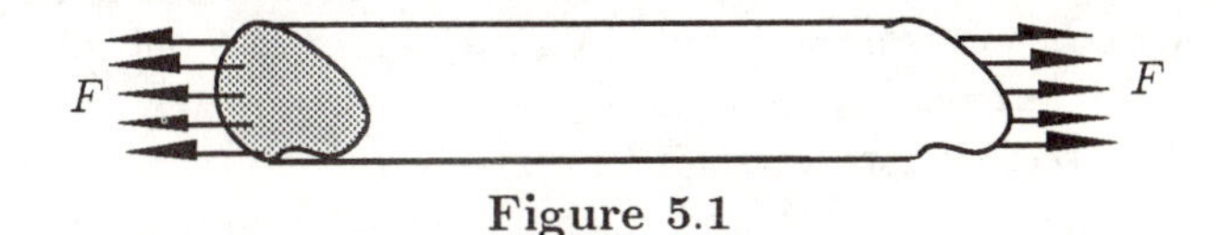

Figure 5.1

The result of applying these forces is to 'stress' the bar (or elastic band). If we touch the exterior of the bar we feel no force. If we cut the bar in the direction of F no change in the appearance of the bar will be observed. However, if we cut the bar transversely at any point then it will immediately begin to tear apart. The forces have been transmitted through the bar to the point of the cut.

We are also familiar with the fact that if the same force F is applied to a bar of larger area of cross section (take a number of elastic bands between your hands) then its effect in deforming the bar is reduced proportionately. Indeed the deformation produced (elongation and thinning) is directly related to the 'normal' stress σ which is defined, for this simple system, as the magnitude of the applied force divided by the cross-sectional area. That is

$$\sigma = \frac{F}{A} \tag{5.1}$$

Now consider the non-uniform bar shown in Figure 5.2

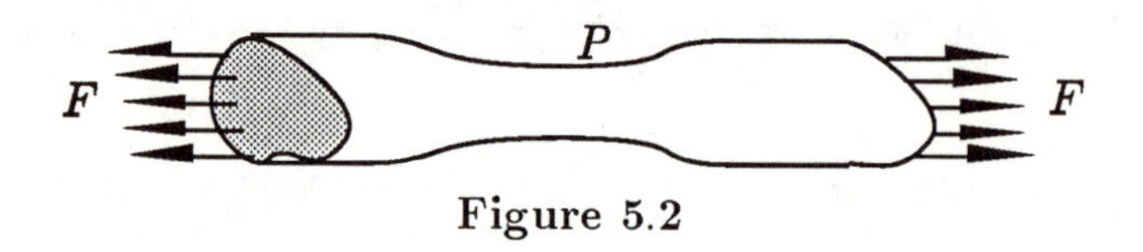

Figure 5.2

It is again a matter of common experience that if the force F is large enough for the bar to rupture then it is very likely that the rupture will occur at a point near P. This is the point at which the normal stress is largest since the same force F is transmitted across each cross section but the area is much reduced in the vicinity of P.

5.2.1 The stress matrix

Consider a continuous body acted upon by forces as shown in Figure 5.3. Within the body we examine a sub-region V which is separated into two parts by a dividing plane S of area A.

The forces exerted by region I on region II are transmitted across S and, by Newton's third law, are equal and opposite to the forces exerted by region II on region I.

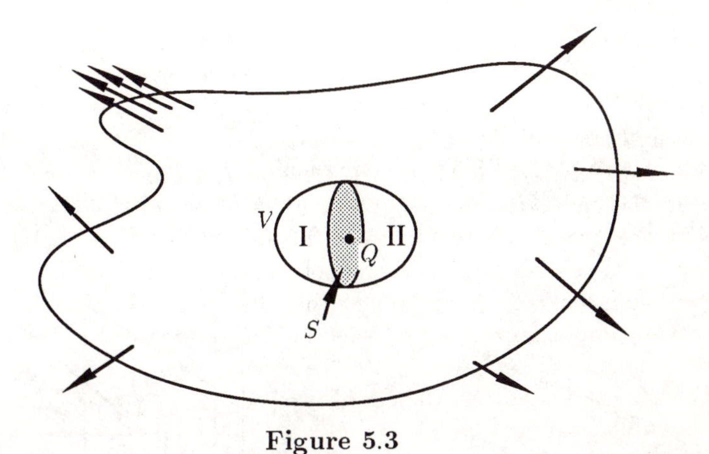

Figure 5.3

Let Q be a point on S. The forces acting in region II may be resolved and are equivalent to a single force

$$\{F\} = \{F_1, F_2, F_3\}^T$$

and a single couple

$$\{M\} = \{M_1, M_2, M_3\}^T$$

acting at Q.

Now imagine shrinking V (and hence S) but always ensuring that Q lies on S. It is a basic assumption of stress analysis that, as the volume V decreases and the resolved force $\{F\}$ and the resolved couple $\{M\}$ also decrease in magnitude, the ratio of force to area tends to a definite limit while the ratio of the couple to area tends to zero.

Mathematically we write

$$\lim_{A\to 0} \frac{\{F\}}{A} = \{T\} \qquad \text{and} \qquad \lim_{A\to 0} \frac{\{M\}}{A} = 0 \tag{5.2}$$

The vector $\{T\}$ is called the **stress vector** at Q and its components have the dimensions of force/area. $\{T\}$ depends not only on the coordinates of Q but also upon the orientation of the plane S. The fundamental result of stress analysis is that there exists, at each point Q of the material, a square matrix $[\sigma]$ independent of the orientation of S and such that

$$\{T\} = [\sigma]\{n\} \tag{5.3}$$

where $\{n\}$ denotes the unit vector normal to S (i.e. $\{n\}^T\{n\} = 1$). The proof of (5.3), which is known as Cauchy's relation, follows from the requirement that the net forces on an elementary region of the body which is in equilibrium should be zero.

The stress vector $\{T\}$ at each point Q within a body is fully defined by (5.3). It is naturally associated with an infinitessimal plane area with unit normal $\{n\}$.

The vector $\{T\}$ has a component along $\{n\}$ which is called the **normal stress** at Q and a component tangential to the plane area called the **shear stress** at Q. See Figure 5.4(a).

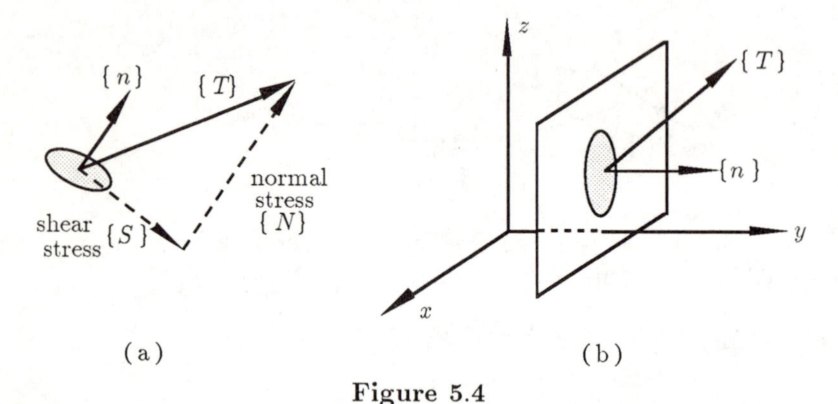

Figure 5.4

By choosing $\{n\}$ to be directed along each of the coordinate axes in turn it is possible to give a direct interpretation to each component of the stress matrix. For example, if we choose, as in Figure 5.4(b), a plane area with unit normal along the y-axis i.e. $\{n\} = \{0, 1, 0\}^T$ then, from (5.3),

$$\{T\} = [\sigma]\{n\} = \begin{Bmatrix} \sigma_{12} \\ \sigma_{22} \\ \sigma_{32} \end{Bmatrix}$$

In this case the normal stress vector say $\{N\}$ (in the y-direction) has magnitude:

$$\{T\}^T\{n\} = \sigma_{22} \qquad \text{so that} \qquad \{N\} = \sigma_{22}\{n\}.$$

Also the shear stress $\{S\}$ in this case is

$$\begin{aligned}\{S\} &= \{T\} - \{N\} \\ &= \{T\} - \sigma_{22}\{n\} \\ &= \begin{Bmatrix} \sigma_{12} \\ \sigma_{22} \\ \sigma_{32} \end{Bmatrix} - \sigma_{22}\begin{Bmatrix} 0 \\ 1 \\ 0 \end{Bmatrix} = \begin{Bmatrix} \sigma_{12} \\ 0 \\ \sigma_{32} \end{Bmatrix}\end{aligned}$$

Repeating the argument for the case $\{n\} = \{1, 0, 0\}^T$ we find

$$\{T\} = \begin{Bmatrix} \sigma_{11} \\ \sigma_{21} \\ \sigma_{31} \end{Bmatrix} \qquad \{T\}^T\{n\} = \sigma_{11} \qquad \{S\} = \begin{Bmatrix} 0 \\ \sigma_{21} \\ \sigma_{31} \end{Bmatrix}$$

whilst for $\{n\} = \{0, 0, 1\}^T$ we obtain

$$\{T\} = \begin{Bmatrix} \sigma_{13} \\ \sigma_{23} \\ \sigma_{33} \end{Bmatrix} \qquad \{T\}^T\{n\} = \sigma_{33} \qquad \{S\} = \begin{Bmatrix} \sigma_{13} \\ \sigma_{23} \\ 0 \end{Bmatrix}$$

We can conclude that the first subscript of σ_{ij} denotes the direction of the component of the stress-vector while the second subscript denotes the direction of the normal to the surface on which it acts (see Figure 5.5).

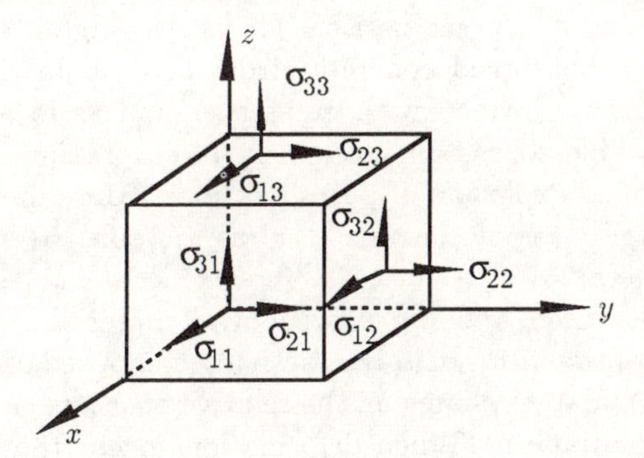

Figure 5.5

The relation, (5.3), between the stress vector and the stress matrix can be deduced using Newton's first law as applied to an infinitessimal region of the body. If this law is applied to an arbitrarily shaped region of the body which has been stressed as a result of the application of body forces $\{B\} = \{B_1, B_2, B_3\}^T$ (such as gravity) to the interior and surface forces $\{T\}$ acting on the boundary then we find that the 'stress' represented by $[\sigma]$ is further constrained.

These 'equilibrium equations' are

$$\begin{aligned} \frac{\partial \sigma_{11}}{\partial x} + \frac{\partial \sigma_{12}}{\partial y} + \frac{\partial \sigma_{13}}{\partial z} + \rho B_1 &= 0 \\ \frac{\partial \sigma_{21}}{\partial x} + \frac{\partial \sigma_{22}}{\partial y} + \frac{\partial \sigma_{23}}{\partial z} + \rho B_2 &= 0 \\ \frac{\partial \sigma_{31}}{\partial x} + \frac{\partial \sigma_{32}}{\partial y} + \frac{\partial \sigma_{33}}{\partial z} + \rho B_3 &= 0 \end{aligned} \qquad (5.4)$$

where ρ denotes the density of the body. (Note that the surface forces $\{T\}$ are absent from these relations). The equations (5.4) relate to the equilibrium of forces. Of course it is also required that the moments be in equilibrium as well. This constrains the stress matrix to be symmetric:

$$[\sigma]^T = [\sigma] \qquad (\text{or } \sigma_{ij} = \sigma_{ji}) \tag{5.5}$$

5.3 The strain matrix

The concept of strain is applicable to any material whether it flows like water or is rigid like iron. In this section we shall be concerned with structural materials which are able to sustain or support large loads. The material may take many geometrical forms: it might, for example, be in the form of a reinforced concrete girder as used in the construction industry. It may be formed with great precision as in an engine block which must be able to sustain extremely large pressure loads resulting from the controlled explosion of petrol with air. More mundanely it may simply be a large mass of concrete used as a stable 'bed' for a sensitive optical experiment.

Consider then a body of structural material. Although often modelled as a continuum the body is actually composed of a large number of particles (crystals). A change in the relative positions between particles is called a **deformation.** When deformations occur the body is said to be **strained.** The strain may result from the application of compressive or tensile forces applied to the surface or from the action of body forces acting throughout the body. However, for many applications the effect of body forces, particularly gravity, can be ignored as their deforming effects can be shown to be small in comparison to the effects of surface stresses. Strain can also be produced by simply heating the body but in this introductory text we shall ignore temperature effects.

In mechanics, bodies are often treated as rigid i.e. all particles are assumed to remain a fixed distance apart. In elasticity theory it is the very fact that bodies are deformable that is of interest. Most structural materials **appear** not to deform unless the loads applied to them are extremely large. In fact, under 'normal' loads all bodies (fixed in position so as not to be movable as a rigid body) do deform but the deformations are extremely small. In small-displacement elasticity theory this natural property of structural materials is exploited.

Of course, we are all familiar with substances like soft putty or rubber which easily deform under the action of normal loads. These, and substances like them, cannot be analysed by the theory outlined here.

To obtain a quantative measure of strain we introduce a

mathematical description of deformation.

Consider the body shown in Figure 5.6. Each particle in the body, such as that at the point A has a specific coordinate position

$$\{p\} = \{x, y, z\}^T$$

with respect to the Cartesian system with an origin at O.

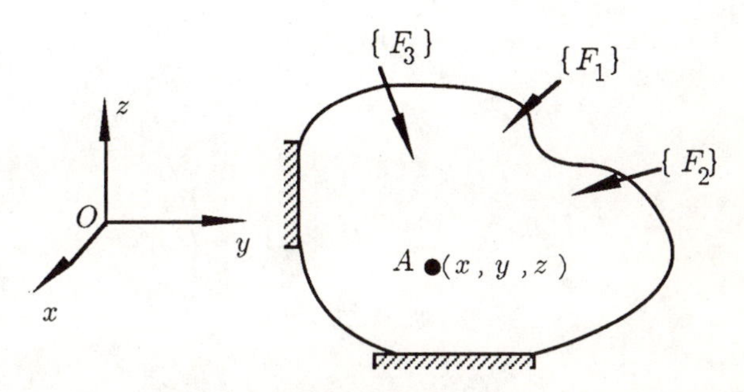

Figure 5.6

Suppose that under the action of applied forces the body is deformed, this resulting in the movement or displacement of the particle at A to a new point A^*. If the coordinates of A^* are $\{p^*\} \equiv \{x^*, y^*, z^*\}^T$ then the displacement of A, denoted by $\{d\} = \{u, v, w\}$ is given by the vector difference

$$\{d\} = \{p^*\} - \{p\} \tag{5.6}$$

As we have noted, for structural materials under normal loads the magnitude of this displacement $(\{d\}^T\{d\})^{1/2}$ is extremely small for all particles.

To obtain a measure of the deformation or strain induced in the body by the applied loads we concentrate on the change in length of an elementary line of particles AB. Suppose this line is deformed to the line A^*B^* after the loads are applied. We shall develop a measure of strain which is appropriate to a very small region within the body and assume that B, and hence B^*, are very close to A and A^* respectively. Anticipating this we represent the coordinates of B as $\{x+\delta x, y+\delta y, z+\delta z\}^T$ and those of B^* as $\{x^*+\delta x^*, y^*+\delta y^*, z^*+\delta z^*\}^T$. See Figure 5.7 where, for clarity, the deformation has been exaggerated.

The lengths of the element before and after deformation are written as δs and δs^* respectively and, using the scalar product to determine the magnitude, we obtain:

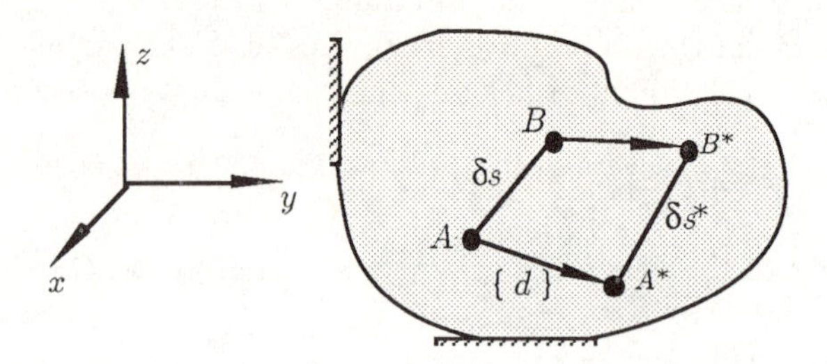

Figure 5.7

$$(\delta s)^2 = \{\delta p\}^T\{\delta p\} = \{\delta x, \delta y, \delta z\}\begin{Bmatrix} \delta x \\ \delta y \\ \delta z \end{Bmatrix}$$
$$= \delta x^2 + \delta y^2 + \delta z^2$$

and

$$(\delta s^*)^2 = \{\delta p^*\}^T\{\delta p^*\}$$
$$= (\delta x^*)^2 + (\delta y^*)^2 + (\delta z^*)^2$$

An obvious measure of strain would be the change in length $(\delta s^* - \delta s)$ of a line of particles in each small region of the body. However as this quantity involves square roots the measure that has been universally adopted is $(\delta s^*)^2 - (\delta s)^2$:

$$(\delta s^*)^2 - (\delta s)^2 = \{\delta p^*\}^T\{\delta p^*\} - \{\delta p\}^T\{\delta p\} \quad \textbf{(5.7)}$$

For a given deformation we shall assume that the final coordinates $\{p^*\}$ of a particle are smooth functions of the original coordinates $\{p\}$.

Since $\{p^*\} = \{d\} + \{p\}$ then, from the chain rule of calculus,

$$\{\delta p^*\} = \frac{\partial}{\partial x}\{p^*\}\delta x + \frac{\partial}{\partial y}\{p^*\}\delta y + \frac{\partial}{\partial z}\{p^*\}\delta z$$
$$= \frac{\partial}{\partial x}\{d\}\delta x + \frac{\partial}{\partial y}\{d\}\delta y + \frac{\partial}{\partial z}\{d\}\delta z + [\mathrm{I}]\{\delta p\}$$

where [I] is the 3×3 identity matrix. This relation is clearly

$$\{\delta p^*\} = \{\delta d\} + [\mathrm{I}]\{\delta p\}$$

Hence, carrying through the algebra implied in (5.7) we find

$$(\delta s^*)^2 - (\delta s)^2 = \{\delta d\}^T\{\delta p\} + \{\delta p\}^T\{\delta d\} \quad \textbf{(5.8)}$$

In (5.8) we have ignored the **non-linear** term $\{\delta d\}^T\{\delta d\}$. We do this in order to produce a **linear theory** of elasticity. The predictions (on deformations and stresses) that are obtained from this linear theory compare well with the deformations and stresses actually measured in structural materials thereby justifying the assumption of linearity.

If the right hand side of (5.8) is expanded (which is left as an exercise for the reader) it will clearly be seen to be a **quadratic form** in the components of $\{\delta p\}$ namely $\{\delta x, \delta y, \delta z\}^T$. Thus (5.8) can be rewritten in the standard form:

$$(\delta s^*)^2 - (\delta s)^2 = 2\{\delta p\}^T[\varepsilon]\{\delta p\} \tag{5.9}$$

in which $[\varepsilon]$ is a square symmetric matrix called the **strain matrix**:

$$[\varepsilon] = \begin{bmatrix} \frac{\partial u}{\partial x} & \frac{1}{2}(\frac{\partial u}{\partial y} + \frac{\partial v}{\partial x}) & \frac{1}{2}(\frac{\partial u}{\partial z} + \frac{\partial w}{\partial x}) \\ \frac{1}{2}(\frac{\partial u}{\partial y} + \frac{\partial v}{\partial x}) & \frac{\partial v}{\partial y} & \frac{1}{2}(\frac{\partial v}{\partial z} + \frac{\partial w}{\partial y}) \\ \frac{1}{2}(\frac{\partial u}{\partial z} + \frac{\partial w}{\partial x}) & \frac{1}{2}(\frac{\partial v}{\partial z} + \frac{\partial w}{\partial y}) & \frac{\partial w}{\partial z} \end{bmatrix} \tag{5.10}$$

Obviously if each component of the strain matrix is zero then, from (5.9), $(\delta s^*)^2 = (\delta s)^2$ implying that the particles of the body remain a fixed distance apart, this being the description of a rigid body.

The diagonal components of $[\varepsilon]$, say $\varepsilon_{11}, \varepsilon_{22}, \varepsilon_{33}$, are called the longitudinal components of strain. The interpretation of each of these components is most easily realized if we choose, in turn, elementary filaments initially aligned along the coordinate axes. For example if the filament of particles is, before deformation, aligned along the x-axis then

$$\{\delta p\} = \{\delta x, 0, 0\}^T$$

and so (5.9) becomes

$$\begin{aligned}(\delta s^*)^2 - (\delta s)^2 &= 2\{\delta x, 0, 0\}[\varepsilon]\begin{Bmatrix} \delta x \\ 0 \\ 0 \end{Bmatrix} \\ &= 2\varepsilon_{11}\delta x^2\end{aligned}$$

that is

$$(\delta s^* - \delta s)(\delta s^* + \delta s) = 2\varepsilon_{11}\delta x^2$$

But $\delta s = \delta x$ in this case and $\delta s^* \approx \delta s$ implying that

$$\frac{\delta s^* - \delta s}{\delta s} = \varepsilon_{11}$$

Hence we can interpret ε_{11} as the change in length per unit length of a line of particles originally along the x-direction. This interpretation will be used in Section 5.8 where the longitudinal deformation of a beam is considered. Note that if $\varepsilon_{11} > 0$ at a point this indicates that straight line segments in the x-direction in the close neighbourhood of the point are stretched and if $\varepsilon_{11} < 0$ at the point then contraction of these segments has occurred.

Similar interpretations follow for ε_{22} and ε_{33}. The off-diagonal components of $[\varepsilon]$ are less easy to interpret. They are connected with shearing effects and are known therefore as the **shear components** of strain. This interpretation is obtained by considering two filaments in one of the coordinate planes which are perpendicular before deformation. After deformation these two elements will no longer be perpendicular but the angle between the two will **close** by an angle ϕ (in radians) which is found to be

$$\phi = 2\varepsilon_{12}$$

(if the plane is in the xy plane.) Similar interpretations apply to ε_{13} and ε_{23}.

5.4 The constitutive equations

The simple tension experiment and the generalized Hooke's law

Perhaps the simplest experiment in structural mechanics is the stretching of a thin wire by the application of a force of magnitude F (see Figure 5.8(a)). For each value of F the extension of the wire from its undeformed length ℓ can be determined. It is found (if F is not too large) that, for structural materials such as steel, the relation between force and extension is linear (see Figure 5.8(b)), that is

$$F = c(\ell^* - \ell) \qquad \textbf{(5.11)}$$

where the constant of proportionality c is called the stiffness of the wire. This simple result is known as **Hooke's law**.

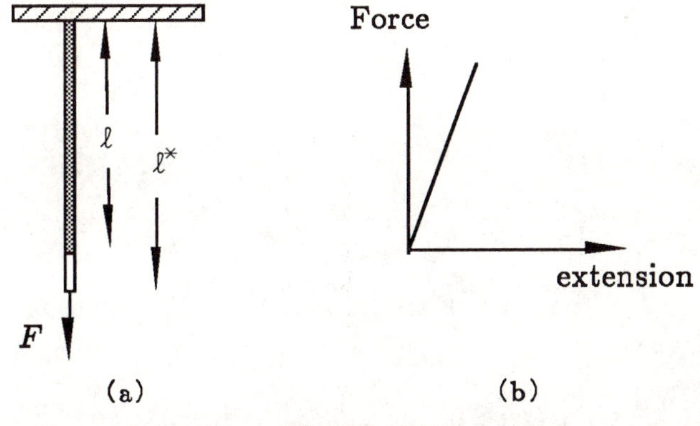

Figure 5.8

If A_0 is the original cross-sectional area of the wire then we can rewrite (5.11) as

$$\frac{F}{A_0} = E\left(\frac{\ell^* - \ell}{\ell}\right) \tag{5.12}$$

where $E = c\ell/A_0$ is known as **Young's modulus** for the material from which the wire has been formed. For engineering materials E is very large, ranging from 7×10^{10} N m^{-2} for aluminium to 2×10^{11} N m^{-2} for structural steel.

In (5.12) F/A_0 is the longitudinal stress and $(\ell^* - \ell)/\ell$ is the longitudinal strain. For this simple configuration, then, the relation between stress and strain is **linear**.

As well as the obvious longitudinal extension in the above experiment the wire is also observed to 'thin', that is the lateral dimensions of the wire contract if F is a tensile force. The lateral 'strain' can be measured and it is found to be proportional to the longitudinal strain (and hence to the longitudinal stress):

$$\text{lateral strain} = -\nu(\text{longitudinal strain}) \tag{5.13}$$

The constant of proportionality ν is known as **Poisson's ratio**. For steel $\nu \approx 0.3$ and for most structural materials it lies between 0.2 and the theoretical maximum value of 0.5.

The simple tension experiment is essentially one-dimensional. To proceed further we must demonstrate a relationship between the stress matrix and the strain matrix for generally shaped bodies in two and three dimensions. For simplicity we shall restrict our attention to linear elastic materials (or, more technically, to linear homogeneous isotropic elastic materials). Fortunately, most structural materials may be so described.

A material is said to be **elastic** if the state of stress at a point is dependent only on the deformation (strain) at that point and not on the history of the deformation. Such a material, when unloaded, will always return to its initial configuration. We shall consider materials in which the relation between stress and strain matrices depends on just two constants and is given by

$$[\sigma] = 2\mu[\varepsilon] + \lambda[\mathrm{I}]tr[\varepsilon] \tag{5.14}$$

Here $tr[\varepsilon]$, called the trace of the matrix $[\varepsilon]$, is the sum of the diagonal components i.e.

$$tr[\varepsilon] = \varepsilon_{11} + \varepsilon_{22} + \varepsilon_{33} \tag{5.15}$$

The two constants λ and μ are known as the **Lamé constants**. They are directly related to the Young's modulus E and the Poisson ratio ν:

$$\lambda = \frac{E\nu}{(1+\nu)(1-2\nu)} \qquad \mu = \frac{E}{2(1+\nu)} \tag{5.16}$$

The relations in (5.14) were first derived by Cauchy in 1828 and are the **constitutive relations** for the theory of linear elasticity. Whilst it can be shown that a general crystalline material is characterised by 21 independent stiffnesses, most structural materials can be fully characterised by just two stiffnesses.

5.5 Plane strain

In this section we restrict our attention to two-dimensional elasticity. There are two basic methods by which a reduction from a full three-dimensional theory of linear elasticity can be carried through.

- **Theory of plane strain**
  Here it is assumed that the displacement vector has only two non-zero components (each of which is a function of only two variables).

- **Theory of generalized plane stress**
  In this case one variable is removed from the equations of linear elasticity by an averaging process.

Each of these theories has applications to important, physically realistic, elastic systems. It is an unexpected bonus that the equations which arise in each are identical in form so allowing the topics to be treated in parallel (solving a problem in plane strain implies the solution of a corresponding problem in generalized plane stress).

We only consider the equations of plane strain and remove the z-variable. That is, we assume that the displacement components $\{u, v, w\}^T$ have the particular form:

$$u = f(x, y) \qquad v = g(x, y) \qquad w = 0 \tag{5.17}$$

This particular choice for the non-zero components of the displacement vector implies (using (5.10)) that the only non-zero strain components are $\varepsilon_{11}, \varepsilon_{12}$ and ε_{22} so that deformations only occur in the plane z =constant; hence the terminology 'plane strain'.

Intuitively, we expect that surface stresses will need to be applied in order to maintain the component w at zero. In a practical situation this may be difficult to arrange and to some extent this reduces the value of the theory based upon (5.17). However there are a number of important structures in which a state of plane strain may be considered to exist. For example, within a thin cross section of a dam or a long pressure vessel, far from the ends, the assumption of the plane strain conditions (5.17) would be a reasonably accurate approximation to the actual deformations of these sections (see Figure 5.9).

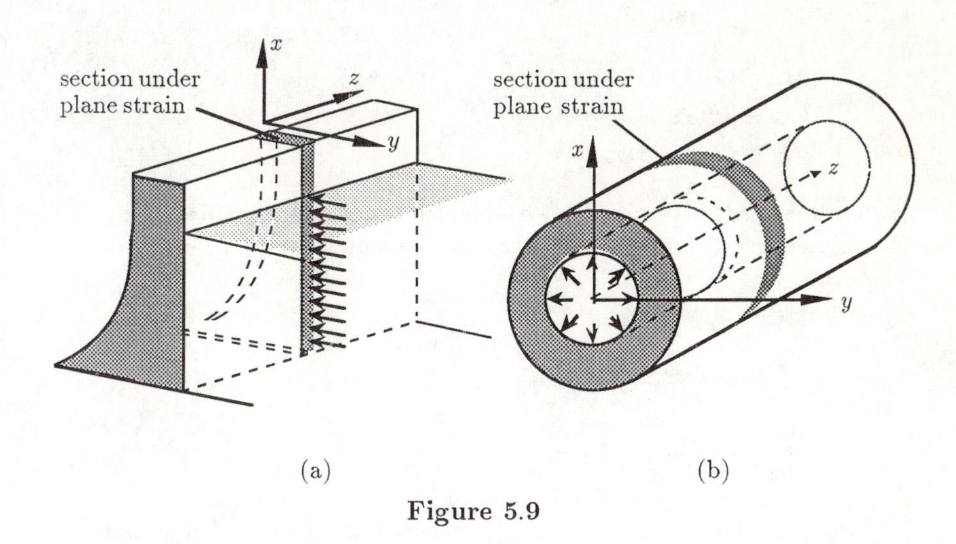

Figure 5.9

In the remainder of this section we will assume that (5.17) applies to a two-dimensional region A (possibly containing holes) which is bounded by a curve (or collection of curves) Γ. On the boundary, deforming planar forces may be applied (see Figure 5.10).

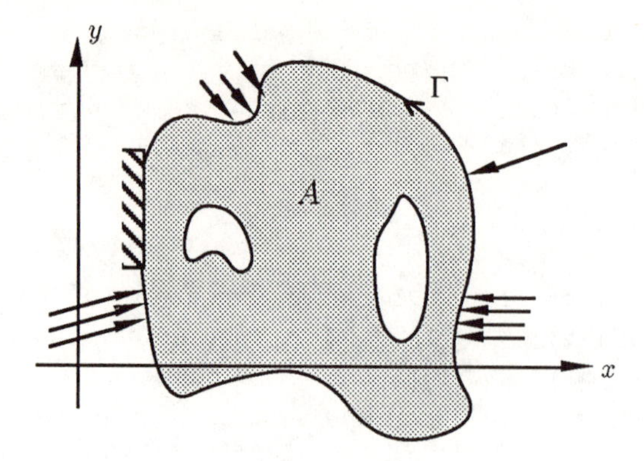

Figure 5.10

The elasticity equations to be solved when plane strain conditions hold are easily obtained from the more general equations (5.4), (5.10) and (5.14). These equations are:

- Strain-displacement equations

$$\varepsilon_{11} = \frac{\partial u}{\partial x} \qquad \varepsilon_{12} = \frac{1}{2}\left(\frac{\partial u}{\partial y} + \frac{\partial v}{\partial x}\right) \qquad \varepsilon_{22} = \frac{\partial v}{\partial y} \tag{5.18}$$

- Constitutive equations

$$\begin{aligned} \sigma_{11} &= (2\mu + \lambda)\varepsilon_{11} + \lambda\varepsilon_{22} \\ \sigma_{12} &= 2\mu\varepsilon_{12} \\ \sigma_{22} &= (2\mu + \lambda)\varepsilon_{22} + \lambda\varepsilon_{11} \end{aligned} \tag{5.19}$$

(Also, necessarily, $\sigma_{33} = \nu(\sigma_{11} + \sigma_{22})$ which results from the demand that the displacement component w in the z-direction should be zero.)

- Equilibrium equations

$$\begin{aligned} \frac{\partial \sigma_{11}}{\partial x} + \frac{\partial \sigma_{12}}{\partial y} + \rho B_1 = 0 \\ \frac{\partial \sigma_{21}}{\partial x} + \frac{\partial \sigma_{22}}{\partial y} + \rho B_2 = 0 \end{aligned} \tag{5.20}$$

The above equations can be written in matrix form. We first define column vectors of displacement, strain and stress:

$$\{d\} \equiv \begin{Bmatrix} u \\ v \end{Bmatrix} \qquad \{s\} \equiv \begin{Bmatrix} \varepsilon_{11} \\ \varepsilon_{22} \\ 2\varepsilon_{12} \end{Bmatrix} \qquad \{\Omega\} \equiv \begin{Bmatrix} \sigma_{11} \\ \sigma_{22} \\ \sigma_{12} \end{Bmatrix} \tag{5.21}$$

(Note the factor of 2 inserted for convenience in the third component of the strain vector $\{s\}$.)

Further, we define the matrix of differential operators:

$$[\partial] \equiv \begin{bmatrix} \dfrac{\partial}{\partial x} & 0 \\ 0 & \dfrac{\partial}{\partial y} \\ \dfrac{\partial}{\partial y} & \dfrac{\partial}{\partial x} \end{bmatrix} \tag{5.22}$$

and, finally, it is convenient to define a matrix of stiffnesses:

$$[C] \equiv \begin{bmatrix} 2\mu + \lambda & \lambda & 0 \\ \lambda & 2\mu + \lambda & 0 \\ 0 & 0 & \mu \end{bmatrix} \tag{5.23}$$

With these definitions we can reformulate (5.18), (5.19) and (5.20) respectively as:

$$\{s\} = [\partial]\{d\} \qquad \{\Omega\} = [C]\{s\} \qquad [\partial]^T\{\Omega\} + \rho\{B\} = 0 \tag{5.24}$$

(The derivation is left as an exercise for the reader.)

The system of differential and algebraic equations (5.24) must be supplemented by boundary conditions. On part of the boundary, Γ_0 say, a stress vector may be applied. On those portions the Cauchy relation (5.3) implies:

$$\begin{aligned} \sigma_{11}n_1 + \sigma_{12}n_2 &= T_1 \\ \sigma_{21}n_1 + \sigma_{22}n_2 &= T_2 \end{aligned} \tag{5.25}$$

in which $\{T\} = \{T_1, T_2\}^T$ is a **given** stress vector on the boundary. Equations (5.25) are called **suppressible** boundary conditions.

On the remaining part (if any) of the boundary, Γ_1 say, the displacement is specified:

$$\{d\} = \{d_0\} \tag{5.26}$$

in which $\{d_0\} = \{u_0, v_0\}^T$ is a **given** displacement (usually zero). Equation (5.26) defines **essential** boundary conditions. (The reason for the terminology 'essential' and 'suppressible' is outlined in Section 5.7 after the introduction of the potential energy concept.)

5.6 Strain energy

When a force $\{F\}$ moves its point of application through a displacement $\{d\}$ the work done is given by $\{d\}^T\{F\}$. When an elastic body is deformed by the application of surface or body forces the work done by these forces is stored in the body and used to ensure that the body recovers its original configuration when the deforming forces are removed.

A simple example is a wire of stiffness c. If the wire is deformed by applying a force then, on removing the force, the wire recovers its original length.

Work stored within a body in this way is called **strain energy**. The strain energy is exactly equal to the work done by the external forces in deforming the body. For the case of a simple wire it is a straightforward matter to determine the strain energy. Suppose a wire of unstressed length ℓ_0 is deformed to a length ℓ. From Hooke's law

$$F = c(\ell - \ell_0)$$

The force required to extend the spring by a further length $\delta\ell$ is $F + \delta F$ where

$$F + \delta F = c(\ell + \delta\ell - \ell_0)$$

that is

$$\delta F = c\delta\ell$$

The work done by this force is

$$\delta W = (F + \delta F)\delta\ell$$

and hence, to a good approximation,

$$\delta W = c(\ell - \ell_0)\delta\ell$$

The total work done in extending the spring from length ℓ_0 to length ℓ^* is therefore

$$\begin{aligned} W &= \int_{\ell_0}^{\ell^*} c(\ell - \ell_0)\, d\ell \\ &= \frac{1}{2}c(\ell^* - \ell_0)^2 \end{aligned}$$

We can rewrite this result in the form

$$W = \tfrac{1}{2}F^*(\ell^* - \ell_0)$$

where F^* is the magnitude of the final force applied to the wire. Thus, for this simple system, the strain energy is equal to one half of the final force multiplied by the final displacement at the point of application of F^*. We also note that if A is the cross-sectional area of the wire then, in terms of stress and strain,

$$W = \frac{1}{2}\frac{F^*}{A}A\ell_0\frac{(\ell^* - \ell_0)}{\ell_0} = \frac{1}{2}\sigma_{11}\varepsilon_{11}V$$

(where V is the volume of the wire). We have assumed here that the wire lies along the x-axis.

It is also useful to express the strain energy entirely in terms of the displacement of the end point $u = (\ell^* - \ell_0)$. We obtain

$$W = \frac{1}{2}c(\ell^* - \ell_0)^2 = \frac{1}{2}c\,u^2$$

and, clearly,

$$\frac{\partial W}{\partial u} = c\,u = c(\ell^* - \ell_0) = F^*$$

This last expression is a special case of the so-called **Castigliano relations**. These are relations (for a general elastic body) between the displacement $\{d\}$ at the point of application of a force $\{F\}$ and the strain energy W:

$$\frac{\partial W}{\partial u} = F_1 \qquad \frac{\partial W}{\partial v} = F_2 \qquad \frac{\partial W}{\partial w} = F_3 \tag{5.27}$$

It is interesting to discuss the form of the strain energy for a more general elastic body. Consider a segment δS of surface on an arbitrarily shaped elastic body which is deformed by the action of surface stresses. We imagine that the final displacement $\{d\}$ at each point is the result of building up the applied stresses from zero to the final value $\{T\}$. The final force on δS is $\{T\}\delta S$. Overall, the displacement of the segment is $\{d\}$ and so we can argue that the work done by the surface stress on this segment is

$$W = \frac{1}{2}\{d\}^T\{T\}\delta S$$

(The factor $\frac{1}{2}$ arises because, as with the wire in the simple tension experiment, the final value of the force $\{T\}\delta S$ does not move its point of application throughout the full magnitude of $\{d\}$.) Repeating this argument for the whole bounding surface of the body we find that the total work done by the surface stresses is given by the surface integral

$$\frac{1}{2}\iint_S \{d\}^T\{T\}\,dS$$

By a similar argument, the total work done by the body forces $\{B\}$ is given by the volume integral

$$\frac{1}{2}\iiint_V \rho\{d\}^T\{B\}\,dV$$

Hence the strain energy W of a body V bounded by a surface S is

$$W = \frac{1}{2}\iint_S \{d\}^T\{T\}\,dS + \frac{1}{2}\iiint_V \rho\{d\}^T\{B\}\,dV \qquad \textbf{(5.28a)}$$

For a body in a plane strain situation, say a plate of area A, the volume integral reduces to a double integral over A and the surface integral becomes a line integral around the boundary Γ (see Figure 5.10). That is, in this case,

$$W = \frac{h}{2}\int_\Gamma \{d\}^T\{T\}\,d\ell + \frac{h}{2}\iint_A \rho\{d\}^T\{B\}\,dx\,dy \qquad \textbf{(5.28b)}$$

where h is the thickness of the plate. (The density term ρ must be interpreted here as a mass per unit area and $\{T\}$ as a force per unit length applied on Γ.)

Now the first integral on the right-hand side of (5.28b) is, using (5.25),

$$\frac{h}{2}\int_\Gamma \{(u\sigma_{11} + v\sigma_{21})n_1 + (u\sigma_{12} + v\sigma_{22})n_2\}\,d\ell$$

But since $\{\frac{dx}{d\ell}, \frac{dy}{d\ell}\}^T$ is a unit tangent vector to Γ then $\{\frac{dy}{d\ell}, -\frac{dx}{d\ell}\}^T$ is a unit normal to Γ and so

$$\frac{h}{2}\int_\Gamma \{d\}^T\{T\}\,d\ell = \frac{h}{2}\int_\Gamma \{(u\sigma_{11} + v\sigma_{21})dy - (u\sigma_{12} + v\sigma_{22})dx\}$$

But, by Green's Theorem in the Plane, a line integral over a closed curve Γ and a double integral over a region A bounded by Γ are related by

$$\int_\Gamma (F dx + G dy) = \iint_A \left(\frac{\partial G}{\partial x} - \frac{\partial F}{\partial y}\right)\,dx\,dy$$

for suitably defined functions G, F.

Hence we have

$$\frac{h}{2}\int_{\Gamma}\{d\}^T\{T\}\,d\ell = \frac{h}{2}\iint_A \{\frac{\partial u}{\partial x}\sigma_{11} + \frac{\partial v}{\partial x}\sigma_{21} + \frac{\partial u}{\partial y}\sigma_{12} + \frac{\partial v}{\partial y}\sigma_{22}\}\,dx\,dy$$
$$+\frac{h}{2}\iint_A \{u\frac{\partial \sigma_{11}}{\partial x} + v\frac{\partial \sigma_{21}}{\partial x} + u\frac{\partial \sigma_{12}}{\partial y} + v\frac{\partial \sigma_{22}}{\partial y}\}\,dx\,dy$$
$$= \frac{h}{2}\iint_A \{s\}^T\{\Omega\}\,dx\,dy + \frac{h}{2}\iint_A \{u\}^T[\partial]^T\{\Omega\}\,dx\,dy \tag{5.29}$$

(The reader should verify (5.29), by simple matrix manipulations using the definitions (5.21).) Now, using the last equation of (5.24) and also (5.29) in (5.28b) we find for a body in plane strain:

$$W = \frac{h}{2}\iint_A \{s\}^T\{\Omega\}\,dx\,dy \tag{5.30a}$$

For a general elastic body it can be shown that the strain energy is

$$W = \frac{1}{2}\sum_{i=1}^{3}\sum_{j=1}^{3}\iiint_V \sigma_{ij}\varepsilon_{ij}\,dV \tag{5.30b}$$

5.7 Potential energy

The strain energy discussed in the previous section has, as we saw, a direct physical interpretation in terms of the work done on a body in order to deform it. The **potential energy** associated with a strained body, on the other hand, is a purely mathematical concept but one which has far reaching applications in elasticity theory, particularly in the development of finite element approximations to elastic problems.

We shall restrict our attention (in the main) to plane strain problems. Consider an elastic body A, bounded by a curve Γ, and let $(\{T\},\{B\},\{d\},\{s\},\{\Omega\})$ be an 'elastic force system' associated with the body. This system satisfies the elasticity equations (5.24), (5.25), (5.26).

We define the potential energy U for plane strain conditions as

$$U = \frac{h}{2}\iint_A \{\tilde{s}\}^T\{\tilde{\Omega}\}\, dxdy - h\int_{\Gamma_0} \{\tilde{d}\}^T\{T\}\, d\ell - h\iint_A \rho\{\tilde{d}\}^T\{B\}\, dxdy \qquad \textbf{(5.31)}$$

The first term in (5.31) is the 'strain energy' (5.30a); the other two terms are the 'work done' due to $\{T\}$ and $\{B\}$ respectively.

In (5.31) the term $\{\tilde{d}\}$ is an arbitrary vector. We can consider it to be an approximation to the actual displacement $\{d\}$ in a given problem, except that it is required to satisfy the boundary conditions (5.26). (This is an 'essential' constraint on $\{\tilde{d}\}$. There is no requirement that the 'force' boundary conditions (5.25) be satisfied. These conditions are 'suppressed'. This explains the terminology 'suppressible' and 'essential' introduced earlier.) The terms $\{\tilde{s}\}$ and $\{\tilde{\Omega}\}$ are related to $\{\tilde{d}\}$ by the usual strain/displacement and stress/strain relations (5.24):

$$\{\tilde{s}\} = [\partial]\{\tilde{d}\} \qquad \{\tilde{\Omega}\} = [C]\{\tilde{s}\} \qquad \textbf{(5.32)}$$

We note that potential energy is a scalar quantity and, for a given elastic problem, its value depends on the choice made for $\{\tilde{d}\}$. It is natural therefore to enquire which vector $\{\tilde{d}\}$, if any, minimizes U. In fact there is a theorem known as the **Theorem of Minimum Potential Energy** which states:

- of all values of U that are obtained by using different vectors $\{\tilde{d}\}$ the minimum value is obtained when $\{\tilde{d}\}$ is the **true** displacement $\{d\}$ (and hence, from (5.32), $\{\tilde{s}\}$ and $\{\tilde{\Omega}\}$ are the true strains and stresses respectively).

The proof of this theorem, while not difficult, is omitted.

5.8 The direct approach to finite elements

It is clear, even from a brief study of the elasticity equations (5.4), (5.10) and (5.14), that the exact determination of the deformation of an elastic body resulting from an applied load is only a realistic possibility if the body and/or loading has an especially simple form. However, in a number of practical situations we do not require the precise detail on stress and strain levels and displacement values that would be available from an exact solution. It follows that, for such situations, we should attempt to approximate the physical system being modelled by simpler systems of equations than those of the exact theory of linear infinitessimal elasticity.

This sort of simplification is used, for example, in the modelling of a slender beam by the well-known Euler–Bernoulli equation. In this approximation it is assumed that the transverse deformation of a slender beam (due to the action of transverse forces) is adequately characterized by the deformation of the centre-line. The deformation of particles not on the centre-line is ignored. As a result of this simplification the full system of linear elasticity equations can be reduced to a single fourth-order linear boundary value problem which can be readily solved.

A still simpler boundary value problem can be obtained by modelling the longitudinal deformation of such a beam due to longitudinal directed forces using an Euler–Bernoulli approach.

Consider then a uniform beam of length ℓ and cross-sectional area A subjected to longitudinal forces P_0, P_ℓ applied at each end (see Figure 5.11). We shall ignore body forces.

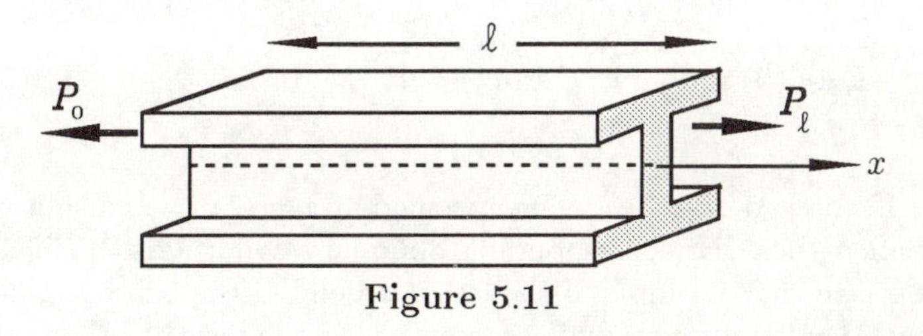

Figure 5.11

In modelling the deformation a reasonable assumption in this case is to imagine that each point on a plane x = constant undergoes the same displacement $u(x)$. That is, we approximate the exact displacement (which, from the exact treatment of the simple tension experiment in Section 5.4, must also include a 'thinning') by:

$$\{d\} = \{u, 0, 0\}^T$$

Consequently the only non-zero component of the strain matrix is

$$\varepsilon_{11} = \frac{du}{dx}$$

Also since the longitudinal stress σ_{11} is related to longitudinal strain as in the simple tension experiment we have

$$\sigma_{11} = E\varepsilon_{11}$$

from which we find the strain energy (5.30b) to be

$$W = \frac{1}{2}\iiint_V \sigma_{11}\varepsilon_{11}\, dV = \frac{1}{2}EA\int_0^\ell \left(\frac{du}{dx}\right)^2 dx$$

The work done by the deforming forces P_0 and P_ℓ is

$$P_0 u(0) + P_\ell u(\ell)$$

Hence the total potential energy for this assumed displacement is

$$U = \frac{1}{2} EA \int_0^\ell \left(\frac{du}{dx}\right)^2 dx - P_0 u(0) - P_\ell u(\ell) \tag{5.33}$$

Now by the theorem of minimum potential energy, U is minimized by the true value of $u(x)$ ('true', that is, to the level of approximation being considered). The problem of finding the form of $u(x)$ is a standard problem in the Calculus of Variations which is outlined in Chapter 8. It is readily shown that $u(x)$ must satisfy the second-order boundary value problem:

$$\frac{d}{dx}\left(EA\frac{du}{dx}\right) = 0 \qquad 0 \leq x \leq \ell \tag{5.34a}$$

$$-EAu'(0) = P_0 \qquad EAu'(\ell) = P_\ell \tag{5.34b}$$

If, instead of forces being specified at $x = 0, x = \ell$, displacements are specified then the corresponding boundary conditions in (5.34b) are inappropriate. For example, if the left-hand end of the beam is held fixed so that $u(0) = 0$ and a stretching force P_ℓ is applied at the other end then (5.34b) must be replaced by

$$u(0) = 0 \qquad EAu'(\ell) = P_\ell \tag{5.34c}$$

In the above analysis, although we are making approximations to the full theory of elasticity, we are nonetheless still modelling the beam in a **continuous** manner in that the basic unknown in the problem (the displacement component u in this case) is modelled by a differential equation. It is now convenient to model the longitudinal deformation in a **discrete** manner, that is, in terms of the measured displacements at each end.

Assuming the cross-section A is constant along the beam (5.34a) can be integrated to give

$$u(x) = c_1 x + c_2 \tag{5.35}$$

where c_1, c_2 are constants. These constants can be expressed in terms of the end-displacements $u(0), u(\ell)$.

Since $u(0) = c_2$ and $u(\ell) = c_1\ell + c_2$ we obtain

$$u(x) = [\frac{u(\ell) - u(0)}{\ell}]x + u(0)$$
$$= u(0)[1 - \frac{x}{\ell}] + u(\ell)[\frac{x}{\ell}] \qquad \textbf{(5.36)}$$

(Note the appearance of the linear shape functions (3.3) and (3.4).)

The relation (5.36) shows that if the end-displacements are known then the displacement throughout the beam can be determined. Also, from (5.34b) and (5.36),

$$P_\ell = EAu'(\ell) = \frac{EA}{\ell}[u(\ell) - u(0)]$$

and

$$-P_0 = EAu'(0) = -\frac{EA}{\ell}[u(\ell) - u(0)]$$

or, in matrix form,

$$\begin{Bmatrix} P_0 \\ P_\ell \end{Bmatrix} = \frac{EA}{\ell}\begin{bmatrix} 1 & -1 \\ -1 & 1 \end{bmatrix}\begin{Bmatrix} u(0) \\ u(\ell) \end{Bmatrix} \qquad \textbf{(5.37)}$$

This relation, (5.37), is the **discrete** model for the longitudinal deformation of a slender beam. The discrete model is, in fact, of far greater use than the continuous model in the analysis of structures comprised of many elements. The reason for this will become clear as we proceed but, even at a superficial level, it is clearly easier to deal with large combinations of matrices than with large numbers of boundary value problems of which one would be required for each element.

An alternative method of deriving (5.37) utilizes the Castigliano relations introduced in Section 5.6. Using (5.36)

$$W = \frac{1}{2}EA\int_0^\ell \left(\frac{du}{dx}\right)^2 dx = \frac{1}{2}EA\int_0^\ell \left\{\frac{u(\ell) - u(0)}{\ell}\right\}^2 dx$$
$$= \frac{1}{2}\frac{EA}{\ell}[u(\ell) - u(0)]^2$$

Hence using the first of (5.27) at $x = 0$ and $x = \ell$:

$$P_0 = \frac{\partial W}{\partial u(0)} = -\frac{EA}{\ell}[u(\ell) - u(0)]$$

and

$$P_\ell = \frac{\partial W}{\partial u(\ell)} = \frac{EA}{\ell}[u(\ell) - u(0)]$$

These relations are equivalent to (5.37).

5.9 Pin-jointed elements

A pin-jointed element is a slender beam for which it is assumed that the predominant deformation is longitudinal and which can only sustain longitudinal forces. The end-points (see Figure 5.12) are universal joints not capable of transmitting torsional couples, bending couples or shear forces.

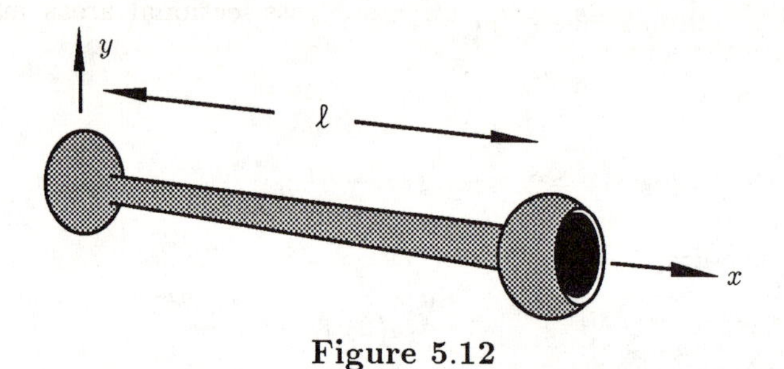

Figure 5.12

A pin-jointed element can only be stretched or compressed. As we have seen in the previous section it is fully characterized by the displacement measured at its end points (or nodes) and by the forces applied there. This information is displayed in Figure 5.13

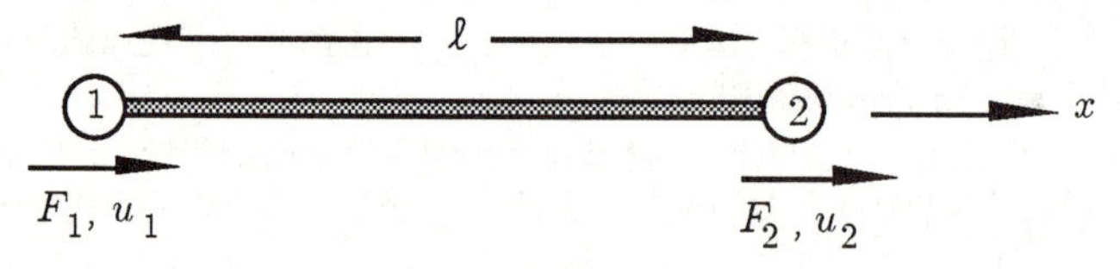

Figure 5.13

Using the notation of Figure 5.13, (5.37) becomes

$$\begin{Bmatrix} F_1 \\ F_2 \end{Bmatrix} = \frac{EA}{\ell} \begin{bmatrix} 1 & -1 \\ -1 & 1 \end{bmatrix} \begin{Bmatrix} u_1 \\ u_2 \end{Bmatrix} \tag{5.38}$$

or

$$\{F\} = [k]\{d\} \tag{5.39}$$

The matrix $[k]$ relating the applied forces to the measured displacements is called the **stiffness matrix** of the pin-jointed element. It is referred to 'local' coordinates (x, y, z). The stiffness matrix can be regarded as the basic building block of structural analysis. The factor EA/ℓ is known as the stiffness of the element. Note that the stiffness matrix in (5.38) is singular and symmetric and has non-negative terms on the leading diagonal. These are properties common to all stiffness matrices.

• Assemblages of pin-jointed elements

We now study the connection of pin-jointed elements together in an assemblage. A simple grouping consisting of two pin-jointed elements connected at a node is shown in Figure 5.14. The elements may be made from different materials, have different cross-sectional areas and be of different lengths and consequently have different stiffnesses

$$k_a = \frac{E_a A_a}{\ell_a} \qquad \text{and} \qquad k_b = \frac{E_b A_b}{\ell_b}$$

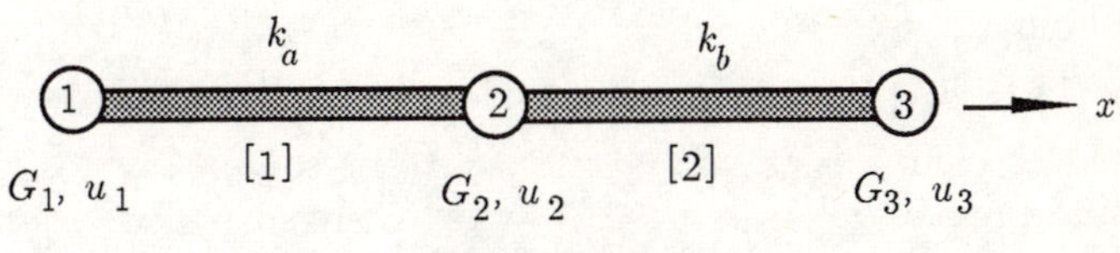

Figure 5.14

We now have two elements labelled [1] and [2] as shown. At each node a force may be applied and a corresponding displacement measured (each directed along the positive x-axis).

The forces G_1, G_2 and G_3 are called the **global forces** acting at the nodes. They are to be distinguished from the **internal forces** developed within each element which are (for a pin-jointed element) related to the displacements measured at each node by (5.38). The global forces give rise to the internal forces. At each node the internal forces within the adjoining elements combine (vectorially) to produce the global force at that node. This behaviour is to be contrasted with the displacement at a node, which is the **same** for every element attached to that node.

The extension of element [1] is $(u_2 - u_1)$ whilst that of element [2] is $(u_3 - u_2)$. Thus the strain energy of the assemblage is

$$W = \frac{1}{2}k_a(u_2 - u_1)^2 + \frac{1}{2}k_b(u_3 - u_2)^2 \qquad \textbf{(5.40)}$$

The Castigliano relations can be used to determine the relation between nodal forces and nodal displacements:

$$G_1 = \frac{\partial W}{\partial u_1} = -k_a(u_2 - u_1) \tag{5.41a}$$

$$G_2 = \frac{\partial W}{\partial u_2} = k_a(u_2 - u_1) - k_b(u_3 - u_2) \tag{5.41b}$$

$$G_3 = \frac{\partial W}{\partial u_3} = k_b(u_3 - u_2) \tag{5.41c}$$

In matrix form:

$$\begin{Bmatrix} G_1 \\ G_2 \\ G_3 \end{Bmatrix} = \begin{bmatrix} k_a & -k_a & 0 \\ -k_a & k_a + k_b & -k_b \\ 0 & -k_b & k_b \end{bmatrix} \begin{Bmatrix} u_1 \\ u_2 \\ u_3 \end{Bmatrix} \tag{5.42}$$

or

$$\{G\} = [K]\{D\} \tag{5.43}$$

in which

$$[K] = \begin{bmatrix} k_a & -k_a & 0 \\ -k_a & k_a + k_b & -k_b \\ 0 & -k_b & k_b \end{bmatrix} \begin{matrix} 1 \\ 2 \\ 3 \end{matrix} \tag{5.44}$$

The matrix $[K]$ connecting the forces to the displacements is called the **overall stiffness matrix** of the structure. We emphasize that the stiffness matrix is meaningful only when written with respect to the force/displacement equation (5.42). If we want to refer to the stiffness matrix alone we should always emphasize (unless it is clear) which nodes each matrix component refers to. That is why we have attached the corresponding displacement node number to each row of $[K]$ in (5.44).

• Superposition

One of the major problems when considering larger assemblages is the construction of the overall stiffness matrix of the structure. It is laborious to have to evaluate the strain energy and then use the Castigliano relations. An alternative procedure may be deduced by noting, from (5.44), that this matrix is formed by simply 'adding in' the local stiffness matrices for each pin-jointed element forming the assemblage. We simply take each pin-joint member in turn and superimpose the components of the local stiffness matrix into the appropriate positions in the overall matrix. (See Section 4.3 for a similar procedure used in the solution of boundary value problems.)

Consider as an example a three-element pin-jointed assemblage with individual element stiffnesses k_a, k_b and k_c. The local element stiffness matrices are

$$[k]^{[1]} = \begin{bmatrix} k_a & -k_a \\ -k_a & k_a \end{bmatrix} \begin{matrix} 1 \\ 2 \end{matrix} \qquad [k]^{[2]} = \begin{bmatrix} k_b & -k_b \\ -k_b & k_b \end{bmatrix} \begin{matrix} 2 \\ 3 \end{matrix}$$

and

$$[k]^{[3]} = \begin{bmatrix} k_c & -k_c \\ -k_c & k_c \end{bmatrix} \begin{matrix} 3 \\ 4 \end{matrix}$$

We expect the overall stiffness matrix to be a 4×4 square matrix. We begin with a 4×4 zero matrix and superimpose $[k]^{[1]}, [k]^{[2]}$ and $[k]^{[3]}$ into this matrix as shown:

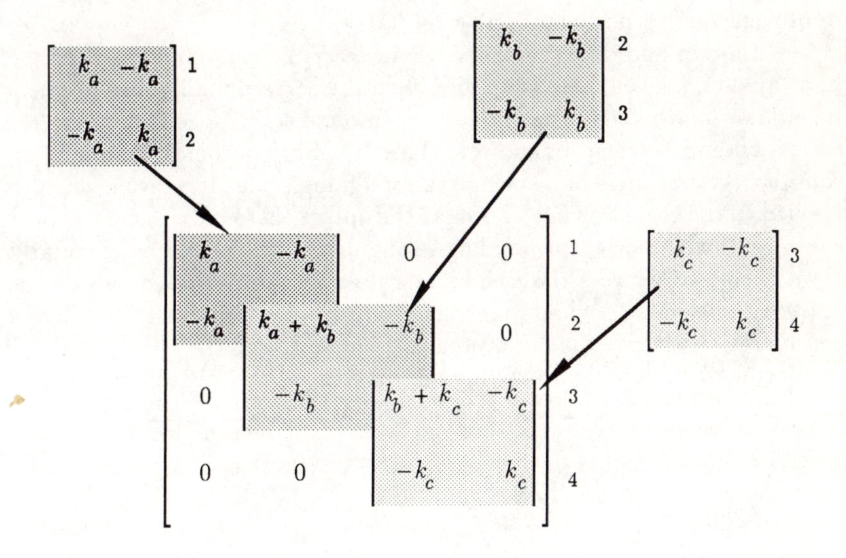

As mentioned earlier the internal forces (5.38) also combine to produce the global forces. This is, in effect, a superposition of the element force vectors.

- **Problem solving**

It has already been noted that the stiffness matrix is singular. This is not surprising because if the stiffness matrix $[K]$ in (5.43) were non-singular we would be able to determine a unique set of displacements $\{D\}$ for a given set of applied forces $\{G\}$ using

$$\{D\} = [K]^{-1}\{G\}$$

But for a pin-jointed element not fixed in some way this would be physically unreasonable as the solution for $\{D\}$ is only unique up to a rigid body motion. Consider, for example, a 'spring' (an example of a very flexible pin-jointed element) which is deformed by stretching. The spring can be moved as a rigid body – changing the nodal displacements of the spring without affecting the deformation induced in it. For one-dimensional problems this implies that to each set of values for $\{D\} = \{u_1, u_2, \ldots, u_n\}^T$ we can always add an arbitrary constant to each component giving $\{D\} = \{u_1 + \alpha, u_2 + \alpha, \ldots, u_n + \alpha\}^T$ without affecting the physical situation in any way. It is only the difference between nodal displacements that is physically significant and not the absolute values. The two-element problem of Figure 5.14 (or indeed any structural problem) may be solved for the unknowns in terms of known quantities only if sufficient physical conditions (boundary conditions) are imposed to prevent a rigid body motion.

Forces and displacements are, in a certain sense, complementary. If, at a node, a given force is applied then the corresponding displacement is unknown whereas if a nodal displacement is specified then the corresponding force is unknown. This 'duality' becomes clearer if we consider the workings of a spring-type weighing scale. If a given weight is applied to the scale it is not known at the outset (before the scale has been calibrated) what the corresponding spring displacement will be. Similarly, if we decide to depress the weighing scale by a given amount we do not know, at the outset, what weight is required to accomplish this. These considerations imply that in systems of equations of the type (5.43) the unknowns of the problem may appear on either side of the equations.

Example 5.1

Consider the pin-jointed system shown in Figure 5.15 where node 3 is held fixed and given forces P_1 and P_2 are applied at nodes 1 and 2 respectively. Determine all the unknowns.

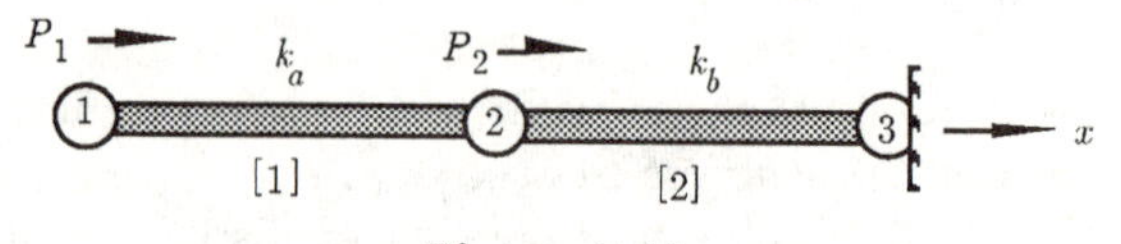

Figure 5.15

Solution

For this problem the unknowns are the displacements u_1, u_2 of nodes 1 and 2 and G_3 the reaction force at node 3. The known quantities are:

$$u_3 = 0 \qquad G_1 = P_1 \qquad G_2 = P_2$$

We have, from (5.43),

$$\begin{Bmatrix} P_1 \\ P_2 \\ G_3 \end{Bmatrix} = \begin{bmatrix} k_a & -k_a & 0 \\ -k_a & k_a + k_b & -k_b \\ 0 & -k_b & k_b \end{bmatrix} \begin{Bmatrix} u_1 \\ u_2 \\ 0 \end{Bmatrix} \tag{5.45}$$

(Note one unknown on the left hand side, the other two on the right.) The three equations contained in (5.45) may be written as

$$\begin{Bmatrix} P_1 \\ P_2 \end{Bmatrix} = \begin{bmatrix} k_a & -k_a \\ -k_a & k_a + k_b \end{bmatrix} \begin{Bmatrix} u_1 \\ u_2 \end{Bmatrix} \tag{5.46a}$$

and

$$G_3 = -k_b u_2 \tag{5.46b}$$

Equation (5.46a) may be solved by simply inverting the coefficient matrix (which is **not** singular) to obtain

$$\begin{Bmatrix} u_1 \\ u_2 \end{Bmatrix} = \frac{1}{k_a k_b} \begin{bmatrix} k_a + k_b & k_a \\ k_a & k_a \end{bmatrix} \begin{Bmatrix} P_1 \\ P_2 \end{Bmatrix}$$

Then, having determined the unknown displacements, we can substitute into (5.46b) to find the unknown reaction force:

$$G_3 = -P_1 - P_2$$

This result confirms that all three nodal forces are in equilibrium.

5.10 Two- and three-dimensional pin-jointed structures

A collection of pin-jointed elements connected together to form a two- or three-dimensional structure is called a **truss**. In Figure 5.16 two such assemblages are shown.

Although the transmission tower in Figure 5.16(b) is not normally composed of pin-jointed elements it can nonetheless be modelled quite accurately as though it were. This is also the case for other structures (bridge supports, roof supports etc.) provided that bending, torsional and shearing stresses are small in comparison with longitudinal stresses.

It is an implicit assumption in the analysis of trusses that the loading on the structure is **applied only at the joints**. In real structures pin-joints are rarely used (particularly in three dimensions) and the connections that are used do transmit moments from one element to another. Thus the material of this section applies only to ideal trusses. The modelling of real structures by a truss with members jointed together with frictionless hinges can be regarded as a first approximation. Often this approach leads to results sufficiently accurate as not to warrant further calculation. If greater accuracy is required the structure could be modelled by beam elements. (These more complicated elements are not considered in detail in this text; however, see Chapter 10.)

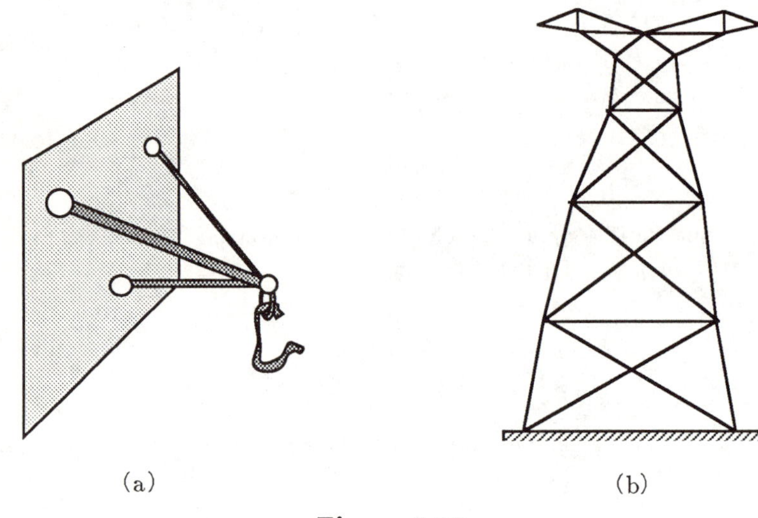

Figure 5.16

In order to develop an expression for the overall stiffness matrix of two-dimensional (and three-dimensional) structures we must firstly construct the stiffness matrix of a pin-jointed element in a more general coordinate system than hitherto. For one-dimensional structures we have chosen a coordinate system (x, y) with the x-axis lying along the axis of each element. However, in a general structure individual elements need not all be aligned in the same direction. We need therefore two distinct coordinate systems: a **global coordinate system** (x', y') chosen in a convenient manner for the complete structure and, for each individual element, a **local coordinate system** (x, y) with the x-axis parallel to the element axis (see Figure 5.17).

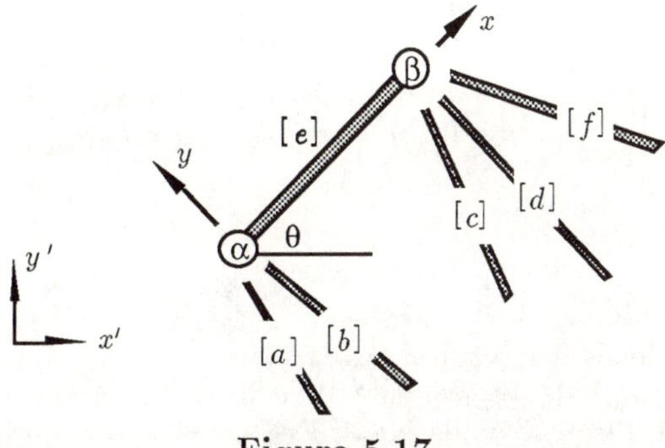

Figure 5.17

Consider the element $[e]$ orientated at an angle θ to the global x'-axis. The element has two nodes (the joints) labelled α, β with local coordinates (x_α, y_α) and (x_β, y_β) respectively. Each node may also be attached to other elements ($[a], [b]$ at node α, and $[c], [d], [f]$ at node β in the figure). In the general case nodes α, β may be subjected to global forces $\{G_\alpha\}$ and $\{G_\beta\}$ respectively, and at each node displacements $\{d_\alpha\}$ and $\{d_\beta\}$ may be measured. See Figure 5.18 where, to aid clarity, we have omitted the other elements adjoining the nodes. Each of these vectors may be referred to either the xy or $x'y'$ system of coordinates.

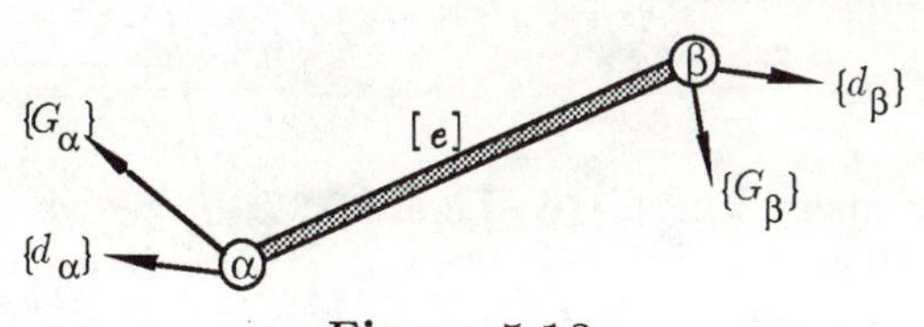

Figure 5.18

At each node the global forces will be distributed between the various elements connected to that node in such a way that the vector sum of the forces developed in each element (the **member** forces) is equal to the global force applied. We denote the internal forces developed within this element as $\{F_\beta\} = \{F_{x\beta}, F_{y\beta}\}^T$ at node β and $\{F_\alpha\} = \{F_{x\alpha}, F_{y\alpha}\}^T$ at node α. The components of these forces with respect to the global coordinate system are written $\{F'_\beta\}$ and $\{F'_\alpha\}$ respectively. It must be emphasized that, for a pin-jointed element, the components of the member forces satisfy the relations

$$F_{x\alpha} = -F_{x\beta} \qquad F_{y\alpha} = F_{y\beta} = 0 \tag{5.47}$$

in the unprimed (local) coordinate system. These characteristics of the pin-jointed element are contained in an amended form of (5.38):

$$\begin{Bmatrix} F_{x\alpha} \\ F_{y\alpha} \\ F_{x\beta} \\ F_{y\beta} \end{Bmatrix} = \frac{EA}{\ell} \begin{bmatrix} 1 & 0 & -1 & 0 \\ 0 & 0 & 0 & 0 \\ -1 & 0 & 1 & 0 \\ 0 & 0 & 0 & 0 \end{bmatrix} \begin{Bmatrix} u_\alpha \\ v_\alpha \\ u_\beta \\ v_\beta \end{Bmatrix} \tag{5.48a}$$

or

$$\{F\} = [k]\{d\} \tag{5.48b}$$

Now since force is a vector quantity the components of $\{F_\alpha\}$ with respect to the two coordinate systems $x'y'$, xy are related by

$$\begin{Bmatrix} F_{x\alpha} \\ F_{y\alpha} \end{Bmatrix} = \begin{bmatrix} \cos\theta & \sin\theta \\ -\sin\theta & \cos\theta \end{bmatrix} \begin{Bmatrix} F'_{x'\alpha} \\ F'_{y'\alpha} \end{Bmatrix}$$

with an identical relation between the components at node β. These relations may be expressed as a single matrix equation

$$\begin{Bmatrix} F_{x\alpha} \\ F_{y\alpha} \\ F_{x\beta} \\ F_{y\beta} \end{Bmatrix} = \begin{bmatrix} c & s & 0 & 0 \\ -s & c & 0 & 0 \\ 0 & 0 & c & s \\ 0 & 0 & -s & c \end{bmatrix} \begin{Bmatrix} F'_{x'\alpha} \\ F'_{y'\alpha} \\ F'_{x'\beta} \\ F'_{y'\beta} \end{Bmatrix} \tag{5.49}$$

where

$$c \equiv \cos\theta = \frac{x_\alpha - x_\beta}{\ell} \qquad s \equiv \sin\theta = \frac{y_\alpha - y_\beta}{\ell}$$

and ℓ is the element length (see Figure 5.17). We can write (5.49) in matrix form:

$$\{F\} = [T]\{F'\} \tag{5.50}$$

Similarly for the displacements since

$$\{d\} = \{u_\alpha, v_\alpha, u_\beta, v_\beta\}^T$$

we have

$$\{d\} = [T]\{d'\} \tag{5.51}$$

Using (5.50) and (5.51) in (5.48b) we obtain

$$[T]\{F'\} = [k][T]\{d'\}$$

Now it is easily verified that $[T]$ is an orthogonal matrix, that is

$$[T][T]^T = [\mathrm{I}] \qquad \text{or} \qquad [T]^T = [T]^{-1} \tag{5.52}$$

Thus

$$\{F'\} = [T]^T[k][T]\{d'\} \tag{5.53}$$

where the matrix $[k'] \equiv [T]^T[k][T]$ is the stiffness matrix of the element referred to the global coordinate system $x'y'$. We find

$$[k'] = \frac{EA}{\ell}\begin{bmatrix} [t] & -[t] \\ -[t] & [t] \end{bmatrix} \begin{matrix} \alpha \\ \beta \end{matrix} \qquad \text{where} \quad [t] = \begin{bmatrix} c^2 & cs \\ cs & s^2 \end{bmatrix} \tag{5.54}$$

(Note that as in (5.44) we have 'attached' node numbers α, β to emphasize to which node each stiffness component refers.)

Equation (5.54) contains the local information for each element. If we develop an expression for the strain energy of the complete structure and again appeal to the Castigliano relations we find that the global forces $\{G\}$ are related to the nodal displacements $\{D\}$ by the relations:

$$\{G\} = [K]\{D\} \tag{5.55}$$

where $[K]$ is the overall stiffness matrix. An examination of the formation of (5.55) by this process shows that, as previously, $[K]$ may be formed directly by superimposing the individual stiffness matrices $[k]$, expressed in global coordinates, for each element.

Example 5.2

For the assemblage of Figure 5.19 determine the deflection of node 2 which is subject to a force P as shown. Both elements are made of the same material and have the same uniform cross-sectional area. The elements are fixed at nodes 1 and 3.

Solution

We construct local stiffness matrices for each element.

- Element [1] nodes 1,2 length ℓ

$$c = \frac{\ell - 0}{\ell} = 1 \qquad s = 0$$

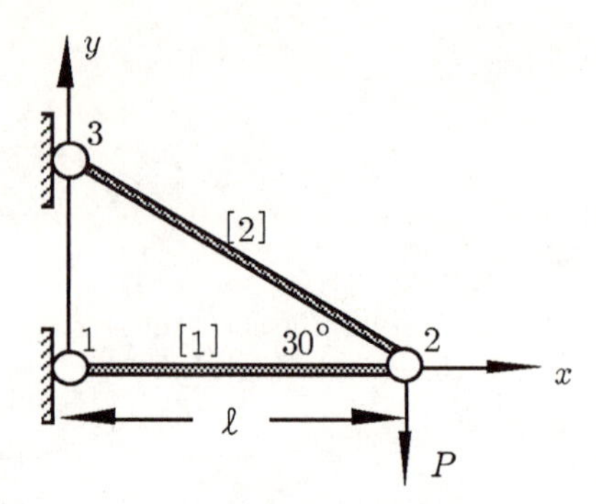

Figure 5.19

Hence by (5.54)

$$[k']^{[1]} = \frac{EA}{\ell}\begin{bmatrix} [t] & -[t] \\ -[t] & [t] \end{bmatrix}\begin{matrix} {\scriptstyle 1} \\ {\scriptstyle 2} \end{matrix} \qquad [t] = \begin{bmatrix} 1 & 0 \\ 0 & 0 \end{bmatrix}$$

- Element [2] nodes 2,3 length $\dfrac{2\ell}{\sqrt{3}}$

$$c = \frac{0-\ell}{2\ell/\sqrt{3}} = -\frac{\sqrt{3}}{2} \qquad s = \frac{\ell/\sqrt{3}-0}{2\ell/\sqrt{3}} = \frac{1}{2}$$

Hence

$$[k']^{[2]} = \frac{EA\sqrt{3}}{2\ell}\begin{bmatrix} [t] & -[t] \\ -[t] & [t] \end{bmatrix}\begin{matrix} {\scriptstyle 2} \\ {\scriptstyle 3} \end{matrix} \qquad [t] = \begin{bmatrix} 3/4 & -\sqrt{3}/4 \\ -\sqrt{3}/4 & 1/4 \end{bmatrix}$$

Hence (5.55) is, for this system,

$$\begin{Bmatrix} G_{x'1} \\ G_{y'1} \\ 0 \\ -P \\ G_{x'3} \\ G_{y'3} \end{Bmatrix} = \frac{EA}{\ell}\begin{bmatrix} 1 & 0 & -1 & 0 & 0 & 0 \\ 0 & 0 & 0 & 0 & 0 & 0 \\ -1 & 0 & 1+\frac{3\sqrt{3}}{8} & -\frac{3}{8} & -\frac{3\sqrt{3}}{8} & \frac{3}{8} \\ 0 & 0 & -\frac{3}{8} & \frac{\sqrt{3}}{8} & \frac{3}{8} & -\frac{\sqrt{3}}{8} \\ 0 & 0 & -\frac{3\sqrt{3}}{8} & \frac{3}{8} & \frac{3\sqrt{3}}{8} & -\frac{3}{8} \\ 0 & 0 & \frac{3}{8} & -\frac{\sqrt{3}}{8} & -\frac{3}{8} & \frac{\sqrt{3}}{8} \end{bmatrix}\begin{Bmatrix} 0 \\ 0 \\ u_2' \\ v_2' \\ 0 \\ 0 \end{Bmatrix} \tag{5.56}$$

leading to the equations (for the unknown displacements)

$$\begin{Bmatrix} 0 \\ -P \end{Bmatrix} = \frac{EA}{\ell}\begin{bmatrix} 1+\frac{3\sqrt{3}}{8} & -\frac{3}{8} \\ -\frac{3}{8} & \frac{\sqrt{3}}{8} \end{bmatrix}\begin{Bmatrix} u_2' \\ v_2' \end{Bmatrix}$$

with solution

$$u'_2 = -\sqrt{3}\frac{\ell P}{EA} \qquad v'_2 = -(\frac{8}{\sqrt{3}} + 3)\frac{\ell P}{EA} \tag{5.57}$$

The reaction forces at nodes 1 and 3 can be obtained from (5.56) and using (5.57). We obtain

$$G_{x'1} = \sqrt{3}P \qquad G_{y'1} = 0$$

$$G_{x'3} = -\sqrt{3}P \qquad G_{y'3} = P$$

As expected the global forces are in equilibrium:

$$G_{x'1} + G_{x'2} + G_{x'3} = \sqrt{3}P + 0 - \sqrt{3}P = 0$$

$$G_{y'1} + G_{y'2} + G_{y'3} = 0 - P + P = 0$$

Extension to three-dimensional trusses is straightforward with each force and each displacement having an extra component – in the z-direction.

Defining $\{F'\}, \{d'\}$ in an obvious way:

$$\{F'\} = \{F'_{x'\alpha}, F'_{y'\alpha}, F'_{z'\alpha}, F'_{x'\beta}, F'_{y'\beta}, F'_{z'\beta}\}^T$$

$$\{d'\} = \{u'_\alpha, v'_\alpha, w'_\alpha, u'_\beta, v'_\beta, w'_\beta\}^T$$

we find

$$\{F'\} = [k']\{d'\}$$

in which

$$[k'] = \frac{EA}{\ell}\begin{bmatrix} [t] & -[t] \\ -[t] & [t] \end{bmatrix} \begin{matrix} \alpha \\ \beta \end{matrix} \tag{5.58}$$

Here $[t]$ is the 3×3 sub-matrix:

$$[t] = \begin{bmatrix} c_1^2 & c_1c_2 & c_1c_3 \\ c_1c_2 & c_2^2 & c_2c_3 \\ c_1c_3 & c_2c_3 & c_3^2 \end{bmatrix} \tag{5.59}$$

where

$$c_1 = \frac{x_\alpha - x_\beta}{\ell} \qquad c_2 = \frac{y_\alpha - y_\beta}{\ell} \qquad c_3 = \frac{z_\alpha - z_\beta}{\ell} \tag{5.60}$$

Example 5.3

The three element pin-jointed truss shown in Figure 5.20 has a force P is applied to node 1 in the direction indicated. Determine the unknown displacements.

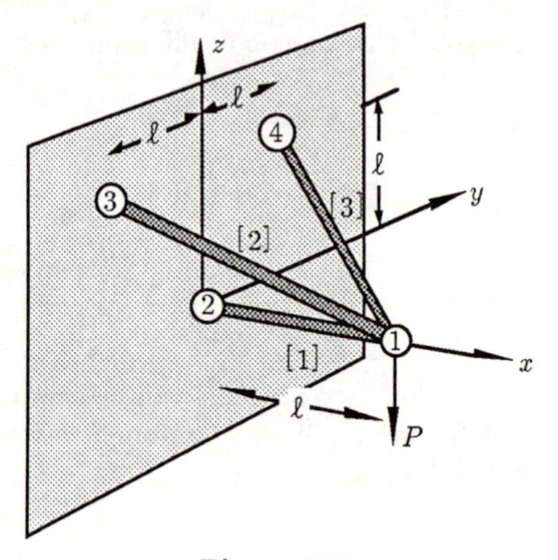

Figure 5.20

Solution

In this problem the nodal coordinates are

$$1:\ (\ell,0,0) \qquad 2:\ (0,0,0) \qquad 3:\ (0,-\ell,\ell) \qquad 4:\ (0,\ell,\ell)$$

We now proceed to determine the element stiffness matrices in global coordinates.

- Element [1]: nodes $\alpha = 1,\ \beta = 2$ length $= \ell$

$$c_1 = -1 \quad c_2 = 0 \quad c_3 = 0$$

Hence

$$[k']^{[1]} = \frac{EA}{\ell}\begin{bmatrix} [t] & -[t] \\ -[t] & [t] \end{bmatrix}\begin{matrix} 1 \\ 2 \end{matrix} \qquad [t] = \begin{bmatrix} 1 & 0 & 0 \\ 0 & 0 & 0 \\ 0 & 0 & 0 \end{bmatrix}$$

- Element [2]: nodes $\alpha = 1,\ \beta = 3$ length $= \sqrt{3}\ell$

$$c_1 = -\frac{1}{\sqrt{3}} \quad c_2 = -\frac{1}{\sqrt{3}} \quad c_3 = \frac{1}{\sqrt{3}}$$

Hence

$$[k']^{[2]} = \frac{EA}{\ell}\begin{bmatrix} [t] & -[t] \\ -[t] & [t] \end{bmatrix}\begin{matrix} 1 \\ 3 \end{matrix} \qquad [t] = \frac{1}{3\sqrt{3}}\begin{bmatrix} 1 & 1 & -1 \\ 1 & 1 & -1 \\ -1 & -1 & 1 \end{bmatrix}$$

- Element [3]: nodes $\alpha = 1,\ \beta = 4$ length $= \sqrt{3}\ell$

$$c_1 = -\frac{1}{\sqrt{3}} \quad c_2 = \frac{1}{\sqrt{3}} \quad c_3 = \frac{1}{\sqrt{3}}$$

Hence

$$[k']^{[3]} = \frac{EA}{\ell}\begin{bmatrix} [t] & -[t] \\ -[t] & [t] \end{bmatrix}\begin{matrix} 1 \\ 4 \end{matrix} \qquad [t] = \frac{1}{3\sqrt{3}}\begin{bmatrix} 1 & -1 & -1 \\ -1 & 1 & 1 \\ -1 & 1 & 1 \end{bmatrix}$$

Superimposing these three element stiffness matrices gives an overall 12×12 stiffness matrix $[K]$ and the force/displacement equations

$$\{G\} = [K]\{D\} \tag{5.61}$$

can now be analysed. The reader should construct this matrix in full. He/she will then appreciate that this technique, although extremely useful, would require a powerful computer when realistic structural problems (which may have hundreds of pin-jointed elements) are being analysed.

The first three equations of (5.61) are

$$\begin{Bmatrix} 0 \\ 0 \\ -P \end{Bmatrix} = \frac{EA}{\ell}\begin{bmatrix} 1 + \frac{2\sqrt{3}}{9} & 0 & -\frac{2\sqrt{3}}{9} \\ 0 & \frac{2\sqrt{3}}{9} & 0 \\ -\frac{2\sqrt{3}}{9} & 0 & \frac{2\sqrt{3}}{9} \end{bmatrix}\begin{Bmatrix} u_1' \\ v_1' \\ w_1' \end{Bmatrix} \tag{5.62}$$

where we have utilized the boundary conditions on displacement

$$(u_\alpha', v_\alpha', w_\alpha') = 0 \qquad \alpha = 2, 3, 4$$

The second equation of (5.62) immediately implies that $v_1' = 0$ (as we expect from considerations of symmetry). The other two equations are easily solved to give

$$u_1' = -\frac{P\ell}{EA} \qquad w_1' = -(1 + \frac{9}{2\sqrt{3}})\frac{P\ell}{EA}$$

In realistic problems this displacement is extremely small. For example, if we use steel rods with a Young's modulus of 2×10^{11} N m^{-2}, cross-sectional area $A = 0.005$ m^2 (radius ≈ 4 cm) and $\ell = 1$ m then for a load of magnitude 10^4 N we find

$$u_1 = -10^{-5} \text{ m} \qquad w_1 = -3.6 \times 10^{-5} \text{ m}$$

5.11 Finite element formulation of plane strain

In the treatment of one-dimensional pin-jointed elements we obtained in (5.36) the variation of the displacement function $u(x)$ as

$$u(x) = u_1\left(1 - \frac{x}{\ell}\right) + u_2\left(\frac{x}{\ell}\right) \tag{5.63}$$

in which u_1, u_2 are the measured displacements at nodes 1 and 2 respectively (see Figure 5.21).

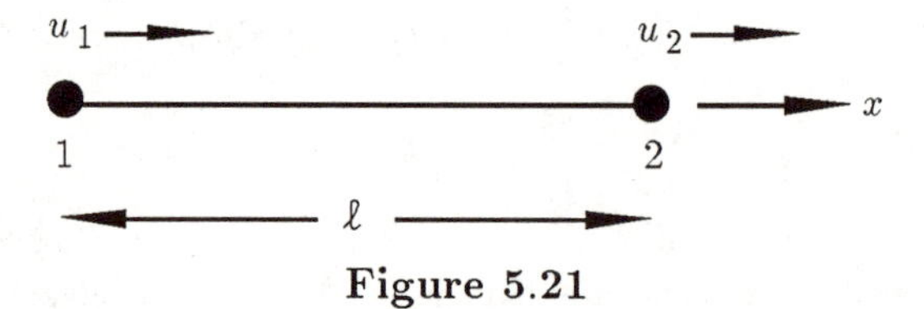

Figure 5.21

In fact, to the level of approximation considered, (5.63) is exact.

We shall attempt to apply a similar approach to the analysis of plane strain problems. For these problems we have two unknown displacement components $u(x, y), v(x, y)$. However, whereas the unknown displacement for a one-dimensional strut satisfies a simple boundary value problem such as (5.34), this is not the case for plane strain deformations. Here, the displacement components u, v satisfy the considerably more complicated system of differential and algebraic equations (5.24) with boundary conditions (5.25) and (5.26). As there is no possibility (in the general case) of obtaining exact solutions to this system we must, necessarily, seek numerical solutions.

These approximate solutions are obtained by utilising the finite element approach in conjunction with the theorem of minimum potential energy. A number of steps are involved:

- We split the domain of the problem into non-overlapping triangular elements (see Figure 5.22). Clearly, however small we choose the triangular elements, there will be an error in representing an irregular region. This is represented by the shading in Figure 5.22. However, this error can be controlled. To reduce its size we simply increase the number of elements.

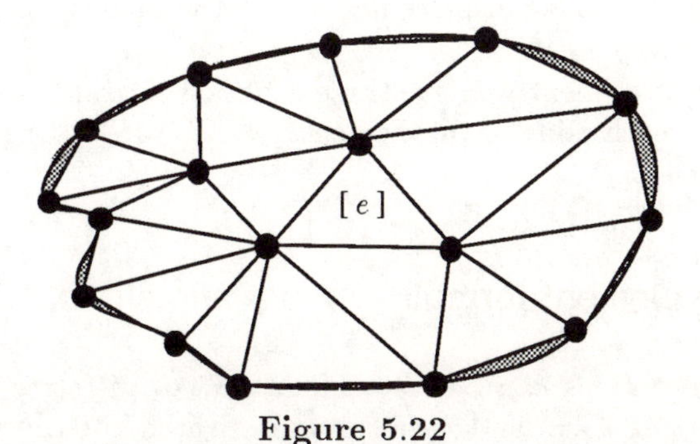

Figure 5.22

• Within each element we assume that both components u and v vary linearly with respect to x and y. That is, within each element we assume

$$\begin{aligned}\tilde{u}(x,y) &= a_0 + a_1 x + a_2 y \\ \tilde{v}(x,y) &= b_0 + b_1 x + b_2 y\end{aligned} \tag{5.64}$$

where a_i, b_i $i = 0, 1, 2$ are constants.
(If the triangles are sufficiently small this approximation to the true variation of u and v within the element will be acceptable.)

• As we move from element to element we demand that the variation of u and v be continuous across common element edges. This will ensure there are no 'tears' or 'breaks' in the material. This is easily arranged as follows.

Assume that there are 'nde' nodes and that the values of $\tilde{u}, \tilde{v}$ at these nodes are denoted by $\tilde{u}_i, \tilde{v}_i$, $i = 1, 2, \ldots, nde$ respectively. Continuity across element boundaries will be guaranteed if the linear variation within each element is chosen in such a way that at each node 'i', with coordinates (x_i, y_i), both $\tilde{u}(x,y)$ and $\tilde{v}(x,y)$ assume the nodal values: i.e.

$$\begin{aligned}\tilde{u}(x_i,y_i) &\equiv a_0 + a_1 x_i + a_2 y_i = \tilde{u}_i \\ \tilde{v}(x_i,y_i) &\equiv b_0 + b_1 x_i + b_2 y_i = \tilde{v}_i\end{aligned} \tag{5.65}$$

The complete approximation now depends only upon the parameters $(\tilde{u}_i, \tilde{v}_i)$ $i = 1, 2, \ldots, nde$. These are chosen so that the potential energy of the configuration is minimized. The approximation will improve (though with substantially increased computation) as the number of elements is increased.

To continue the analysis we shall consider a particular element $[e]$ with nodes 1,2,3. Our attention will be focused primarily on the u-component of the nodal displacements, but the analysis applies equally well to the v-component.

For this element we have three equations arising from the first of (5.65), one for each node of the element:

$$\begin{aligned} \tilde{u}_1 &= a_0 + a_1 x_1 + a_2 y_1 \\ \tilde{u}_2 &= a_0 + a_1 x_2 + a_2 y_2 \\ \tilde{u}_3 &= a_0 + a_1 x_3 + a_2 y_3 \end{aligned} \tag{5.66}$$

These three equations can be solved for a_0, a_1, a_2 in terms of the nodal function values $\tilde{u}_1, \tilde{u}_2, \tilde{u}_3$ and the results substituted into the first of (5.64). We find:

$$\begin{aligned} \tilde{u}(x,y) &= \tilde{u}_1 L_1(x,y) + \tilde{u}_2 L_2(x,y) + \tilde{u}_3 L_3(x,y) \\ &= \sum_{i=1}^{3} \tilde{u}_i L_i(x,y) \end{aligned} \tag{5.67}$$

in which the functions L_1, L_2, L_3 are most easily expressed using determinants:

$$L_1 = \frac{1}{D}\begin{vmatrix} 1 & x & y \\ 1 & x_2 & y_2 \\ 1 & x_3 & y_3 \end{vmatrix} \qquad L_2 = \frac{1}{D}\begin{vmatrix} 1 & x_1 & y_1 \\ 1 & x & y \\ 1 & x_3 & y_3 \end{vmatrix} \qquad L_3 = \frac{1}{D}\begin{vmatrix} 1 & x_1 & y_1 \\ 1 & x_2 & y_2 \\ 1 & x & y \end{vmatrix} \tag{5.68}$$

where

$$D = \begin{vmatrix} 1 & x_1 & y_1 \\ 1 & x_2 & y_2 \\ 1 & x_3 & y_3 \end{vmatrix} = 2 \times \text{area of triangle } [e]$$

The shape functions L_1, L_2 and L_3 are the two-dimensional equivalent of the functions $(1 - x/\ell)$, (x/ℓ) which occur in (5.63).

The three-noded triangular element is often called a constant strain element as, within each element, each of the strain components is constant.

Example 5.4

For the simple rectangular region shown in Figure 5.23(a) determine the shape functions appropriate to each triangular element. Verify that $u(x, y)$ is continuous across the common element boundary.

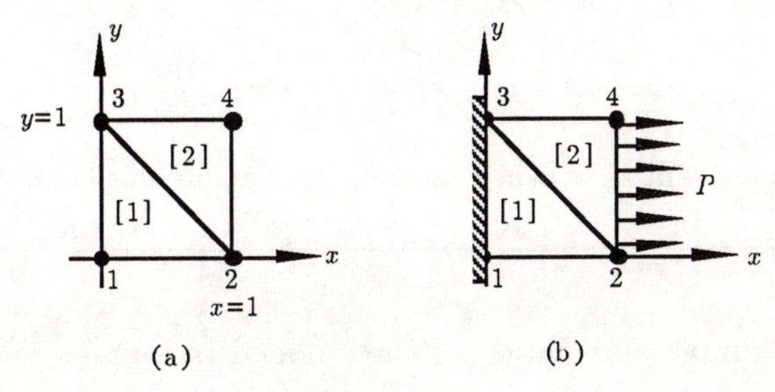

Figure 5.23

Solution

The nodal coordinates are:

$1: (0,0) \quad 2: (1,0) \quad 3: (0,1) \quad 4: (1,1)$

- Element [1]: local nodes 1,2,3 global nodes 1,2,3

$$L_1 = \begin{vmatrix} 1 & x & y \\ 1 & 1 & 0 \\ 1 & 0 & 1 \end{vmatrix} = 1 - x - y$$

Similarly $L_2 = x, \; L_3 = y$

- Element [2]: local nodes 1,2,3 global nodes 2,4,3

$$L_1 = \begin{vmatrix} 1 & x & y \\ 1 & 1 & 1 \\ 1 & 0 & 1 \end{vmatrix} = 1 - y$$

Similarly $L_2 = -1 + x + y, \; L_3 = 1 - x$

Hence, within element [1]

$$\tilde{u}^{[1]}(x, y) = \tilde{u}_1(1 - x - y) + \tilde{u}_2 x + \tilde{u}_3 y \tag{5.69}$$

and within element [2]

$$\tilde{u}^{[2]}(x, y) = \tilde{u}_2(1 - y) + \tilde{u}_4(-1 + x + y) + \tilde{u}_3(1 - x) \tag{5.70}$$

Along the common element boundary (whose equation is $1 - x - y = 0$) we obtain

$$\tilde{u}^{[1]}(x, y) = \tilde{u}_2 x + \tilde{u}_3(1 - x)$$

$$\tilde{u}^{[2]}(x, y) = \tilde{u}_2(1 - (1 - x)) + \tilde{u}_3(1 - x) = \tilde{u}^{[1]}(x, y)$$

hence verifying that $\tilde{u}(x, y)$ is continuous across this boundary.

A relatively simple application of three-noded triangular elements is the problem of the stretching of a thin plate (of thickness h) as shown in Figure 5.23(b). As indicated in the figure we shall model the plate using two elements (chosen for convenience as in Example 5.4). The boundary conditions imply that

$$\tilde{u}_1 = \tilde{u}_3 = 0, \qquad \tilde{v}_1 = \tilde{v}_3 = 0$$

Thus there are four unknowns viz. the displacement components at nodes 2 and 4: $(\tilde{u}_2, \tilde{v}_2)$ and $(\tilde{u}_4, \tilde{v}_4)$. Our starting point for the finite element analysis is the potential energy (5.31). If we ignore the effects of body forces then

$$U = \frac{h}{2} \iint_A \{\tilde{s}\}^T \{\tilde{\Omega}\}\, dx dy - h \int_\Gamma \{\tilde{d}\}^T \{T\}\, d\ell \tag{5.71}$$

The boundary conditions on stress show that

$$\{T\} = \begin{Bmatrix} P \\ 0 \end{Bmatrix} \text{ on } x = 1 \qquad \{T\} = \begin{Bmatrix} 0 \\ 0 \end{Bmatrix} \text{ on } y = 0 \text{ and } y = 1 \tag{5.72}$$

and so

$$\int_\Gamma \{\tilde{d}\}^T \{T\}\, d\ell = \int_{x=1} \tilde{u} P\, dy \tag{5.73}$$

Using (5.70) with $x = 1$

$$\begin{aligned} \int_{x=1} \tilde{u} P\, dy &= \int_0^1 \{\tilde{u}_2(1 - y) + \tilde{u}_4 y\} P\, dy \\ &= \frac{1}{2}\{\tilde{u}_2 + \tilde{u}_4\} P \end{aligned} \tag{5.74}$$

Also, for the double integral in (5.71)

$$\iint_A \{\tilde{s}\}^T\{\tilde{\Omega}\}\, dxdy = \iint_{[1]} \{\tilde{s}\}^T\{\tilde{\Omega}\}\, dxdy + \iint_{[2]} \{\tilde{s}\}^T\{\tilde{\Omega}\}\, dxdy$$

in which, within element [1],

$$\{\tilde{s}\} = [\partial]\{\tilde{d}\} = \begin{Bmatrix} \tilde{u}_2 \\ 0 \\ \tilde{v}_2 \end{Bmatrix}$$

where we have used (5.69) for $\tilde{u}(x,y)$ and the equivalent expression for $\tilde{v}(x,y)$.

Hence

$$\{\tilde{\Omega}\} = [C]\{\tilde{s}\} = \begin{Bmatrix} (2\mu+\lambda)\tilde{u}_2 \\ \lambda\tilde{u}_2 \\ \mu\tilde{v}_2 \end{Bmatrix}$$

and so

$$\{\tilde{s}\}^T\{\tilde{\Omega}\} = (2\mu+\lambda)\tilde{u}_2^2 + \mu\tilde{v}_2^2 \tag{5.75}$$

A similar construction is carried out for element [2] giving

$$\{\tilde{s}\} = \begin{Bmatrix} \tilde{u}_4 \\ \tilde{v}_4 - \tilde{v}_2 \\ -\tilde{u}_2 + \tilde{u}_4 + \tilde{v}_4 \end{Bmatrix}$$

$$\{\tilde{\Omega}\} = \begin{Bmatrix} (2\mu+\lambda)\tilde{u}_4 + \lambda(\tilde{v}_4 - \tilde{v}_2) \\ \lambda\tilde{u}_4 + (2\mu+\lambda)(\tilde{v}_4 - \tilde{v}_2) \\ \mu(-\tilde{u}_2 + \tilde{u}_4 + \tilde{v}_4) \end{Bmatrix}$$

and so

$$\begin{aligned}\{\tilde{s}\}^T\{\tilde{\Omega}\} =& (2\mu+\lambda)\tilde{u}_4^2 + 2\lambda(\tilde{v}_4 - \tilde{v}_2)\tilde{u}_4 \\ &+ (2\mu+\lambda)(\tilde{v}_4 - \tilde{v}_2)^2 + \mu(-\tilde{u}_2 + \tilde{u}_4 + \tilde{v}_4)^2\end{aligned} \tag{5.76}$$

Since (5.75) and (5.76) are constant the double integrals in U are easily evaluated. The final expression is

$$U = \frac{h}{4}\{(2\mu+\lambda)\tilde{u}_2^2 + \mu\tilde{v}_2^2 + (2\mu+\lambda)\tilde{u}_4^2 + 2\lambda\tilde{u}_4(\tilde{v}_4 - \tilde{u}_2) + (2\mu+\lambda)(\tilde{v}_4 - \tilde{v}_2)^2 + \mu(-\tilde{u}_2 + \tilde{u}_4 + \tilde{v}_4)^2\} - \frac{h}{2}(\tilde{u}_2 + \tilde{u}_4)P \tag{5.77}$$

This expression is now minimized by choosing values for $(\tilde{u}_2, \tilde{v}_2)$ and $(\tilde{u}_4, \tilde{v}_4)$ such that

$$\frac{\partial U}{\partial \tilde{u}_2} = 0 \qquad \frac{\partial U}{\partial \tilde{u}_4} = 0 \qquad \frac{\partial U}{\partial \tilde{v}_2} = 0 \qquad \frac{\partial U}{\partial \tilde{v}_4} = 0 \tag{5.78}$$

This process leads to the following system of linear equations:

$$\begin{bmatrix} 6\mu+2\lambda & 0 & -2\mu & -2\mu \\ 0 & 6\mu+2\lambda & -2\lambda & -4\mu-2\lambda \\ -2\mu & -2\lambda & 6\mu+2\lambda & 2\mu+2\lambda \\ -2\mu & -4\mu-2\lambda & 2\mu+2\lambda & 6\mu+2\lambda \end{bmatrix} \begin{Bmatrix} \tilde{u}_2 \\ \tilde{v}_2 \\ \tilde{u}_4 \\ \tilde{v}_4 \end{Bmatrix} = \begin{Bmatrix} 2P \\ 0 \\ 2P \\ 0 \end{Bmatrix} \tag{5.79}$$

From (5.79) the solution for the unknown displacements may be obtained, and hence the strain and stress levels throughout each element found.

In realistic plane strain problems large numbers of elements will be necessary if high accuracy is required for displacement values. Indeed more sophisticated elements than the simple three-noded triangular element should be considered. This is particularly true for irregular shaped regions where elements capable of approximating curved boundaries more closely than can straight-edged elements need to be used. Also, a more methodical approach than that carried out above would need to be applied, particularly for large scale problems. These and related matters are considered in Chapters 6 and 7. In the last section of Chapter 7 we continue our study of elasticity and, in particular, derive the matrix formulation of the finite element model of plane strain.

EXERCISES

5.1 A uniform bar of cross-sectional area A has its axis aligned along the x-axis and is subjected to equal and opposite forces F as shown in Figure Q5.1. If

$$\sigma_{11} = \frac{F}{A} \qquad \text{other } \sigma_{ij} = 0$$

determine the stress vector at any point on a plane at an angle α to the axis of the bar. Determine the magnitudes of the normal and shear stresses on this plane and verify that the total force on the plane is F.

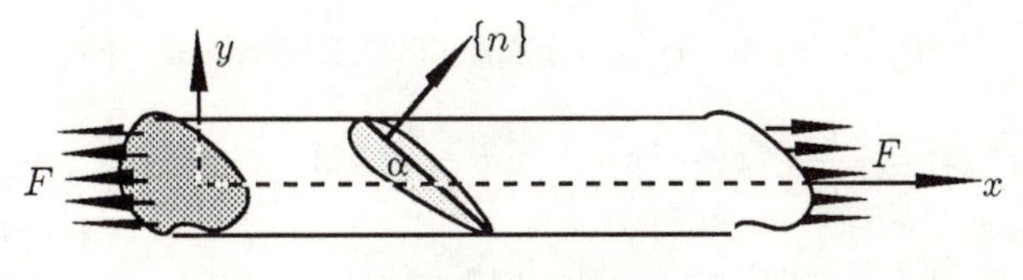

Figure Q5.1

5.2 A square plate is acted on by a shearing stress τ on each of its edges as shown in Figure Q5.2. Determine the tensile and shearing stresses on planar elements at Q and P. Assume that the stress matrix components are constant throughout the plate.

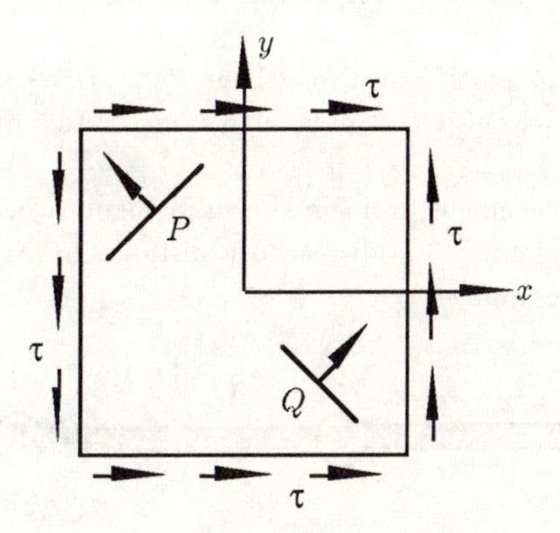

Figure Q5.2

5.3 The state of stress at a point P in a body is defined by

$$\sigma_{11} = 200 \quad \sigma_{12} = 400 \quad \sigma_{13} = 300 \quad \sigma_{22} = 0 \quad \sigma_{23} = 0 \quad \sigma_{33} = -100$$

Determine the stress vector acting on a plane with normal $\{1, 2, 2\}^T$ passing through P. Find also the magnitudes of the normal and shear stresses on the plane.

5.4 In the torsion problem of a circular cylinder aligned along the z-axis and which is rigidly fixed at $(0, 0, 0)$ it may be shown that the displacement vector is

$$\{d\} = \{-\tau yz, \tau xz, 0\}^T$$

where τ is a constant.

Determine the strain matrix components and show that plane sections, originally perpendicular to the axis of the cylinder, suffer no distortion apart from a rotation about the axis.

5.5 Determine the strain matrix for each of the following displacement vectors:

$$\text{(a)} \quad u = -\frac{\nu}{E}\rho g(xz - hx) \qquad v = -\frac{\nu}{E}\rho g(yz - hy)$$

$$w = -\frac{1}{E}\rho g\{(\frac{z^2}{2} - hz) + \frac{1}{2}(x^2 + y^2)$$

(This describes the displacement of a cylinder deforming under gravity.)

$$\text{(b)} \quad u = -\frac{\nu M}{EI}xy \qquad v = \frac{M}{EI}(-\nu y^2 + \frac{\nu x^2 - z^2}{2}) \qquad w = \frac{M}{EI}yz$$

(This is the displacement of a beam deformed by pure couples.)

5.6 A wire has an undeformed length of 1 m. Determine the longitudinal strain in the wire if it is stretched to a length 1.05 m.

5.7 Solve the two-element problem shown in Figure Q5.7, in which node 1 is held fixed and node 3 is displaced a distance u. Node 2 is free with no external force applied.

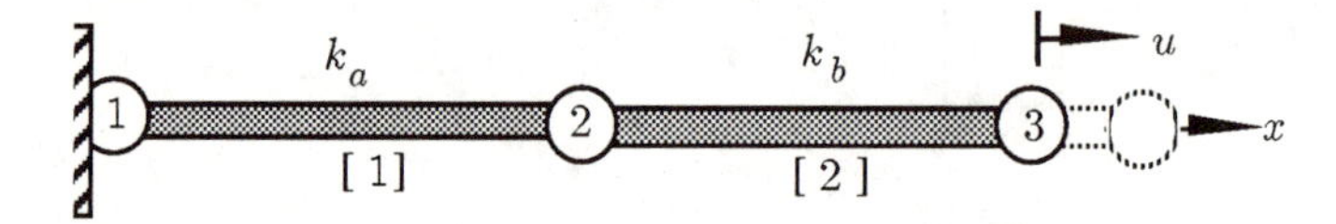

Figure Q5.7

5.8 (a) Obtain the stiffness matrix for the assemblage in Figure Q5.8 (i) using the Castigliano relation (ii) by superposition.

(b) If nodes 1 and 4 are held fixed calculate the unknown displacements, reactions and member forces, assuming that applied loads act at nodes 2 and 3 in the positive x-direction.

(c) If node 2 of the assemblage is subjected to a displacement $u_2 = 1$, and assuming nodes 1 and 4 are held fixed, determine u_3, G_2 and the unknown reaction forces.

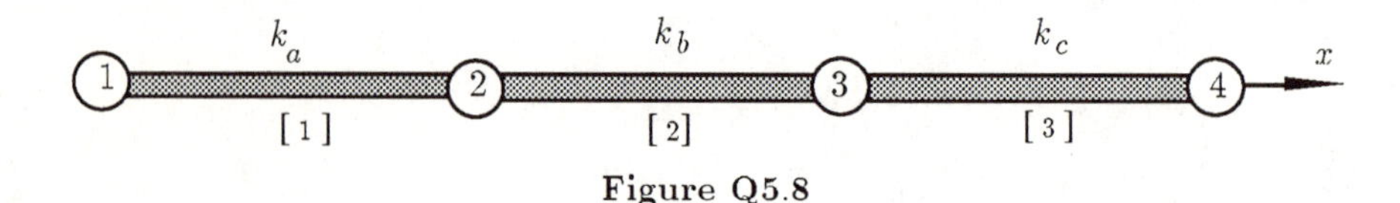

Figure Q5.8

5.9 For the assemblage of pin-jointed elements shown in Figure Q5.9 and the stiffnesses, applied forces and boundary conditions indicated calculate the displacements at nodes 2 and 3.

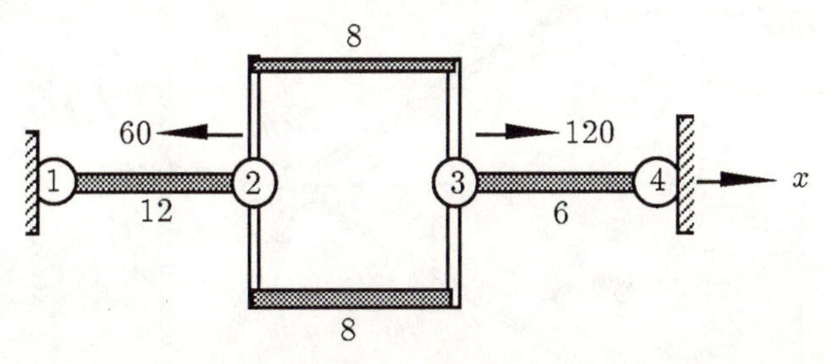

Figure Q5.9

5.10 The three element pin-jointed structure shown in Figure Q5.10 is loaded at node 1. Each element has the same cross-sectional area A and the same Young's modulus E.

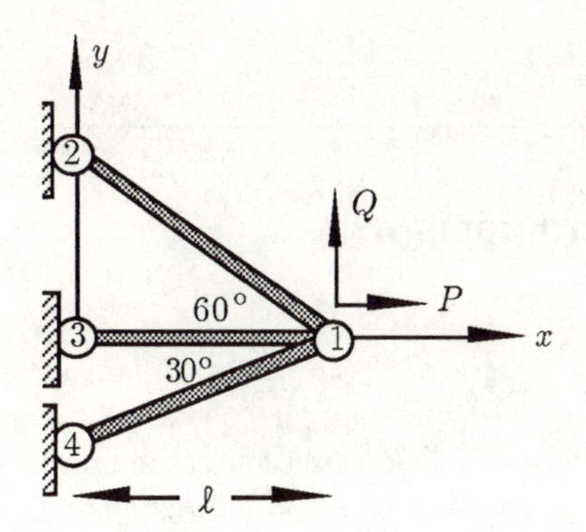

Figure Q5.10

Use the stiffness matrix method to show that the displacement obtained at node 1 is

$$u_1 = \frac{\ell}{3EA(\sqrt{3}+1)}\left\{(\sqrt{3}+3)P + (\sqrt{3}-3)Q\right\}$$

$$v_1 = \frac{\ell}{3EA(\sqrt{3}+1)}\left\{(\sqrt{3}-3)P + (3\sqrt{3}+9)Q\right\}$$

5.11 A seven-element truss (see Figure Q5.11) is loaded at nodes 2 and 3. All elements have the same cross-sectional area and Young's modulus. Calculate all the nodal displacements. (*Hint*: utilize the geometrical symmetry in the structure to reduce the number of unknowns to manageable proportions.)

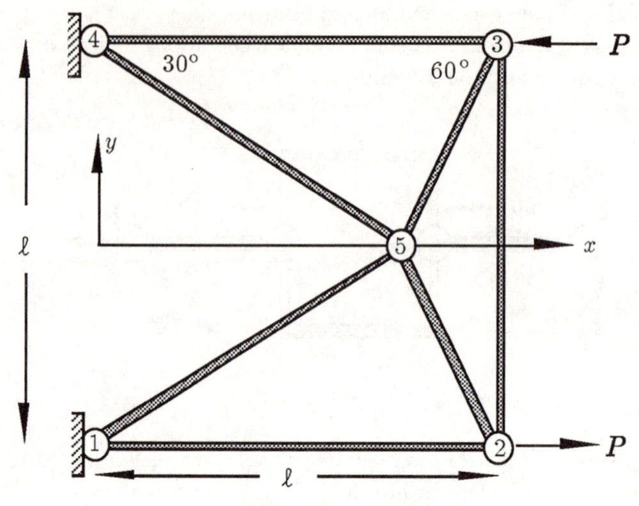

Figure Q5.11

SUPPLEMENT

5S.1 Computer programs

In this supplement two programs in the Fortran77 language are given. The first program can be used to solve a one-dimensional assembly of springs or pin-jointed elements. The second program is an extension of the first to analyse assemblages of two-dimensional pin-jointed elements. Both programs contain brief explanatory comments.

Program 1

Program to solve assemblage of springs by the stiffness method

```
dimension nenn(15,2),stiff(15),ostiff(15,15),sb(15,15),force(15)
dimension noded(15),nodef(15),displ(15),val(15)
integer hbw
```

Input of basic information.

```
print*, 'how many nodes are there'
read(1,*)nde
```

```
      print*,'how many elements are there'
      read(1,*)nel
      print*,'type in element no. and associated node no.s'
      read(1,*)(kel,nenn(kel,1),nenn(kel,2),i=1,nel)
      print*,'type in element no. and associated stiffnesses'
      read(1,*)(kel,stiff(kel),i=1,nel)
```

Initialize stiffness matrix at zero.

```
      do 2 i=1,nde
      do 1 j=1,nde
      ostiff(i,j)=0.
1     continue
2     continue
```

Construct global stiffness matrix.

```
      do 3 k=1,nel
      ii=nenn(k,1)
      jj=nenn(k,2)
      ostiff(ii,ii)=ostiff(ii,ii)+stiff(k)
      ostiff(ii,jj)=ostiff(ii,jj)-stiff(k)
      ostiff(jj,ii)=ostiff(jj,ii)-stiff(k)
      ostiff(jj,jj)=ostiff(jj,jj)+stiff(k)
3     continue
```

Obtain information on prescribed displacements and forces.

```
      print*,
      print*,'how many nodes have prescribed displacements?'
      read(1,*)nbn
      print*,'type in node no.s and associated displacements'
      read(1,*)(noded(i),val(noded(i)),i=1,nbn)
      print*,'what are node no.s at which forces are applied'
      read(1,*)(nodef(i),force(nodef(i)),i=1,(nde-nbn))
```

Begin solution procedure. In contrast with the direct approach used in the program described in the supplement to Chapter 4, we use the Payne–Irons method of implementing the displacement boundary conditions. In this method each diagonal component of the overall stiffness matrix which corresponds to a node on which a displacement boundary condition is imposed is multiplied by a large number and the corresponding term in the force vector amended accordingly. For example if node k has boundary condition $u_k = c$ then the kth equation would be (after amendment)

$$K_{k1}u_1 + \ldots + K_{kk}u_k 10^{12} + \ldots + K_{k,nde}u_{nde} = K_{kk}10^{12}c$$

Since the diagonal term on the left-hand side is now considerably larger than the remaining terms this equation gives, to a good approximation, $u_k = c$ as required.

```
      do 7 i=1,nbn
      force(noded(i))=ostiff(noded(i),noded(i))*val(noded(i))*1.0e+12
      ostiff(noded(i),noded(i))=ostiff(noded(i),noded(i))*1.0e+12
7     continue
```

Find band width (see Chapter 9 if more detail is required).

```
      hbw=abs(nenn(1,2)-nenn(1,1))
      do 8 n=1,nel
      nic1=nenn(n,1)
      nic2=nenn(n,2)
      if(abs(nic1-nic2).gt.hbw)hbw=abs(nic1-nic2)
8     continue
      hbw=hbw+1
```

Form banded matrix.

```
      do 10 iii=1,nde
      do 9 k=1,hbw
      sb(iii,k)=ostiff(iii,k+iii-1)
9     continue
10    continue
```

Solve the amended system of equations, now written in banded form, using the solution routine *band*. The solution is contained in the array *displ(i)* $i = 1, \ldots, nde$.

```
      call band(sb,force,nde,hbw,displ)
      print*,
      print*,'the values of the nodal displacements are :'
      write(1,11)(displ(i),i=1,nde)
11    format(1h ,/,4f12.4)
```

```
print*,
print*,
stop
end
```

The code for subroutine *band* can be found in the supplement to Chapter 4.

```
subroutine band(sk,ff,nde,hbw,displ)
```

Program 2

Program to solve assemblage of truss elements by the stiffness method

```
common/b1/nenn(15,2),xnode(15),ynode(15),e(15),ar(15)
dimension ostiff(15,15),sb(15,15),f(15),stiff(4,4)
dimension noded(15),displ(15),val(15,2),mode(15,2)
integer hbw
```

Input of basic information.

```
print*,'how many nodes are there'
read(1,*)nde
neqn=2*nde
print*,'how many elements are there'
read(1,*)nel
print*,'type in element no. and associated node no.s'
read(1,*)(kel,nenn(kel,1),nenn(kel,2),i=1,nel)
print*,'type in node no. then x and y coordinates of each node'
read(1,*)(kel,xnode(kel),ynode(kel),i=1,nde)
print*,'type in element no. ,youngs modulus and area'
print*,'of each element'
read(1,*)(kel,e(kel),ar(kel),i=1,nel)
```

Initialize stiffness matrix at zero.

```
      do 2 i=1,neqn
      do 1 j=1,neqn
      ostiff(i,j)=0.
1     continue
```

```
2       continue
```

Construct the overall stiffness matrix by superimposing local stiffness matrices (held in subroutine *strut*2) taking due account of node numbers.

```
        do 110 k=1,nel
        i=nenn(k,1)
        j=nenn(k,2)
        call strut2(k,stiff)
        do 111 iro=1,4
        do 112 ico=1,4
        if(iro.lt.3) irow=2*(i-1)+iro
        if(iro.ge.3) irow=2*(j-1)+iro-2
        if(ico.lt.3) icol=2*(i-1)+ico
        if(ico.ge.3) icol=2*(j-1)+ico-2
        ostiff(irow,icol)=ostiff(irow,icol)+stiff(iro,ico)
112     continue
111     continue
110     continue
```

Input displacement and force boundary conditions.

```
        print*,
        print*,'how many nodes have prescribed displacements'
        read(1,*)nbn
        print*,'input node no.  ,then the type of constraint of each'
        print*,'node. (1 if constrained in a coordinate direction'
        print*,'and 0 if free. then the value of the constraint'
        do 140 i=1,nbn
        read(1,*)noded(i),(mode(noded(i),j),j=1,2),
      1(val(noded(i),j),j=1,2)
140     continue
        print*,'what are node no.s with associated x,y forces'
        do 99 i=1,nde-nbn
        read(1,*)kel,f(2*kel-1),f(2*kel)
99      continue
```

Prepare to solve.

```
        do 210 i=1,nbn
        do 210 j=1,2
```

```
      nx=noded(i)
      if(mode(nx,j).eq.0)goto 210
      j1=2*nx-2+j
      f(j1)=ostiff(j1,j1)*val(nx,j)*1.0e+12
      ostiff(j1,j1)=ostiff(j1,j1)*1.0e+12
210   continue
```

Find band width.

```
      hbw=abs(nenn(1,2)-nenn(1,1))
      do 8 n=1,nel
      nic1=nenn(n,1)
      nic2=nenn(n,2)
      if(abs(nic1-nic2).gt.hbw)hbw=abs(nic1-nic2)
8     continue
      hbw=2*(hbw+1)
```

Form banded matrix.

```
      do 10 iii=1,neqn
      do 9 k=1,hbw
      sb(iii,k)=ostiff(iii,k+iii-1)
9     continue 10    continue
```

Solve equations and output solution in array *displ(i)*, $i = 1, \ldots, neqn$.

```
      call band(sb,f,neqn,hbw,displ)
      print*,
      print*,'the values of the nodal displacements are :'
      write(1,11)(displ(i),i=1,neqn)
11    format(1h ,/,4g12.4)
      print*,
      stop
      end
```

The code for subroutine *band* can be found in the supplement to Chapter 4.

```
      subroutine band(sk,ff,nde,hbw,displ)
```

Subroutine *strut*2 constructs local element stiffness matrices.

```
      subroutine strut2(m,stiff)
      common/b1/nenn(15,2),x(15),y(15),e(15),ar(15)
      dimension xl(15),stiff(4,4)
      i=nenn(m,1)
      j=nenn(m,2)
      xl(m)=sqrt((x(j)-x(i))**2+(y(j)-y(i))**2)
      a1=(x(j)-x(i))
      a2=(y(j)-y(i))
      coeff=e(m)*ar(m)/xl(m)**3
      stiff(1,1)=coeff*a1*a1
      stiff(1,2)=coeff*a1*a2
      stiff(1,3)=-stiff(1,1)
      stiff(1,4)=-stiff(1,2)
      stiff(2,1)=stiff(1,2)
      stiff(2,2)=coeff*a2*a2
      stiff(2,3)=-stiff(2,1)
      stiff(2,4)=-stiff(2,2)
      do 490 ice=1,4
      stiff(3,ice)=-stiff(1,ice)
490   stiff(4,ice)=-stiff(2,ice)
      return
      end
```

6

Finite Element Approximation of Line and Double Integrals

PREVIEW

This chapter examines the application of shape functions and elements to the numerical evaluation of integrals in one- and two-dimensions. We introduce both triangular and quadrilateral elements and the associated shape functions and also show, using appropriate coordinate transformations, how curved regions may be approximated.

6.1 Introduction

A certain amount of preparatory work is needed in order to facilitate the transition from one-dimensional to two-dimensional boundary value problems which involve partial differential equations defined over a plane region. The background knowledge required to understand the formulation and solution of two dimensional problems is not perhaps as familiar to engineers and scientists as it is to specialist mathematicians.

It is the purpose of this chapter, therefore, to cover those mathematical topics which are necessary in order to understand the extension of the finite element method to two dimensions. It is possible to obtain a false impression of the finite element method if it is applied only to one-dimensional problems. In fact for one-dimensional boundary value problems there are a number of rival numerical procedures which, in many respects (particularly the amount of algebra involved) are a great deal easier to apply than the finite element method. The advantages of the finite element technique over other alternatives are more fully appreciated in two- or three-dimensional situations. Problems involving plate bending, plane strain, heat conduction over a plate, and similar problems in other areas of engineering which are very difficult to analyse using other techniques are readily solved using the finite element method. This is particularly true for boundary value problems defined over curved regions.

Much of the mathematical detail for two-dimensional problems is concerned with the evaluation of

(a) integrals along curves lying in the xy plane (line integrals)
(b) integrals over areas in the xy plane (double integrals)

Our immediate aim in the next few sections is to show that such integrals, no matter how complicated, can be treated in a general way by a process of standardisation which involves the use of finite elements and shape functions. Once they are in standard form the integrals can be evaluated to any desired degree of accuracy by the method of Gaussian quadrature.

6.2 Ordinary integrals

An integral of a function, $f(x)$, of a single variable over the region $a \leq x \leq b$ (see Figure 6.1) is written $\int_a^b f(x)\,dx$ and represents, as is well known, the area under the curve $y = f(x)$ between $x = a$ and $x = b$.

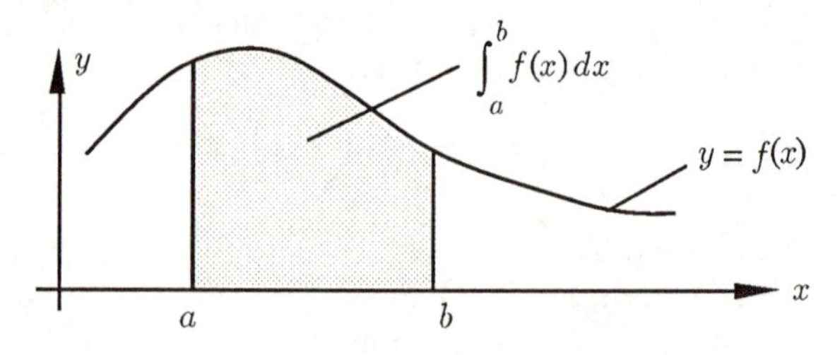

Figure 6.1

Unless the integrand $f(x)$ takes especially simple forms it is unlikely that analytical evaluation of the integral is possible and numerical approximation is required.

As we have already indicated in Section 3.9, in order to use the Gaussian quadrature technique of numerical integration the region $a \leq x \leq b$ must be transformed into a standard region $-1 \leq \xi \leq 1$. This transformation is easily achieved by the change of variable (3.9) which with an obvious change of notation is

$$x = \frac{(b-a)}{2}\xi + \frac{(b+a)}{2} \tag{6.1}$$

When changing variable in an integral three distinct operations must be carried out.

- The range of integration must be amended so as to be appropriate to the new variable. Using (6.1) the new range is indeed $-1 \leq \xi \leq 1$ since when $\xi = -1$ then $x = a$ and when $\xi = +1$ we have $x = b$.

- The integrand must be re-expressed in terms of the new variable. In this case $f(x)$ is replaced by the function

$$f\left(\frac{(b-a)}{2}\xi + \frac{(b+a)}{2}\right)$$

- The increment dx must be replaced by an equivalent increment with respect to the new variable ξ. In this case, using (6.1),

$$\begin{aligned} dx &= \left(\frac{dx}{d\xi}\right) d\xi \\ &= \frac{(b-a)}{2}\, d\xi \end{aligned}$$

Thus we can write (see Figure 6.2)

$$\int_a^b f(x)\, dx = \int_{-1}^{1} h(\xi)\, d\xi \tag{6.2}$$

where

$$h(\xi) = f\left(\frac{(b-a)}{2}\xi + \frac{(b+a)}{2}\right)\frac{(b-a)}{2} \tag{6.3}$$

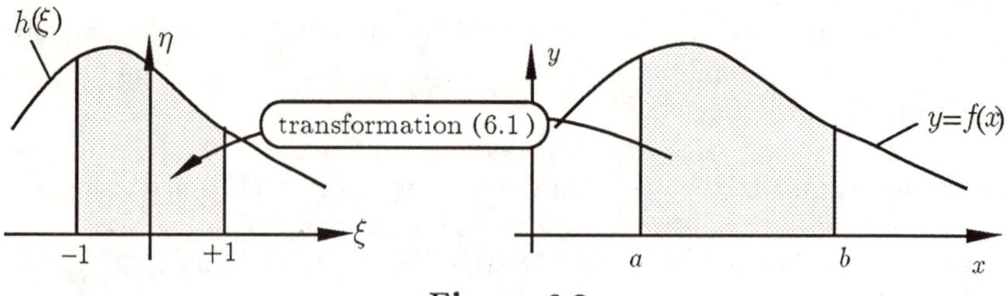

Figure 6.2

The integral on the right-hand side of (6.2) can be evaluated approximately using an nth-order Gaussian quadrature rule:

$$\int_{-1}^{1} h(\xi)\, d\xi \approx \sum_{i=1}^{n} w_i\, h(\xi_i) \tag{6.4}$$

in which ξ_i are the sample points and w_i the corresponding weighting factors. The higher the value of n the greater will be the resulting accuracy of the approximation. We have already, in Table 3.3 Section 3.9, given values of ξ_i and w_i for the cases $n = 2, 3, ..., 6$.

6.3 Line integrals using quadratic elements

As we shall see in Chapter 7 line integrals arise naturally in the finite element solution of two-dimensional boundary value problems. Although such integrals may take many forms, a common way of expressing them is

$$\int_C \{f(x,y)\, dx + g(x,y)dy\}$$

where C is a curve in the xy plane described in a particular sense and $f(x,y)$ and $g(x,y)$ are given functions of the two variables x and y. Note that x and y are not independent as they are constrained by the curve C. See Figure 6.3 for the case where C is a non-closed curve.

The interpretation to be given to a line integral is not of importance in the present discussion. Our interest lies in how such an integral might be evaluated for given functions $f(x,y), g(x,y)$ and a given curve C.

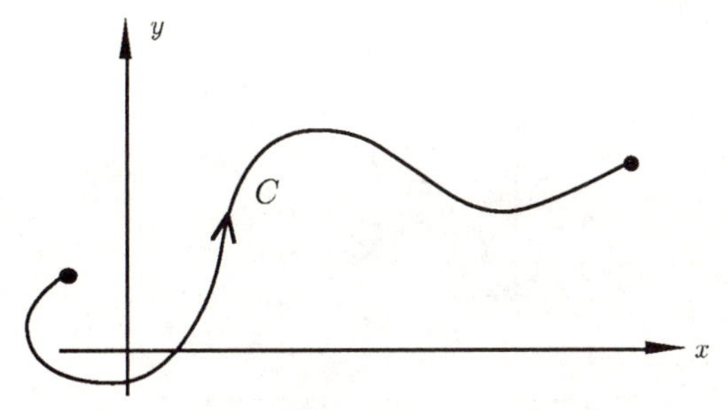

Figure 6.3

Of course, it is presumed that the precise definition of the curve C is known. Normally the equation of C is expressed in **parametric form:**

$$x = x(t) \qquad y = y(t) \qquad t_0 \leq t \leq t_1 \tag{6.5}$$

in which $t = t_0$ specifies the starting point of C and $t = t_1$ defines the end point. For example, for the segment of the circle shown in Figure 6.4 one possible parametric definition is

$$x = 3\cos t \qquad y = 3\sin t \qquad \frac{\pi}{4} \leq t \leq \frac{\pi}{2}$$

It is the parametric description of the curve C that leads directly to a technique for evaluating a line integral viz. by transforming it into an ordinary integral.

Consider a curve C parametrized as in (6.5). Along this curve dx/dt and dy/dt can be determined directly from the parametric equations and so

$$\int_C \{f(x,y)\,dx + g(x,y)\,dy\} = \int_C \left\{f(x(t),y(t))\,\frac{dx}{dt} + g(x(t),y(t))\frac{dy}{dt}\right\} dt$$

But the integral on the right-hand side is just an ordinary integral with respect to a single variable t. Since, along C, t varies from its initial value t_0 to its final value t_1, we can write

$$\int_C \{f(x,y)\,dx + g(x,y)\,dy\} = \int_{t_0}^{t_1} k(t)dt \tag{6.6}$$

where $\quad k(t) = f(x(t),y(t))\dfrac{dx}{dt} + g(x(t),y(t))\dfrac{dy}{dt}$

Of course, once a line integral is expressed in this way it can be evaluated (either analytically or numerically) in the same manner as any other ordinary integral.

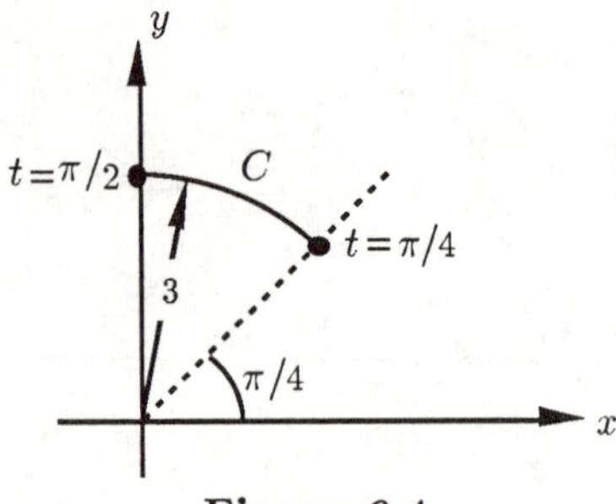

Figure 6.4

The parametric description (6.5) of a curve is particularily useful if it can be arranged to transform the whole of C into a single section from t_0 to t_1 of the t-axis (see Figure 6.5). In this case, a further elementary transformation can always be made so that the curve C is transformed into the standard region $-1 \leq \xi \leq +1$.

Unfortunately, for general curves this is rarely possible. An alternative approach, and one which has direct significance for the finite element method, is to segment C into n discrete sections $C_{[1]}, C_{[2]}, \ldots, C_{[n]}$. There is a an obvious relation between line integrals taken over these segments and the line integral taken over the complete curve:

$$\int_C (f dx + g dy) = \int_{C_{[1]}} (f dx + g dy) + \ldots + \int_{C_{[n]}} (f dx + g dy) \qquad \textbf{(6.7)}$$

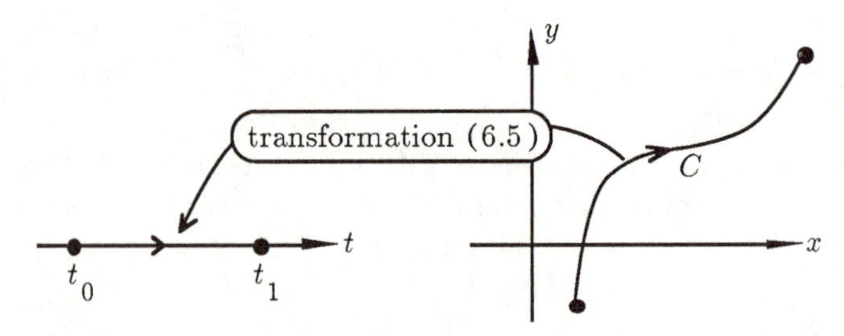

Figure 6.5

It is intuitively clear that, as long as the segments are sufficiently small, each of them may be approximated by a single parametric description. Then, by evaluating each elemental line integral in turn, we can determine an approximation to the line integral taken over the full curve C. As long as the segments are not highly curved it is possible to approximate each by a low order polynomial curve. We shall choose second-order polynomials (parabolae) and then, as might be expected, the quadratic shape functions $Q_i(\xi)$, $i = 1, 2, 3$ play a significant role. There are of course other possible approximations. Use of the linear shape functions $L_i(\xi)$, $i = 1, 2$, would imply segmenting the curve into straight-line sections which is too crude an approximation unless the segments are small. Better approximations can be achieved using higher order polynomials but with a corresponding increase in complexity. It is found, in practice, that the use of parabolae is (generally) a happy balance.

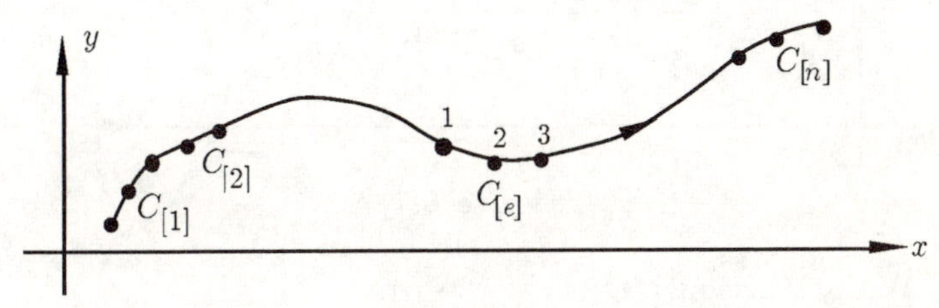

Figure 6.6

In Figure 6.6 the curve C has been approximated by n parabolic segments. On the segment $C_{[e]}$ we choose three points – the two end points (x_1, y_1) and (x_3, y_3) and an interior point (x_2, y_2).

A transformation is now used to map this segment into the **standard segment** $-1 \leq \xi \leq 1$. A suitable transformation for the x-variable is

$$x = x_1 \frac{1}{2}\xi(\xi - 1) + x_2(1 - \xi^2) + x_3 \frac{1}{2}\xi(\xi + 1)$$

since, as is easily checked, the points $\xi = -1, 0, 1$ correspond, respectively, to x_1, x_2, x_3. We recognize, from (3.16), that the coefficients of x_1, x_2, x_3 are the quadratic shape functions $Q_1(\xi), Q_2(\xi)$ and $Q_3(\xi)$. Hence we can write

$$x = x(\xi) \equiv \sum_{i=1}^{3} x_i \, Q_i(\xi) \tag{6.8a}$$

and in a similar way for the y-variable

$$y = y(\xi) \equiv \sum_{i=1}^{3} y_i \, Q_i(\xi) \tag{6.8b}$$

The relations in (6.8) are the parametric equations of a curve passing through the points (x_1, y_1), (x_2, y_2), (x_3, y_3). In fact this curve is a parabola (or possibly a straight line if the three given points lie along a straight line).

We emphasize that the segment $C_{[e]}$ joining (x_1, y_1) to (x_3, y_3) and passing through (x_2, y_2) is approximated by the parabola (6.8). See Figure 6.7 where the relevant portion of the curve has been magnified in the interests of clarity.

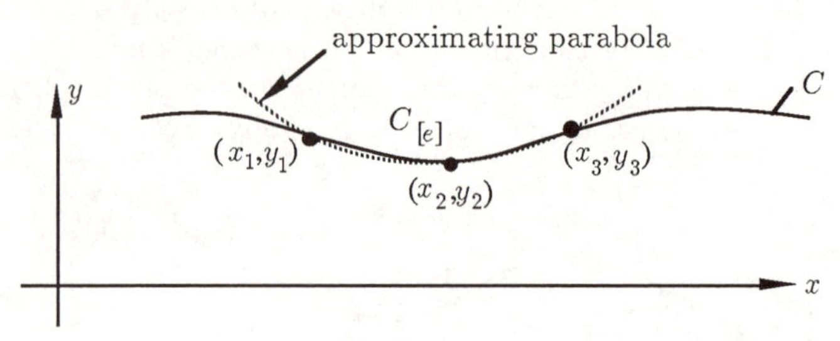

Figure 6.7

With this segmentation we can write

$$\int_C (f(x,y)dx + g(x,y)dy) \approx \sum_{e=1}^{n} \left\{ \int_{-1}^{1} h_{[e]}(\xi)\, d\xi \right\} \tag{6.9}$$

in which

$$h_{[e]}(\xi) \equiv f(x,y)\frac{dx}{d\xi} + g(x,y)\frac{dy}{d\xi}$$

and from (6.8) (within segment [e])

$$\begin{aligned} \frac{dx}{d\xi} &= x_1 \frac{1}{2}(2\xi - 1) + x_2(-2\xi) + x_3 \frac{1}{2}(2\xi + 1) \\ \frac{dy}{d\xi} &= y_1 \frac{1}{2}(2\xi - 1) + y_2(-2\xi) + y_3 \frac{1}{2}(2\xi + 1) \end{aligned} \tag{6.10}$$

We now illustrate these ideas with a numerical evaluation of a simple line integral.

Example 6.1

Using two elements evaluate the line integral $\int_C (ydx + dy)$ in which C is the quarter circle of radius 1 in the first quadrant, described in the sense shown in Figure 6.8(a).

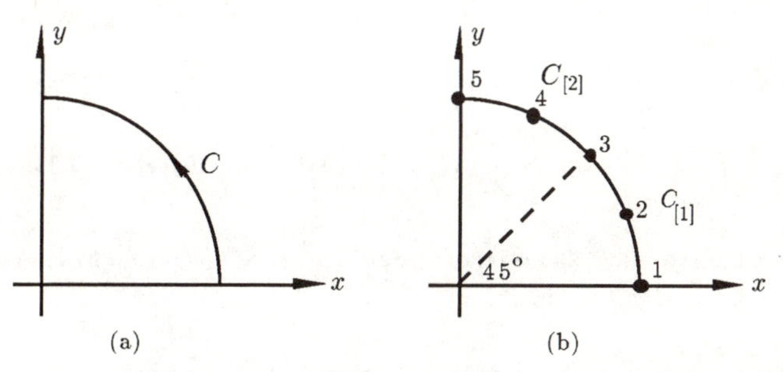

Figure 6.8

Solution

We will approximate C by two quadratic elements, see Figure 6.8(b). The nodal coordinates are

$1:(1,0)$	$4:(0.38268, 0.92388)$
$2:(0.92388, 0.38268)$	$5:(0,1)$
$3:(0.70711, 0.70711)$	

Now, from (6.7),

$$\int_C (ydx + dy) = \int_{C_{[1]}} (ydx + dy) + \int_{C_{[2]}} (ydx + dy)$$

Transforming element [1] into the standard element, we have from (6.10)

$$\frac{dx}{d\xi} = \frac{1}{2}(2\xi - 1) + 0.92388(-2\xi) + \frac{0.70711}{2}(2\xi + 1)$$

$$\frac{dy}{d\xi} = 0 + 0.38268(-2\xi) + \frac{0.70711}{2}(2\xi + 1)$$

from which we obtain (using (6.8b))

$$y\frac{dx}{d\xi} + \frac{dy}{d\xi} = \xi^3(0.00409) + \xi^2(-0.04546) + \xi(-0.16385) + 0.29752$$

Hence

$$\int_{C_{[1]}} (ydx + dy) \approx \int_{-1}^{1} \{\xi^2(-0.04546) + 0.29752\} d\xi$$

$$= 0.56472$$

in which we have ignored odd powers of ξ as they integrate to zero.

Similarly for element [2], we have from (6.10)

$$\frac{dx}{d\xi} = 0.70711\left(\frac{1}{2}\right)(2\xi - 1) + 0.38268(-2\xi)$$

$$\frac{dy}{d\xi} = 0.70711\left(\frac{1}{2}\right)(2\xi - 1) + 0.92388(-2\xi) + \frac{1}{2}(2\xi + 1)$$

giving (using (6.8b))

$$y\frac{dx}{d\xi} + \frac{dy}{d\xi} = \xi^3(0.00409) + \xi^2(0.01633) + \xi(-0.24624) - 0.18020$$

and so

$$\int_{C_{[2]}} (y dx + dy) \approx \int_{-1}^{1} \{\xi^2(0.01633) - 0.18020\} d\xi = -0.34950$$

in which we have again ignored odd powers of ξ. Our approximation to the line integral is therefore

$$\int_C (y dx + dy) \approx 0.56472 - 0.34950 = 0.2152 \text{ to 4 d.p.}$$

The exact value of the given line integral is easily obtained here. An exact parametrization of the whole curve C is of course

$$x = \cos t \qquad y = \sin t \qquad 0 \leq t \leq \tfrac{\pi}{2}$$

$$\therefore \quad \frac{dx}{dt} = -\sin t \qquad \frac{dy}{dt} = \cos t$$

and so

$$\int_C (y dx + dy) = \int_0^{\pi/2} (-\sin^2 t + \cos t)\, dt$$

$$= \int_0^{\pi/2} [\frac{(\cos 2t - 1)}{2} + \cos t]\, dt = 0.2146$$

The line integral in this example is not path independent. We have changed the circular path into two segments, each of which is parabolic. However, the parabolic segments are good approximations to the circular segments which, in part, accounts for the good approximation achieved here.

The important point to note about the use of shape functions and the segmentation procedure is that it can be applied to any line integral no matter how complicated.

In the example above we have evaluated the final integrals analytically, due to the simple form of the integrand and in the interests of clarity. In more realistic examples the integrals on the right hand side of (6.9) are determined numerically using Gaussian quadrature.

6.4 Double integrals using triangular and quadrilateral elements

A double integral is an integral of a function $f(x, y)$ of two variables taken over a two-dimensional region A of the xy plane. Such an integral is generally denoted by

$$\iint_A f(x, y)\, dx\, dy$$

and it can be interpreted as the volume V under the surface $z = f(x, y)$ above the region A (see Figure 6.9).

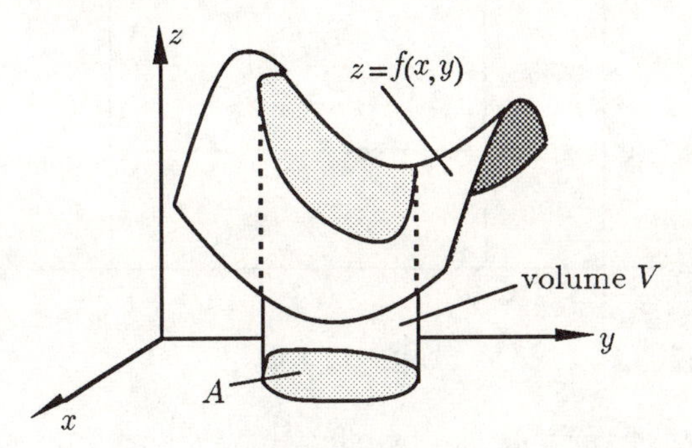

Figure 6.9

A double integral is normally evaluated by expressing it as a so-called **repeated** integral:

$$\iint_A f(x, y)\, dx\, dy = \int_a^b \left\{ \int_{y=h_1(x)}^{y=h_2(x)} f(x, y)\, dy \right\} dx \tag{6.11}$$

in which $a, b,\ h_1(x), h_2(x)$ are defined in Figure 6.10.

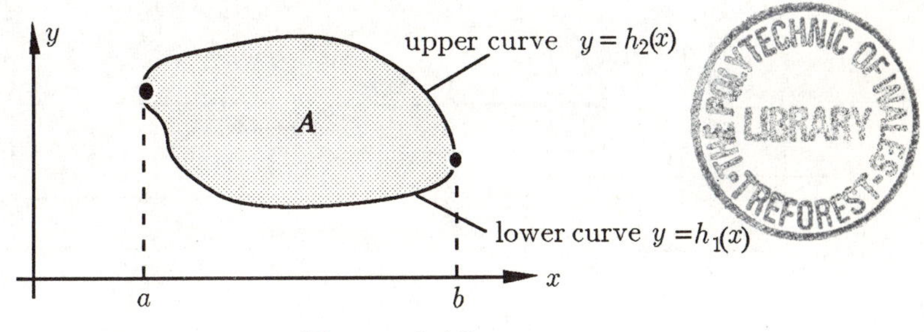

Figure 6.10

In (6.11) the inner integral, with respect to the y-variable, is performed first, treating x as a constant and inserting the limits in the usual way. Then the integral with respect to x is performed, again in the usual way.

If the region A is rectangular then the double integral becomes

$$\iint_A f(x,y)\,dx\,dy = \int_a^b \left\{ \int_\alpha^\beta f(x,y)\,dy \right\} dx$$

which is particularly simple since the limits on the y-integral are now constants and not functions of x (see Figure 6.11).

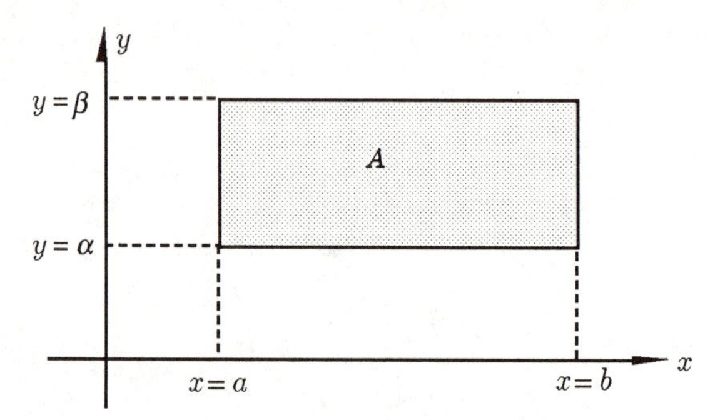

Figure 6.11

Simple changes of variable can be effected to transform such a rectangle in both position and size into the **standard square**

$$-1 \le \xi \le 1 \qquad -1 \le \eta \le 1$$

with respect to two new variables ξ, η. The transformation is

$$x = \frac{(b-a)}{2}\xi + \frac{(a+b)}{2} \qquad y = \frac{(\beta-\alpha)}{2}\eta + \frac{(\alpha+\beta)}{2} \tag{6.12}$$

See Figure 6.12. Compare (6.12) with (6.1) in one dimension.

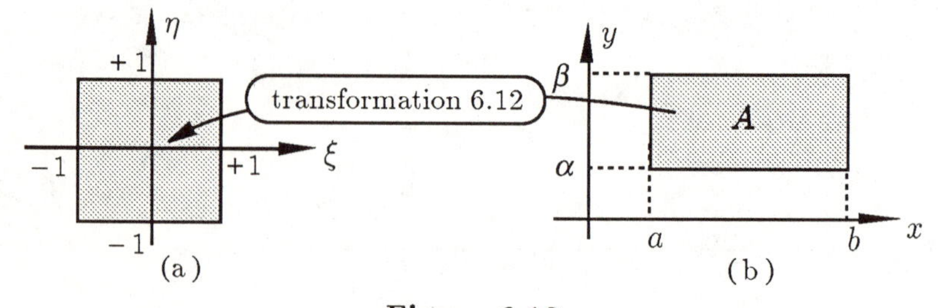

Figure 6.12

Double integrals over other, more difficult, regions than a

rectangle can often be simplified by transforming the region into a standard region which may be the standard square shown or a **standard triangle** which we will take as the triangle with vertices $(-1,-1)$, $(-1,1)$, $(1,-1)$ i.e. the left hand lower half of the standard square. Examples of the use of the standard triangle will be outlined later.

Such transformations are carried out using a suitable change of variables of the form:

$$x = x(\xi, \eta) \qquad y = y(\xi, \eta) \tag{6.13}$$

With respect to these 'standard' variables, ξ and η, the double integral can then be written (for a transformation into the standard square)

$$\iint_A f(x,y)\,dx\,dy = \int_{-1}^{1}\int_{-1}^{1} h(\xi,\eta)\,d\xi\,d\eta \tag{6.14}$$

where

$$h(\xi,\eta) = f(x(\xi,\eta),\, y(\xi,\eta))\,|J| \tag{6.15a}$$

The quantity J is defined by

$$J = \begin{vmatrix} \dfrac{\partial x}{\partial \xi} & \dfrac{\partial y}{\partial \xi} \\ \dfrac{\partial x}{\partial \eta} & \dfrac{\partial y}{\partial \eta} \end{vmatrix} \tag{6.15b}$$

and is called the **Jacobian** of the transformation. The transformation (6.14) of the double integral is the two-dimensional equivalent of the one-dimensional transformation (6.2).

Example 6.2

By transforming the annular region A between the two quarter circles shown in Figure 6.13 into the standard square evaluate the integral $\iint_A (x^2+y^2)\,dx\,dy$.

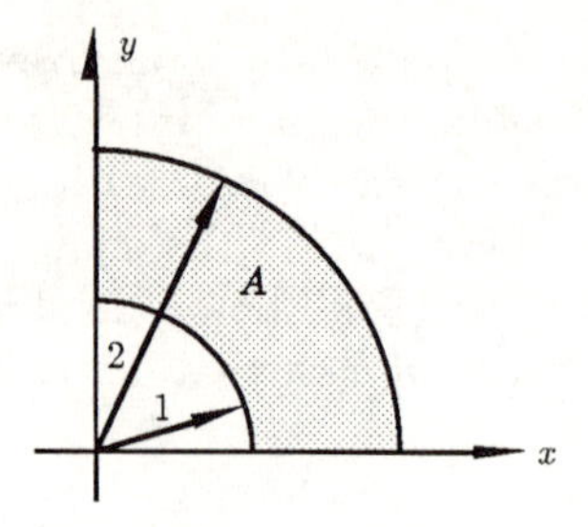

Figure 6.13

Solution

Firstly we use the well-known change of variables:

$$x = r\cos\theta \qquad y = r\sin\theta$$

This transformation maps the given annulus into a rectangular region in the '$r\theta$' plane (see Figure 6.14).

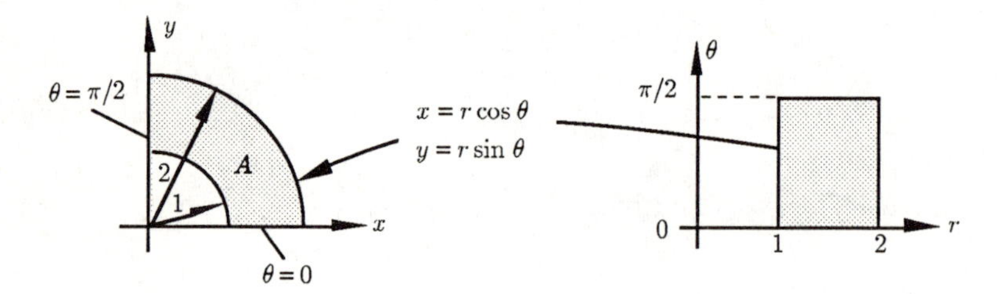

Figure 6.14

Using (6.12) but with (r, θ) instead of (x, y) the further change of variables:

$$r = \frac{1}{2}\xi + \frac{3}{2} \qquad \theta = \frac{\pi}{4}\eta + \frac{\pi}{4}$$

transforms the rectangle into the standard square $-1 \leq \xi, \eta \leq +1$. Combining the two changes of variables it follows that

$$x = \left(\frac{1}{2}\xi + \frac{3}{2}\right)\cos\left(\frac{\pi}{4}\eta + \frac{\pi}{4}\right)$$
$$y = \left(\frac{1}{2}\xi + \frac{3}{2}\right)\sin\left(\frac{\pi}{4}\eta + \frac{\pi}{4}\right) \qquad \textbf{(6.16)}$$

will transform the annulus A directly into the standard square in a single operation.

The Jacobian in this case is

$$J = \begin{vmatrix} \dfrac{\partial x}{\partial \xi} & \dfrac{\partial y}{\partial \xi} \\ \dfrac{\partial x}{\partial \eta} & \dfrac{\partial y}{\partial \eta} \end{vmatrix}$$

$$= \begin{vmatrix} \frac{1}{2}\cos\left(\frac{\pi\eta}{4}+\frac{\pi}{4}\right) & \frac{1}{2}\sin\left(\frac{\pi\eta}{4}+\frac{\pi}{4}\right) \\ -\frac{\pi}{4}\left(\frac{\xi}{2}+\frac{3}{2}\right)\sin\left(\frac{\pi\eta}{4}+\frac{\pi}{4}\right) & \frac{\pi}{4}\left(\frac{\xi}{2}+\frac{3}{2}\right)\cos\left(\frac{\pi\eta}{4}+\frac{\pi}{4}\right) \end{vmatrix}$$

$$= \frac{\pi}{8}\left(\frac{\xi}{2}+\frac{3}{2}\right)$$

Also, from (6.16) $x^2 + y^2 = \left(\frac{1}{2}\xi + \frac{3}{2}\right)^2$. Hence, using (6.14), we have

$$\iint_A (x^2+y^2)\,dx\,dy = \int_{-1}^{1}\int_{-1}^{1}\left(\frac{1}{2}\xi+\frac{3}{2}\right)^3 \frac{\pi}{8}\,d\xi\,d\eta$$

$$= \frac{\pi}{8}\int_{-1}^{1} d\eta \int_{-1}^{1}\left(\frac{1}{2}\xi+\frac{3}{2}\right)^3 d\xi$$

$$= \frac{\pi}{8}(2)\left[\frac{1}{2}\left(\frac{1}{2}\xi+\frac{3}{2}\right)^4\right]_{-1}^{1}$$

$$= \frac{\pi}{4}\left(8-\frac{1}{2}\right) = \frac{15\pi}{8}$$

The operation of transforming to a standard region, whether the square or the triangle, is important because standard numerical procedures can then be used to effect an approximation to the double integral. In particular, for integration over the standard square, we can use the two-dimensional Gaussian quadrature method which takes the form:

$$\int_{-1}^{1}\int_{-1}^{1} h(\xi,\eta)\,d\xi\,d\eta \approx \sum_{i=1}^{n}\sum_{j=1}^{n} w_i\,w_j\,h(\xi_i,\,\eta_j) \qquad \textbf{(6.17)}$$

in which the (ξ_i, η_i) and the w_i are respectively the sample points and weighting factors appropriate to a nth order method. (Compare the one-dimensional form (6.4).) Normally, and particularly when n is large or when $h(\xi, \eta)$ is complicated, such a calculation would be done with the help of a computer. Similar approximation techniques are available for double integrals defined over standard triangles.

In Example 6.2 the whole of the given region of integration was transformed to a standard region in one operation (6.16). This is rarely possible in practical examples of double integrals which may well be defined over irregular shaped regions, not necessarily bounded by straight lines or circles.

To evaluate such integrals we adopt a strategy similar to that used for the treatment of line integrals over arbitrarily shaped curves, viz. we segment the given region A into elements $A_{[1]}, A_{[2]} \ldots A_{[n]}$ and effect a transformation of each element into a standard region.

The success of this strategy follows from the property of double integrals:

$$\iint_A f(x,y)dx\,dy = \iint_{A_{[1]}} f(x,y)dx\,dy + \iint_{A_{[2]}} f(x,y)dx\,dy + \ldots + \iint_{A_{[n]}} f(x,y)dx\,dy$$

We shall see that this segmentation can always be carried through to any desired degree of approximation. Perhaps the simplest way to do this is to use triangular elements. We must firstly, however, show how to transform an arbitrary triangle into the standard triangle.

Example 6.3

Show that the transformation

$$x = x_1\left(-\frac{1}{2}(\xi+\eta)\right) + x_2\left(\frac{1}{2}(1+\xi)\right) + x_3\left(\frac{1}{2}(1+\eta)\right)$$

$$y = y_1\left(-\frac{1}{2}(\xi+\eta)\right) + y_2\left(\frac{1}{2}(1+\xi)\right) + y_3\left(\frac{1}{2}(1+\eta)\right) \qquad \textbf{(6.18)}$$

maps a triangle with vertices (x_i, y_i), $i = 1, 2, 3$, into the standard triangle in the $\xi\eta$ plane.

Solution

The relation between (x, y) and (ξ, η) as defined in (6.18) is linear so that straight lines in the $\xi\eta$ plane will remain straight lines in the xy plane.

Consider, for example, the side

$$\xi = -1, \quad -1 \le \eta \le +1$$

of the standard triangle. We obtain, using (6.18),

$$x = x_1\left(-\frac{1}{2}(\eta - 1)\right) + x_3\left(\frac{1}{2}(1+\eta)\right)$$
$$= \eta\frac{(x_3 - x_1)}{2} + \frac{(x_3 + x_1)}{2}$$

and similarly

$$y = \eta\frac{(y_3 - y_1)}{2} + \frac{(y_3 + y_1)}{2}$$

But these are the parametric equations of a straight line passing through the points (x_1, y_1) (when $\eta = -1$) and (x_3, y_3) (when $\eta = +1$).

The other sides of the standard triangle are treated similarly as the reader should verify (see Figure 6.15).

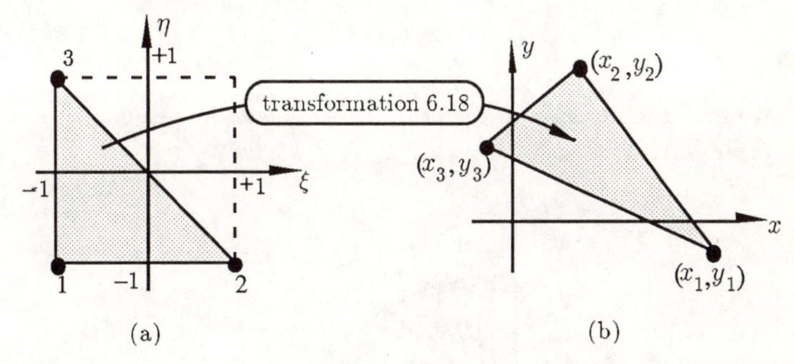

Figure 6.15

Note that the functions used in the transformation (6.18):

$$T_1(\xi, \eta) = -\frac{1}{2}(\xi + \eta)$$

$$T_2(\xi, \eta) = \frac{1}{2}(1 + \xi)$$

$$T_3(\xi, \eta) = \frac{1}{2}(1 + \eta) \qquad \textbf{(6.19)}$$

are called **linear shape functions** for a three-noded triangle. Using them we can write (6.18) as

$$x = \sum_{i=1}^{3} x_i\, T_i(\xi, \eta) \quad \text{and} \quad y = \sum_{i=1}^{3} y_i\, T_i(\xi, \eta)$$

The node numbers 1,2,3 used in the standard triangle are called **local node numbers** and are taken in the order shown in Figure 6.15(a).

Example 6.4

Determine an approximation to the double integral $\iint_A (x+y)dx\,dy$ where A is the octant of the circle centre origin radius 2 shown in Figure 6.16.

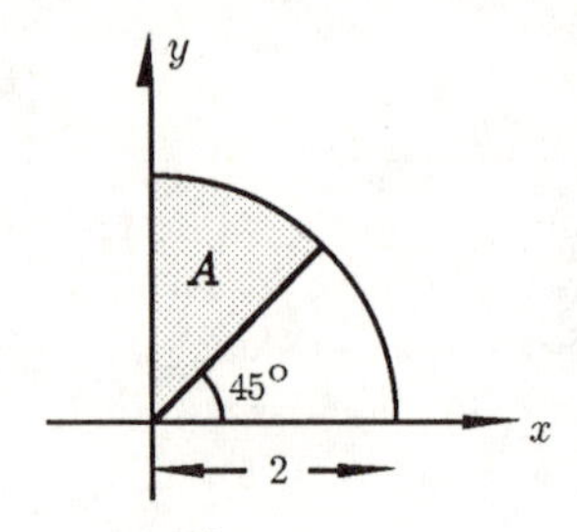

Figure 6.16

Solution

We shall use a triangular segmentation of A with four elements as shown in Figure 6.17.

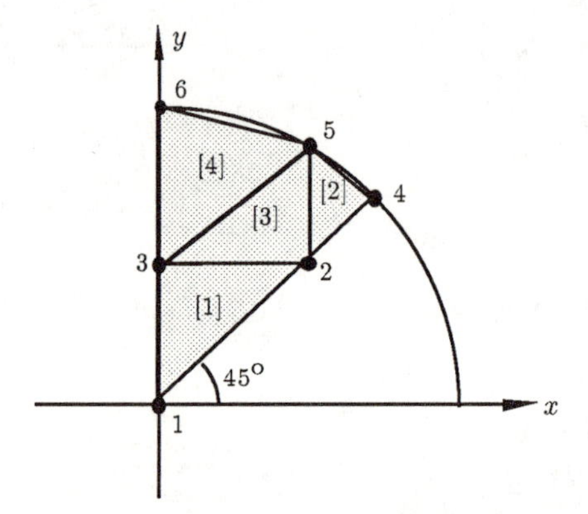

Figure 6.17

Here node 2 is chosen to have coordinates $(1,1)$. The other nodal coordinates follow directly from the diagram.

The following easily obtained results concerning double integrals taken over the 'standard triangle' ST of Figure 6.15(a) will be useful in the calculations below:

$$\iint_{ST} 1\,d\xi\,d\eta = 2 \qquad \iint_{ST} \xi\,d\xi\,d\eta = -\frac{2}{3} \qquad \iint_{ST} \eta\,d\xi\,d\eta = -\frac{2}{3} \tag{6.20}$$

We now map each triangular element in turn into the standard triangle using (6.18). The integral over each element $[e]$ is obtained from the equivalent of (6.14) viz.

$$\iint_{[e]} f(x,y)dx\,dy = \iint_{ST} h(\xi,\eta)\,d\xi\,d\eta \qquad \textbf{(6.21)}$$

with $h(\xi,\eta)$ defined in (6.15a).

- **Element [1]**

With nodes taken in the order 1,2,3, we have using (6.18) for the transformation of this element into the standard triangle

$$x = \frac{1}{2}(1+\xi) \qquad y = \frac{1}{2}(1+\xi) + \frac{1}{2}(1+\eta)$$

Hence the Jacobian in this case is

$$J = \begin{vmatrix} \dfrac{\partial x}{\partial \xi} & \dfrac{\partial y}{\partial \xi} \\ \dfrac{\partial x}{\partial \eta} & \dfrac{\partial y}{\partial \eta} \end{vmatrix} = \begin{vmatrix} \dfrac{1}{2} & \dfrac{1}{2} \\ 0 & \dfrac{1}{2} \end{vmatrix} = \frac{1}{4}$$

and so (6.21) gives

$$\iint_{[1]} (x+y)dx\,dy = \iint_{ST} \left(\frac{3}{2} + \xi + \frac{\eta}{2}\right)\frac{1}{4}\,d\xi\,d\eta$$

$$= \left[3 - \frac{2}{3} - \frac{1}{3}\right]\frac{1}{4} = \frac{1}{2}$$

using (6.20)

- **Element [2]**

With nodes taken in the order 2,4,5 (to 'match' the local nodes 1,2,3) we have, again using (6.18),

$$x = \xi\,\frac{(\sqrt{2}-1)}{2} + \frac{(\sqrt{2}+1)}{2}$$

$$y = \xi\frac{(\sqrt{2}-1)}{2} + \eta\,\frac{(\sqrt{3}-1)}{2} + \frac{\sqrt{3}+\sqrt{2}}{2}$$

from which

$$J = \tfrac{1}{4}(1 - \sqrt{2} - \sqrt{3} + \sqrt{6})$$

Hence the integral over element [2] becomes

$$\iint_{ST} \left\{\xi\,(\sqrt{2}-1) + \eta\,\frac{(\sqrt{3}-1)}{2} + \sqrt{2} + \frac{\sqrt{3}}{2} + \frac{1}{2}\right\} J\,d\xi\,d\eta$$

The integral is readily evaluated using (6.20) and is found to be

$$\tfrac{1}{3}(\sqrt{2}+\sqrt{3}-2)$$

Similar calculations for elements [3] and [4], which the reader is urged to carry out, give for the respective integrals

$$\iint_{[3]}(x+y)dx\,dy = \iint_{ST}\left\{\xi\frac{(\sqrt{3}-1)}{2} - \frac{\eta}{2} + \frac{\sqrt{3}}{2} + 1\right\}\frac{(\sqrt{3}-1)}{4}d\xi\,d\eta$$

$$= \frac{1}{6}(-1+3\sqrt{3})$$

and

$$\iint_{[4]}(x+y)dx\,dy = \iint_{ST}\left\{\xi\frac{\sqrt{3}}{2} + \frac{\eta}{2} + \frac{\sqrt{3}}{2} + \frac{3}{2}\right\}\frac{1}{4}\,d\xi\,d\eta$$

$$= \frac{1}{6}(\sqrt{3}+4)$$

Adding the contributions from each element we obtain

$$\iint_{A}(x+y)dx\,dy \approx \frac{1}{2} + \frac{1}{3}(\sqrt{2}+\sqrt{3}-2) + \frac{1}{6}(-1+3\sqrt{3}) + \frac{1}{6}(\sqrt{3}+4)$$

$$= 2.53678$$

The exact value of the double integral is easily shown to be $8/3 = 2.6666$. The accuracy of our approximation is thus by no means remarkable but is certainly acceptable when we realize that the region A has been modelled by only four linear triangular elements. Greater accuracy can be achieved either by using more elements or by using a distribution of the elements which more closely 'covers' the region A.

- **Quadrilateral elements**

As well as triangular elements with straight edges we can use quadrilateral elements with straight edges to 'mesh' a region of interest. It is easily verified (and the reader should carry through this exercise) that the transformation which takes a standard square with vertices

$$1:\ (-1,-1) \quad 2:\ (1,-1) \quad 3:\ (1,1) \quad 4:\ (-1,1)$$

into a general straight-edged quadrilateral with vertices at (x_i, y_i), $i = 1, 2, 3, 4$, is

$$x = \sum_{i=1}^{4} x_i L_i(\xi, \eta) \qquad y = \sum_{i=1}^{4} y_i L_i(\xi, \eta) \tag{6.22}$$

See Figure 6.18. The transformation (6.12) was, in fact, an example of (6.22). In (6.22) the functions used are

$$\begin{aligned} L_1(\xi,\eta) &= \frac{1}{4}(\xi - 1)(\eta - 1) \quad & L_2(\xi,\eta) &= -\frac{1}{4}(\xi + 1)(\eta - 1) \\ L_3(\xi,\eta) &= \frac{1}{4}(\xi + 1)(\eta + 1) \quad & L_4(\xi,\eta) &= -\frac{1}{4}(\xi - 1)(\eta + 1) \end{aligned} \tag{6.23}$$

These functions $L_i(\xi, \eta)$ are referred to as **bi-linear shape functions** since the behaviour of each is **linear** along the edges of the standard region.

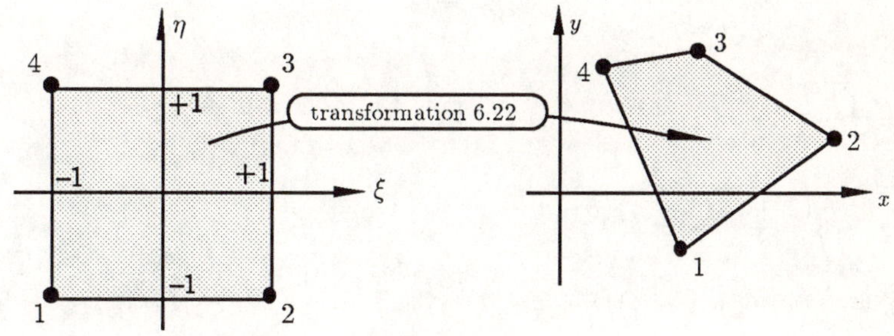

Figure 6.18

A useful exercise for the reader would be to rework Example 6.4 using quadrilateral elements. We suggest that the segmentation shown in Figure 6.19 be used.

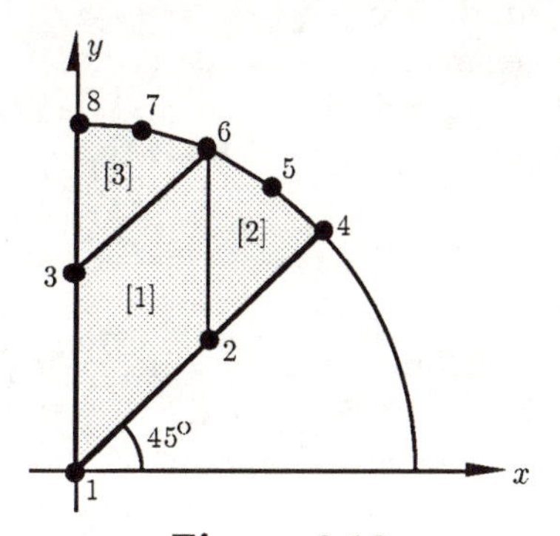

Figure 6.19

6.5 Double integrals using curved elements

Thus far in the modelling of two-dimensional regions we have used straight edged elements. This is clearly unsatisfactory for the accurate approximation of regions with curved boundaries. We have seen, in our discussion on line integrals, that curves can be well approximated by the use of one-dimensional quadratic elements together with a coordinate transformation. There are two possible two-dimensional extensions of the one-dimensional quadratic element – a six noded triangular element and an eight noded quadrilateral element. We shall consider only the latter.

The coordinate transformation used to transform a standard eight noded square element into a general (curved) quadrilateral with 'nodes' at (x_i, y_i), $i = 1, \ldots, 8$, can be shown to be

$$x = \sum_{i=1}^{8} x_i \, Q_i(\xi, \eta) \qquad y = \sum_{i=1}^{8} y_i \, Q_i(\xi, \eta) \tag{6.24}$$

where

$$\begin{aligned}
Q_1 &= -\frac{1}{4}(1-\xi)(1-\eta)(\xi+\eta+1) & Q_2 &= \frac{1}{2}(1-\xi^2)(1-\eta) \\
Q_3 &= \frac{1}{4}(1+\xi)(1-\eta)(\xi-\eta-1) & Q_4 &= \frac{1}{2}(1-\eta^2)(1+\xi) \\
Q_5 &= \frac{1}{4}(1+\xi)(1+\eta)(\xi+\eta-1) & Q_6 &= \frac{1}{2}(1-\xi^2)(1+\eta) \\
Q_7 &= -\frac{1}{4}(1-\xi)(1+\eta)(\xi-\eta+1) & Q_8 &= \frac{1}{2}(1-\eta^2)(1-\xi)
\end{aligned} \tag{6.25}$$

Although these functions contain cubic terms they are still known as the **bi-quadratic shape functions** for an eight-noded quadrilateral element since they reduce to **quadratic** functions of ξ or η along the edges of the standard square. For example on the side of the square with nodes 1, 2 and 3 (see Figure 6.20(a)) we have $\eta = -1$ and the shape functions (6.25) become zero except for Q_1, Q_2, Q_3 which become

$$Q_1(\xi, -1) = \frac{1}{2}\xi(\xi-1) \quad Q_2(\xi, -1) = 1 - \xi^2 \quad Q_3(\xi, -1) = \frac{1}{2}\xi(\xi+1)$$

These are the familiar one-dimensional quadratic shape functions. Similar results are found for the other three edges.

As might now be expected the transformation (6.24) takes the standard square into a quadrilateral, each edge of which is a parabola. This can be verified by examining the effect of the transformation on each edge of the square.

For example, for the edge $\xi = +1$ with nodes 3, 4, 5 (6.24) becomes

$$x = x_3 \tfrac{1}{2}\eta(\eta - 1) + x_4(1 - \eta^2) + x_5 \tfrac{1}{2}\eta(\eta + 1)$$

$$y = y_3 \tfrac{1}{2}\eta(\eta - 1) + y_4(1 - \eta^2) + y_5 \tfrac{1}{2}\eta(\eta + 1)$$

These are the parametric equations of a parabola (since both x and y are quadratic in the parameter η). The parabola passes through the points (x_3, y_3) (when $\eta = -1$), (x_4, y_4) (when $\eta = 0$) and (x_5, y_5) (when $\eta = +1$). We observe that the parabola may degenerate to a straight line when the three points (x_i, y_i), $i = 3, 4, 5$, lie on a straight line.

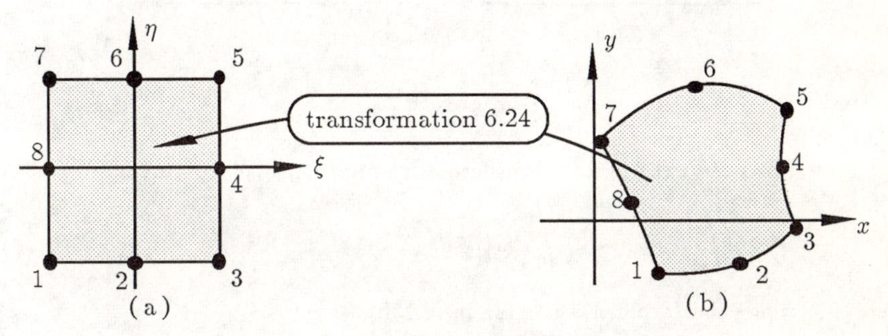

Figure 6.20

The complete transformation performed by (6.24) is shown in Figure (6.20). The importance of (6.24) lies in the fact that we now know how to transform a curved quadrilateral (bounded by parabolae or straight lines) into the standard square. Such transformations are used to considerable advantage in the finite element method. (A fuller discussion of the derivation and properties of (6.25) and other two-dimensional shape functions and elements is given in the Appendix.)

We now illustrate the use of curved elements in the evaluation of a double integral.

Example 6.5

Evaluate the double integral $\iint_A dx\, dy$ where A in the annular region shown in Figure 6.21(a) between quarter circles of radii 1 and 3.

Solution

We are of course being asked to calculate the area of the annulus. To keep matters as simple as possible we will use a single eight-noded element to approximate the region with nodes positioned in an obvious way as detailed in Figure 6.21(b). Specifically the nodal coordinates are:

$1:(1,0)$ $\quad 2:(2,0)$ $\quad 3:(3,0)$ $\quad 4:(3\frac{\sqrt{2}}{2},3\frac{\sqrt{2}}{2})$

$5:(0,3)$ $\quad 6:(0,2)$ $\quad 7:(0,1)$ $\quad 8:(\frac{\sqrt{2}}{2},\frac{\sqrt{2}}{2})$

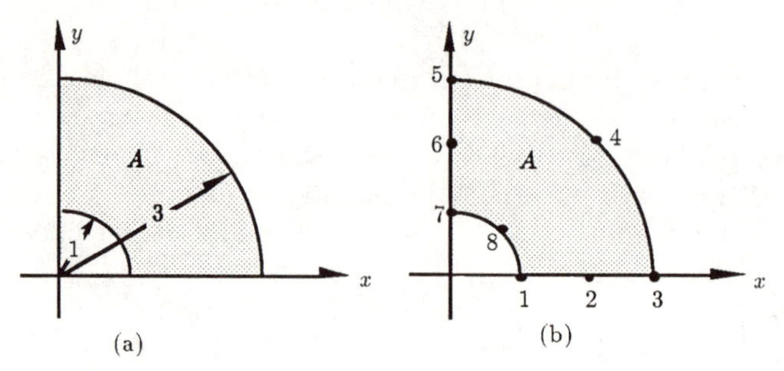

Figure 6.21

Now, using (6.14) for a transformation into the standard square,

$$\iint_A dx\,dy = \int_{-1}^{1}\int_{-1}^{1} |J|\,d\xi\,d\eta \tag{6.26}$$

where the Jacobian J is given in (6.15b).

From the transforming equations (6.24) we have

$$\frac{\partial x}{\partial \xi} = \sum_{i=1}^{8} x_i \frac{\partial Q_i}{\partial \xi} \quad , \quad \frac{\partial y}{\partial \xi} = \sum_{i=1}^{8} y_i \frac{\partial Q_i}{\partial \xi}$$

and similarly for $\partial x/\partial\eta$ and $\partial y/\partial\eta$. The partial derivatives of the shape functions are easily found from (6.25):

$$\begin{aligned}
\frac{\partial Q_1}{\partial \xi} &= \frac{1}{4}(1-\eta)(2\xi+\eta) & \frac{\partial Q_2}{\partial \xi} &= -\xi(1-\eta)\\
\frac{\partial Q_3}{\partial \xi} &= \frac{1}{4}(1-\eta)(2\xi-\eta) & \frac{\partial Q_4}{\partial \xi} &= \frac{1}{2}(1-\eta^2)\\
\frac{\partial Q_5}{\partial \xi} &= \frac{1}{4}(1+\eta)(2\xi+\eta) & \frac{\partial Q_6}{\partial \xi} &= -\xi(1+\eta)\\
\frac{\partial Q_7}{\partial \xi} &= \frac{1}{4}(1+\eta)(2\xi-\eta) & \frac{\partial Q_8}{\partial \xi} &= -\frac{1}{2}(1-\eta^2)
\end{aligned} \tag{6.27}$$

Similarly

$$\begin{aligned}
\frac{\partial Q_1}{\partial \eta} &= \frac{1}{4}(1-\xi)(2\eta+\xi) & \frac{\partial Q_2}{\partial \eta} &= -\frac{1}{2}(1-\xi^2) \\
\frac{\partial Q_3}{\partial \eta} &= -\frac{1}{4}(1+\xi)(\xi-2\eta) & \frac{\partial Q_4}{\partial \eta} &= -\eta(1+\xi) \\
\frac{\partial Q_5}{\partial \eta} &= \frac{1}{4}(1+\xi)(2\eta+\xi) & \frac{\partial Q_6}{\partial \eta} &= \frac{1}{2}(1-\xi^2) \\
\frac{\partial Q_7}{\partial \eta} &= \frac{1}{4}(1-\xi)(2\eta-\xi) & \frac{\partial Q_8}{\partial \eta} &= -\eta(1-\xi)
\end{aligned} \tag{6.28}$$

It follows, after a little algebra, that in this specific example

$$\frac{\partial x}{\partial \xi} = \eta^2\left(\frac{1-\sqrt{2}}{2}\right) + \eta\left(-\frac{1}{2}\right) + \frac{\sqrt{2}}{2}$$

$$\frac{\partial x}{\partial \eta} = \xi\eta(1-\sqrt{2}) + \xi\left(-\frac{1}{2}\right) + \eta(2-2\sqrt{2}) - 1$$

$$\frac{\partial y}{\partial \xi} = \eta^2\left(\frac{1-\sqrt{2}}{2}\right) + \eta\left(\frac{1}{2}\right) + \frac{\sqrt{2}}{2}$$

$$\frac{\partial y}{\partial \eta} = \eta\xi(1-\sqrt{2}) + \xi\left(\frac{1}{2}\right) + \eta(2-2\sqrt{2}) + 1$$

Hence

$$J = \begin{vmatrix} \eta^2\left(\frac{1-\sqrt{2}}{2}\right) + \eta\left(-\frac{1}{2}\right) + \frac{\sqrt{2}}{2} & \eta^2\left(\frac{1-\sqrt{2}}{2}\right) + \eta\left(\frac{1}{2}\right) + \frac{\sqrt{2}}{2} \\ \begin{array}{c}\xi\eta(1-\sqrt{2}) + \xi\left(-\frac{1}{2}\right) \\ +\eta(2-2\sqrt{2}) - 1\end{array} & \begin{array}{c}\eta\xi(1-\sqrt{2}) + \xi\left(\frac{1}{2}\right) \\ +\eta(2-2\sqrt{2}) + 1\end{array} \end{vmatrix}$$

Noting that in the integral on the right-hand side of (6.26) only terms involving **even** powers of ξ, η and $\xi\eta$ will produce a non-zero value we obtain

$$\begin{aligned}
\iint_A dx\,dy &\approx \int_{-1}^{1}\int_{-1}^{1}\left\{\eta^2(-1+\sqrt{2}) + \sqrt{2}\right\} d\xi\, d\eta \\
&= (-1+\sqrt{2})\left(\frac{2}{3}\right)2 + \sqrt{2}\,(4) = 6.20913
\end{aligned}$$

This compares well with the exact solution which is $\pi(3^2 - 1^2)/4 = 2\pi = 6.28318$.

Of course, in more realistic examples more elements are likely to be used and the calculations performed on a computer.

What we hope the reader has appreciated in the last four sections is that the use of a finite element mesh to evaluate line integrals or double integrals defined over general regions can have distinct advantages over analytical techniques. The reader should not make the mistake of being too influenced by the amount of numerical work that is involved. Almost all of the work that we have done by hand can be automated and the calculations carried through on a computer.

6.6 The approximation of surfaces

As one further use of finite element meshes and two-dimensional shape functions we briefly consider their application to the modelling of surfaces.

A surface is often defined in explicit Cartesian form by an equation of the form

$$z = f(x, y) \qquad (x, y) \text{ within } A \tag{6.29}$$

(see Figure 6.22 (a)).

For example, the equation of the upper half of the sphere centre at the origin and of unit radius (Figure 6.22 (b)) has equation

$$z = \sqrt{1 - x^2 - y^2} \tag{6.30}$$

In some circumstances, particularly in the finite element method, it is desirable to replace the exact equation which defines the surface by an approximation. One way of doing this is to segment the domain A into a finite element mesh using triangular or quadrilateral elements (or a combination of both). Only a crude approximation results if three-noded triangular elements is used and a better one is obtained by using eight-noded quadrilateral elements. However, for diagrammatic clarity (Figure 6.23) we have actually highlighted a three-noded triangular element $[e]$.

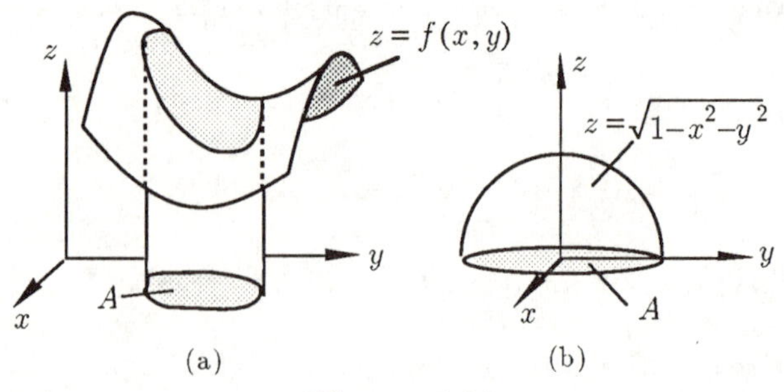

Figure 6.22

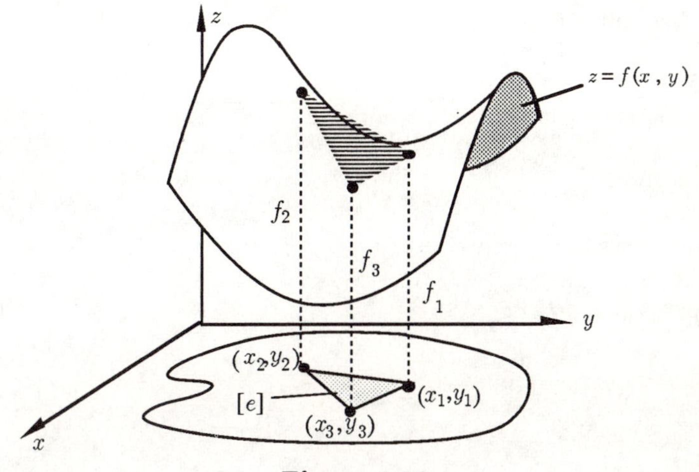

Figure 6.23

The value of $f(x, y)$ is determined at each corner of this element:

$$f_i = f(x_i, y_i) \quad i = 1, 2, 3 \tag{6.31}$$

and then the actual surface lying above element $[e]$ is approximated by the surface segment:

$$z = \tilde{f}(x, y) = \sum_{i=1}^{3} f_i \, T_i(\xi, \eta) \tag{6.32}$$

which is the shaded plane triangular area shown in Figure 6.23. In (6.32)

$$x = \sum_{i=1}^{3} x_i \, T_i(\xi, \eta) \quad \text{and} \quad y = \sum_{i=1}^{3} y_i \, T_i(\xi, \eta)$$

where the $T_i(\xi, \eta)$ are the shape functions (6.19) appropriate to the three-noded triangular element.

Equation (6.32) can be viewed as a transformation that maps the standard triangle (cf. Example 6.4) into the shaded triangle which approximates the surface above element $[e]$.

Conversely, (6.32) can be viewed as the transformation which takes the shaded triangle (a piece of approximated surface) into the standard triangle. This approach is of great significance when dealing with the numerical evaluation of surface integrals. In the following example we illustrate this application.

Example 6.6

Determine an approximation to the surface integral $I = \iint_S 1\,dS$ in which S is the surface of the hemisphere $z = \sqrt{1 - x^2 - y^2}$.

Solution

Using symmetry, we realize that

$$I = 4 \iint\limits_{S_1} 1\,dS \tag{6.33}$$

(see Figure 6.24(a)).

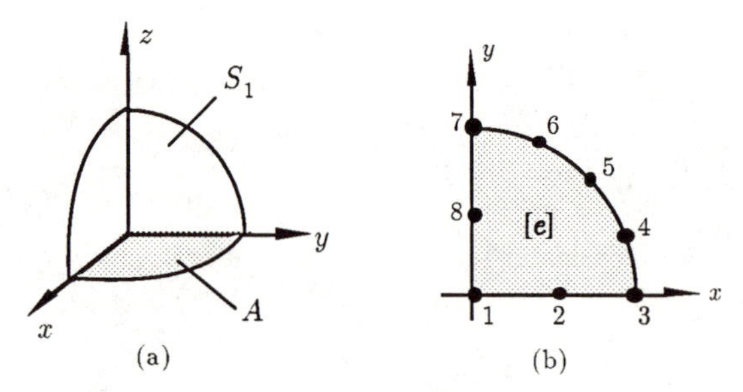

Figure 6.24

As a crude approximation to the quadrant A we use a single eight-noded quadrilateral element with nodal coordinates as follows:

$1 : (0, 0)$	$2 : (0.5, 0)$	$3 : (1, 0)$
$4 : (0.9238, 0.3827)$	$5 : (0.7071, 0.7071)$	$6 : (0.3827, 0.9238)$
$7 : (0, 1)$	$8 : (0, 0.5)$	

Sampling the surface S_1 (with equation $z = \sqrt{1 - x^2 - y^2}$) at each of these points we find:

$f_1 = 1$	$f_2 = 0.8660$	$f_3 = 0$
$f_4 = 0$	$f_5 = 0$	$f_6 = 0$
$f_7 = 0$	$f_8 = 0.8660$	

Therefore, from a variant of (6.32) applicable to eight-noded quadrilaterals the equation of the quarter hemisphere S_1 can be approximated by:

$$\tilde{f}(x, y) = f_1 Q_1(\xi, \eta) + f_2 Q_2(\xi, \eta) + f_8 Q_8(\xi, \eta)$$

Hence, using (6.25),

$$\tilde{f}(x,y) = -\frac{1}{4}(1-\xi)(1-\eta)(1+\eta+\xi) + 0.8660\left[\frac{1}{2}(1-\xi^2)(1-\eta) + \frac{1}{2}(1-\eta^2)(1-\xi)\right]$$
$$= (1-\xi)(1-\eta)(0.6160+0.1830\xi+0.1830\eta) \qquad (6.34)$$

and similarly using (6.24)

$$x = x_2Q_2(\xi,\eta) + x_3Q_3(\xi,\eta) + x_4Q_4(\xi,\eta) + x_5Q_5(\xi,\eta) + x_6Q_6(\xi,\eta)$$

$$y = y_4Q_4(\xi,\eta) + y_5Q_5(\xi,\eta) + y_6Q_6(\xi,\eta) + y_7Q_7(\xi,\eta) + y_8Q_8(\xi,\eta) \qquad (6.35)$$

To evaluate the surface integral we transform it into a double integral by projecting S_1 onto, for example, the xy plane. The element of surface area dS on S_1 can then be expressed in terms of an element of area $dx\,dy$ on the xy plane according to

$$dS = \frac{dx\,dy}{|\hat{\mathbf{n}} \cdot \hat{\mathbf{k}}|}$$

where $\hat{\mathbf{n}}$ is a unit vector normal to S and $\hat{\mathbf{k}}$ is the unit vector normal to the xy plane. For this example:

$$\hat{\mathbf{n}} = x\hat{\mathbf{i}} + y\hat{\mathbf{j}} + z\hat{\mathbf{k}}$$

so that

$$\hat{\mathbf{n}} \cdot \hat{\mathbf{k}} = z = f(x,y) = \sqrt{1-x^2-y^2}$$

Hence

$$\iint_{S_1} dS = \iint_{[e]} 1\,\frac{dx\,dy}{f(x,y)}$$

Now we play the usual trick: transform $[e]$ to the standard square and use the approximation to $f(x,y)$ to give

$$\iint_{S_1} dS = \int_{-1}^{1}\int_{-1}^{1} \frac{|J|}{\tilde{f}(x,y)}\,d\xi\,d\eta$$

in which

$$|J| = \begin{vmatrix} \dfrac{\partial x}{\partial \xi} & \dfrac{\partial y}{\partial \xi} \\ \dfrac{\partial x}{\partial \eta} & \dfrac{\partial y}{\partial \eta} \end{vmatrix}$$

and $\tilde{f}(x, y)$ is given by (6.34) in terms of (ξ, η) variables.

Rather than carry through the algebra we acknowledge that the integral of $|J|/\tilde{f}(x, y)$ over the standard square needs to be evaluated numerically. Choosing a Gaussian quadrature sixth order method (see (6.17) and Table 3.3) we need to sample the function

$$h(\xi, \eta) = \frac{|J|}{\tilde{f}(x, y)}$$

at 36 sample points, which would be virtually impossible to do by hand without incurring errors. Hence we wrote a short computer program to obtain

$$\iint_{S_1} 1\, dS \approx 1.57412$$

taken over the positive quadrant. This value, considering we have used only a single element approximation, compares very favourably with the exact answer of $\pi/2$ or 1.57079.

It would be a useful exercise for the reader to carry through the computations outlined above to verify the final approximation. We suggest that he/she follow our example and relegate the calculations to a computer.

In this section we have shown how to approximate a surface in a piecewise manner using elements and shape functions in tandem with appropriate coordinate transformations. It is this approach that is used (implicitly) in the finite element solution of two dimensional boundary value problems. In such problems the unknown function $f(x, y)$ (the solution to the boundary value problem) is a function of two variables and so may be viewed as a **surface** $z = f(x, y)$. The surface is 'patched' as described in this section: the only difference being that the nodal function values of $f(x, y)$ (cf. (6.31)) are **unknowns**. It is the purpose of the finite element method to determine these unknown nodal function values and so construct an approximation to the exact solution.

EXERCISES

6.1 (a) Deduce the shape functions (6.19) for a three-noded triangle directly by assuming a linear approximation

$$\tilde{u}(\xi, \eta) = \alpha_0 + \alpha_1 \xi + \alpha_2 \eta$$

over the standard triangle and solving for α_0, α_1 and α_2.
(b) Verify that these shape functions enjoy the usual properties of shape functions:

$$T_i\{\xi, \eta\} = \begin{cases} 1 & \text{at node } i \\ 0 & \text{at the other two nodes} \end{cases}$$

6.2 Verify the results (6.20) for integrals taken over the standard triangle.

6.3 A certain three-noded triangular element has nodes with coordinates as follows

$$1: \ (0.13, 0.01) \qquad 2: \ (0.25, 0.06) \qquad 3: \ (0.13, 0.13)$$

(i) Deduce a transformation that will map this element into the standard triangle.
(ii) A finite element analysis has given the following nodal values:

$$\tilde{u}_1 = 190 \qquad \tilde{u}_2 = 160 \qquad \tilde{u}_3 = 185$$

Deduce $\tilde{u}$ at the point $x = 0.2$, $y = 0.06$ and deduce the (constant) values of $\partial \tilde{u}/\partial x$ and $\partial \tilde{u}/\partial y$ within the element.

6.4 (a) Deduce the bilinear shape functions (6.23) directly by assuming an approximation:

$$\tilde{u}(\xi, \eta) = \alpha_0 + \alpha_1 \xi + \alpha_2 \eta + \alpha_3 \xi \eta$$

over the standard square and solving for α_0, α_1, α_2 and α_3.
(b) Deduce these shape functions more quickly by an obvious two-dimensional equivalent of the method discussed in Section 3.6, i.e. assume

$$L_i(\xi, \eta) = F_i(\xi, \eta) G_i(\xi, \eta)$$

and use the required properties of shape functions.
(c) Verify that the two-dimensional shape functions (6.23) reduce either to zero or to one-dimensional linear shape functions along each edge of the standard square.

6.5 Verify that (6.22) maps the standard square into a straight-edged quadrilateral with vertices at (x_i, y_i), $i = 1, 2, 3, 4$.

6.6 A certain four-noded rectangular element has nodal coordinates as follows:

$$1: (0.31, 0.18) \quad 2: (0.38, 0.18) \quad 3: (0.38, 0.25) \quad 4: (0.31, 0.25)$$

(i) Deduce a transformation that will map this element into the standard square.
(ii) A finite element analysis has given the following nodal values:

$$\tilde{u}_1 = 115 \quad \tilde{u}_2 = 85 \quad \tilde{u}_3 = 76 \quad \tilde{u}_4 = 105$$

Deduce $\tilde{u}$ at the point $x = 0.35,\ y = 0.22$.

6.7 Determine an approximation to the line integral

$$\int_C \{(3x - 4y)dx + (4x + 2y)dy\}$$

where C is the closed path shown in Figure Q6.7 made up of the semi-ellipse C_1 and a straight-line base C_2. Use a linear element for the straight line and a single quadratic element to approximate the elliptical portion C_1.

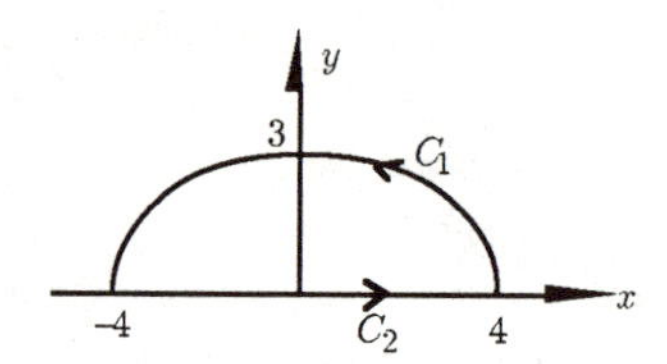

Figure Q6.7

6.8 Repeat Q6.7 by approximating the semi-ellipse with two quadratic elements.

6.9 Use a single quadratic element to determine approximations to the line integrals

$$\text{(a)} \quad \int_C (2x\,dx + dy) \qquad \text{(b)} \quad \int_C (dx + 2x\,dy)$$

where C is the segment of the circle:

$$x = \cos\theta \quad y = \sin\theta \quad 0 \le \theta \le \frac{\pi}{4}$$

Compare with the exact results.

6.10 A curve connecting the 'nodes' $(x_i,\ y_i)$ $i = 1, 2, 3, 4$ can be approximated by a parametric cubic curve defined by

$$x = \sum_{i=1}^{4} x_i C_i(\xi) \qquad y = \sum_{i=1}^{4} y_i C_i(\xi)$$

in which the shape functions $C_i(\xi)$ are the cubic shape functions:

$$C_1(\xi) = -\frac{9}{16}(\frac{1}{3}+\xi)(\frac{1}{3}-\xi)(1-\xi) \qquad C_2(\xi) = \frac{27}{16}(1+\xi)(\frac{1}{3}-\xi)(1-\xi)$$

$$C_3(\xi) = \frac{27}{16}(1+\xi)(\frac{1}{3}+\xi)(1-\xi) \qquad C_4(\xi) = -\frac{9}{16}(1+\xi)(\frac{1}{3}+\xi)(\frac{1}{3}-\xi)$$

See also Q3.12 of Chapter 3. This transformation maps the given nodes (x_i, y_i) into standard nodes $\xi_1 = -1$, $\xi_2 = -\frac{1}{3}$, $\xi_3 = \frac{1}{3}$ and $\xi_4 = 1$. Verify that the following relations are satisfied by these shape functions:

(i) $C_i(\xi) = \begin{cases} 1 & \text{at node } i \\ 0 & \text{at other element nodes} \end{cases}$ (ii) $\sum_{i=1}^{4} C_i(\xi) = 1$

(iii) $\sum_{i=1}^{4} \xi_i C_i(\xi) = \xi$ (iv) $\sum_{i=1}^{4} \xi_i^2 C_i(\xi) = \xi^2$ (v) $\sum_{i=1}^{4} \xi_i^3 C_i(\xi) = \xi^3$

6.11 Repeat Example 6.1 using the single cubic element introduced in Q6.10.

6.12 Use a third-order Gaussian quadrature rule to approximate the double integral

$$\iint_D x^2 \cos y \, dx dy$$

where D is the standard square.

6.13 Using the change of variable:

$$x = (\frac{\xi}{2}+\frac{3}{2})\cos(\frac{\pi\eta}{4}+\frac{\pi}{4}) \qquad y = (\frac{\xi}{2}+\frac{3}{2})\sin(\frac{\pi\eta}{4}+\frac{\pi}{4})$$

first used in Example 6.2 and a three-point Gaussian quadrature rule approximate the integral

$$\iint_A e^{\sqrt{x^2+y^2}} dx dy$$

where A is defined as in Figure 6.13.

6.14 Rework Example 6.4 using the distribution of elements shown in Figure Q6.14 with nodes:

$$1: (0,0) \qquad 2: (\frac{1}{\sqrt{2}}, \frac{1}{\sqrt{2}}) \qquad 3: (0,1) \qquad 4: (\frac{2}{\sqrt{2}}, \frac{2}{\sqrt{2}})$$

$5: (0.76537, 1.84776) \qquad 6: (0, 2)$

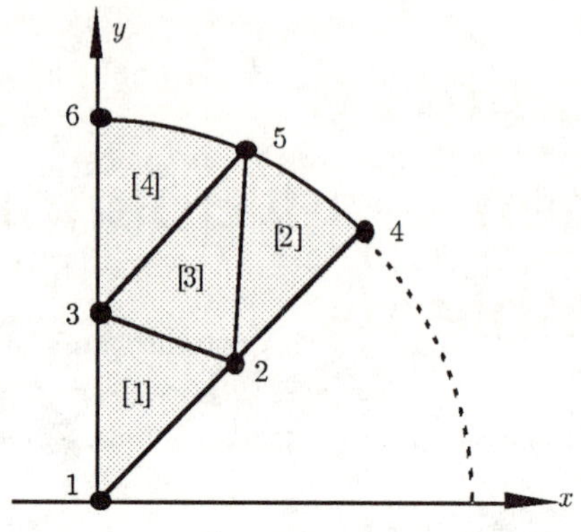

Figure Q6.14

6.15 In Q6.14 replace elements [1] and [3] by a single 4-noded element and re-evaluate the double integral over this element using a 2-point Gaussian quadrature rule.

6.16 Re-work Example 6.6 on the estimation of the surface area of one octant of the sphere $z = \sqrt{1 - x^2 - y^2}$ using 4 three-noded triangular elements as shown in Figure Q6.16.

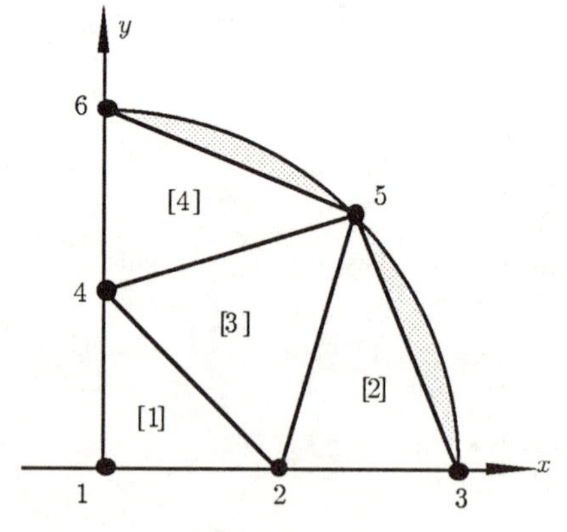

Figure Q6.16

7

Finite Element Solution of Two-dimensional Boundary Value Problems

PREVIEW

In this chapter we consider the application of the finite element technique to two-dimensional linear boundary value problems. We use four-noded quadrilateral elements and three-noded triangular elements. A matrix formulation is also developed and applied to problems in heat conduction and in the analysis of torsion.

7.1 Introduction

The techniques studied in Chapters 2 and 4 for finding approximate solutions to one-dimensional boundary value problems can be extended to two dimensions but some complications arise on moving from one to two dimensions. For example, a segment of the real axis bounded by two points (the 'domain' of a one-dimensional boundary value problem) will now be a region (possibly irregularly shaped) bounded by a curve. Such a curve consists, of course, of an infinite number of points. The ordinary derivatives occurring in one-dimensional boundary value problems become partial derivatives and ordinary integrals generalize to double integrals. Also, as we shall see, the 'boundary terms' arising from integration by parts in the modified Galerkin method now become line integrals taken over the boundary of the two-dimensional domain.

We are also faced, in two dimensions, with choosing an appropriate system of coordinates so that the boundary value problem is expressed in the simplest way. For some two-dimensional regions the choice is obvious: for rectangular shapes we would use Cartesian coordinates while for regions bounded by circles and radial lines polar coordinates might be the most appropriate. For irregularly shaped regions the so-called (n, s) coordinates are useful. Without being too precise at this stage the 's-coordinate' is a measure of arc length along the domain boundary whilst the 'n-coordinate' is a measure of distance perpendicular to the boundary. Further discussion of (n, s) coordinates will be given in Chapter 8.

Generally, we shall adhere to a Cartesian coordinate system as this is conceptually the simplest. However, in common with other texts and with general practice, we may find it convenient (particularly with respect to the boundary conditions) to use a mixture of coordinates – usually Cartesian and (n, s) coordinates.

A typical example of a two-dimensional boundary value problem arises when modelling the conduction of heat over a thin plate. If the plate is made from an inhomogeneous material, it can be shown (see Supplement 2 to Chapter 4) that if $u(x, y)$ is the steady-state temperature then the partial differential equation:

$$\frac{\partial}{\partial x}\left(k_x(x,y)\frac{\partial u}{\partial x}\right) + \frac{\partial}{\partial y}\left(k_y(x,y)\frac{\partial u}{\partial y}\right) + h(x,y) = 0$$

holds within the plate. Here k_x, k_y (both possibly functions of x and y) are the thermal conductivities of the material in the x and y directions respectively and $h(x, y)$ (a given function) represents the rate of internal heat generation. If $k_x = k_y = k$ say, a constant, then the governing equation becomes simply

$$\frac{\partial^2 u}{\partial x^2} + \frac{\partial^2 u}{\partial y^2} + \frac{h(x,y)}{k} = 0$$

which is the two-dimensional version of the Poisson equation. If the heat generation function $h(x,y) \equiv 0$ then we obtain simply

$$\frac{\partial^2 u}{\partial x^2} + \frac{\partial^2 u}{\partial y^2} = 0$$

which is the two-dimensional Laplace equation.

The boundary conditions to be imposed in a heat conduction problem depend on the precise problem being analysed – for example, part of the boundary may be insulated indicating that the rate of heat flow and hence the temperature gradient normal to the boundary is zero ($\partial u/\partial n = 0$). Other parts of the boundary may be kept at a fixed temperature ($u =$ constant).

Examples of two-dimensional boundary value problems occur in many other areas of engineering and physics such as fluid flow, electromagnetism, the analysis of torsion in solid mechanics and the modelling of gravitational attraction.

In this chapter we consider linear second-order two-dimensional boundary value problems i.e. the highest derivative appearing in the differential equation is second order. The three equations above are typical examples. Fourth-order boundary value problems, both in one and two dimensions, are the subject of Chapter 10.

7.2 The Galerkin formulation in two dimensions

We shall begin our study of two-dimensional problems by using the weighted residual technique (and ultimately finite elements) to solve a relatively simple problem viz. Laplace's equation in a rectangular plate A, part of whose boundary is held at a fixed temperature while the remainder is subjected to heating (see Figure 7.1).

If u denotes the temperature at any point of the plate then, as we have seen above, the mathematical model of this physical situation takes the form of a two-dimensional boundary value problem:

$$\frac{\partial^2 u}{\partial x^2} + \frac{\partial^2 u}{\partial y^2} = 0 \qquad (x,y) \text{ within } A \tag{7.1a}$$

$$u = k^* \quad \text{on} \quad C^* \qquad x = 0, \quad y = +2 \quad \text{and} \quad y = -2 \tag{7.1b}$$

$$\frac{\partial u}{\partial x} = k \quad \text{on} \quad C \qquad x = 2 \tag{7.1c}$$

Although we have given a particular physical interpretation to (7.1) this interpretation is largely irrelevant as far as the numerical solution is concerned. We are simply solving a two-dimensional boundary value problem.

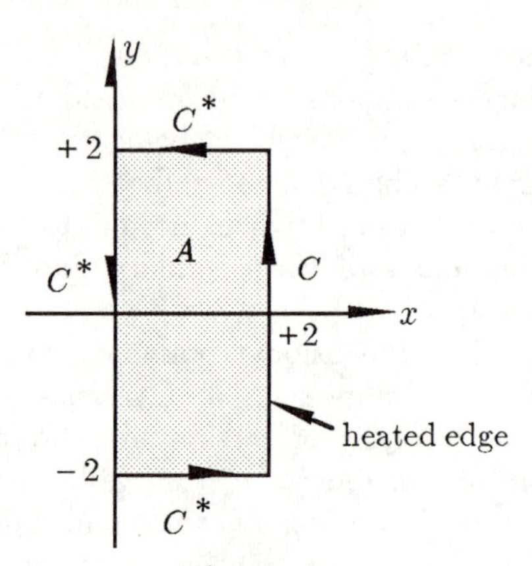

Figure 7.1

Our first step in a weighted residual solution to this problem is to generalise the Galerkin formulation already used for one-dimensional boundary value problems in Chapter 2.

We let $\tilde{u}(x, y)$ denote an approximation to the exact solution of (7.1). The approximation is assumed to take the particular form:

$$\tilde{u}(x, y) = \theta_0(x, y) + \sum_{i=1}^{n} a_i \, \theta_i(x, y) \tag{7.2}$$

(Compare (2.2) for the corresponding one-dimensional formulation.) The function $\theta_0(x, y)$ is chosen to satisfy **all** the boundary conditions of the problem and $\theta_i(x, y)$, $i = 1, 2, \ldots, n$ (the so-called coordinate functions) are chosen to satisfy the corresponding homogeneous form of the boundary conditions. In other words

$$\theta_0(x, y) = k^* \text{ on } C^* \text{ and } \frac{\partial \theta_0}{\partial x} = k \text{ on } C \tag{7.3}$$

whilst

$$\theta_i(x,y) = 0 \text{ on } C^* \text{ and } \frac{\partial \theta_i}{\partial x} = 0 \text{ on } C \quad i = 1, 2, \ldots, n \tag{7.4}$$

It follows from (7.3) and (7.4) that $\tilde{u}(x,y)$ satisfies **all** the boundary conditions of the problem regardless of the values of the parameters a_i. The functions $\theta_0(x,y)$, $\theta_i(x,y)$, $i = 1, 2, \ldots, n$, are chosen by the user so that once values for the parameters a_i are determined an approximation $\tilde{u}$ is obtained from (7.2). Because $\tilde{u}$ is an approximation to u it will not satisfy the partial differential equation (7.1). Indeed

$$\frac{\partial^2 \tilde{u}}{\partial x^2} + \frac{\partial^2 \tilde{u}}{\partial y^2} = \epsilon(x,y) \tag{7.5}$$

where $\epsilon(x,y)$ is the residual function. In the Galerkin formulation the parameters a_i, $i = 1, 2, \ldots, n$, are chosen to satisfy the n residual equations:

$$\iint_A \left\{ \frac{\partial^2 \tilde{u}}{\partial x^2} + \frac{\partial^2 \tilde{u}}{\partial y^2} \right\} \theta_i(x,y) dx\, dy = 0 \qquad i = 1, 2, \ldots, n \tag{7.6}$$

(Compare (2.3) for one-dimensional boundary value problems.)

As in the one-dimensional Galerkin approach we shall amend equations (7.6)

(i) in order to reduce the order of the partial derivative terms in the integrand
(ii) in order to incorporate the derivative boundary conditions.

The mathematical theorem which aids this reformulation is called **Green's Theorem in the Plane.** This theorem states that if A is a region in the xy plane bounded by a closed curve Γ and if $F(x,y)$ and $G(x,y)$ are suitably 'smooth' functions then:

$$\iint_A \left\{ \frac{\partial G}{\partial x} - \frac{\partial F}{\partial y} \right\} dx\, dy = \oint_\Gamma (F\, dx + G\, dy) \tag{7.7}$$

In order to utilize this result in (7.6) we choose

$$F = -\frac{\partial \tilde{u}}{\partial y} \theta_i \quad \text{and} \quad G = \frac{\partial \tilde{u}}{\partial x} \theta_i$$

so that, from (7.7)

$$\iint_A \left\{ \frac{\partial^2 \tilde{u}}{\partial x^2}\theta_i + \frac{\partial \theta_i}{\partial x}\frac{\partial \tilde{u}}{\partial x} + \frac{\partial^2 \tilde{u}}{\partial y^2}\theta_i + \frac{\partial \theta_i}{\partial y}\frac{\partial \tilde{u}}{\partial y} \right\} dx\,dy$$

$$= \oint_\Gamma \theta_i \left\{ -\frac{\partial \tilde{u}}{\partial y}\,dx + \frac{\partial \tilde{u}}{\partial x}dy \right\}$$

Hence

$$\iint_A \left\{ \frac{\partial^2 \tilde{u}}{\partial x^2} + \frac{\partial^2 \tilde{u}}{\partial y^2} \right\} \theta_i\,dx\,dy = -\iint_A \left\{ \frac{\partial \theta_i}{\partial x}\frac{\partial \tilde{u}}{\partial x} + \frac{\partial \theta_i}{\partial y}\frac{\partial \tilde{u}}{\partial y} \right\} dx\,dy + \oint_\Gamma \theta_i \left\{ -\frac{\partial \tilde{u}}{\partial y}\,dx + \frac{\partial \tilde{u}}{\partial x}\,dy \right\} \qquad \textbf{(7.8)}$$

We can now begin to impose some of the boundary conditions for the particular problem (7.1). Since $\theta_i(x,y) = 0$ on C^* and $\partial\tilde{u}/\partial x = k$ on C we must have

$$\oint_\Gamma \theta_i \left\{ -\frac{\partial \tilde{u}}{\partial y}\,dx + \frac{\partial \tilde{u}}{\partial x}dy \right\} = \int_C k\,\theta_i\,dy \qquad \textbf{(7.9)}$$

where we have also put $dx = 0$ on C.

Hence, using (7.8) and (7.9), (7.6) becomes

$$-\iint_A \left\{ \frac{\partial \theta_i}{\partial x}\frac{\partial \tilde{u}}{\partial x} + \frac{\partial \theta_i}{\partial y}\frac{\partial \tilde{u}}{\partial y} \right\} dx\,dy + \int_C k\,\theta_i\,dy = 0 \qquad i = 1, 2, \cdots, n \qquad \textbf{(7.10)}$$

We may consider (7.10) as the basis of a new Galerkin technique – the **modified Galerkin method**. In this method $\tilde{u}$ need only satisfy the non-derivative (or essential) boundary conditions of the problem. This is not too surprising since we have already used the derivative boundary condition to obtain (7.10). To be precise we use (7.10) together with (7.2) and demand only that $\tilde{u}$ satisfies the boundary condition $\tilde{u} = k^*$ on C^*.

Example 7.1

Determine a two-parameter approximation to the boundary value problem (7.1) with $k^* = 0$. Use the modified Galerkin equations (7.10).

Solution

From the symmetry of the problem we expect the solution to be an even function of y. We choose:

$$\tilde{u}(x,y) = x(y^2-4)(a_1+a_2x)$$

which clearly satisfies the non-derivative (essential) boundary conditions. In the notation of (7.2) this implies

$$\theta_1(x,y) = x(y^2-4), \qquad \theta_2(x,y) = x^2(y^2-4)$$

(We have taken $\theta_0(x,y) = 0$ because the non-derivative boundary conditions, with $k^* = 0$, are homogeneous.)

In this case it is an easy matter to confirm using (7.2) and (7.10) that the two modified Galerkin equations are :

$$-\iint_A \left\{(y^2-4)\ [(y^2-4)(a_1+2a_2x)] + 2xy\ [(2xy)(a_1+a_2x)]\right\} dxdy$$
$$+\int_{-2}^{2} 2k(y^2-4)\ dy = 0$$

$$-\iint_A \left\{2x(y^2-4)[(y^2-4)(a_1+2a_2x)] + 2x^2y[(2xy)(a_1+a_2x)]\right\} dxdy$$
$$+\int_{-2}^{2} 4k(y^2-4)\ dy = 0$$

which give, after straightforward evaluation and simplification,

$$\begin{aligned} 88a_1 + 156a_2 &= -15k \\ 78a_1 + 176a_2 &= -15k \end{aligned}$$

with solution

$$a_1 = -0.090k \qquad a_2 = -0.045k$$

Therefore our two-parameter approximation is

$$\tilde{u}(x,y) = x(y^2-4)(-0.090 - 0.045x)k \tag{7.11}$$

The exact solution to this boundary value problem may be obtained by a separation of variables technique. As a crude indication of the accuracy obtained in (7.11) we note that $\tilde{u}(2,0) = 1.44k$ whereas the exact solution at this point is, to two decimal places, $u(2,0) = 1.35k$.

7.3 Roof function approach

We will continue in this section to use the boundary value problem (7.1) as our example. The first step is to segment the domain, the rectangular region A, into a number of elements. The type of element to use (triangular or quadrilateral) and the order, for example 3- or 6-noded triangles, 4- or 8-noded quadrilaterals is largely at our behest. Obviously, if the domain has a curved boundary we would prefer to use elements which closely model the boundary.

We will solve (7.1) by hand, using a very crude mesh consisting of two four-noded rectangular elements (with, therefore, just one unknown nodal value to determine).

• Initial approximation

The elements and nodes are numbered as in Figure 7.2 so that $\tilde{u}_4$ is the single unknown nodal value.

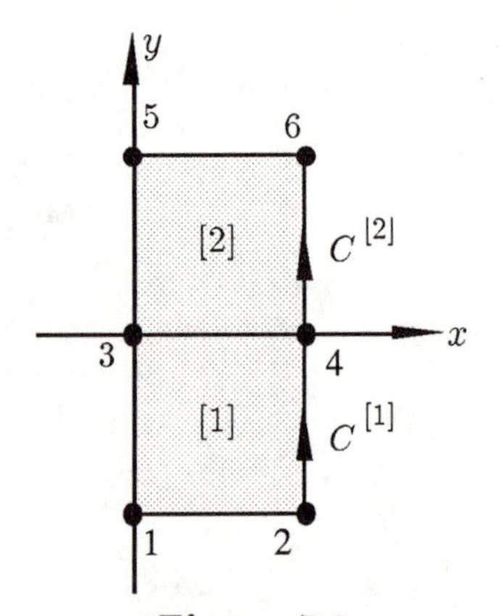

Figure 7.2

The solution of a two-dimensional boundary value problem is a function of two variables; geometrically it represents a surface. In obtaining approximations to the boundary value problem we are, in effect, obtaining an approximation to the solution surface. In the finite element technique the approximate solution surface comprises discrete segments with a separate segment for each element. What we do is to approximate the true surface $z = u(x, y)$ within each element, by the surface segment:

$$\tilde{u}(x,y) = \sum_{i=1}^{4} \tilde{u}_i^{[e]} \, L_i\,(\xi,\eta) \tag{7.12a}$$

where

$$x = \sum_{i=1}^{4} x_i^{[e]} \, L_i\,(\xi,\eta) \qquad y = \sum_{i=1}^{4} y_i^{[e]} \, L_i\,(\xi,\eta) \tag{7.12b}$$

(Compare (3.9) in one dimension.) Here $(x_i^{[e]}, y_i^{[e]})$, $i = 1, 2, 3, 4$, are the nodal coordinates in element $[e]$ and $\tilde{u}_i^{[e]}$ are the nodal function values of $\tilde{u}(x, y)$ (see Section 6.6). Also, $L_i(\xi, \eta)$, $i = 1, 2, 3, 4$, are the bi-linear shape functions (6.23) appropriate to a four-noded quadrilateral element:

$$\begin{aligned} L_1 &= \frac{1}{4}(1-\eta)(1-\xi) \qquad & L_2 &= \frac{1}{4}(1-\eta)(1+\xi) \\ L_3 &= \frac{1}{4}(1+\eta)(1+\xi) \qquad & L_4 &= \frac{1}{4}(1+\eta)(1-\xi) \end{aligned} \tag{7.13}$$

with derivatives

$$\begin{aligned} \frac{\partial L_1}{\partial \xi} &= -\frac{1}{4}\,(1-\eta) \qquad & \frac{\partial L_1}{\partial \eta} &= -\frac{1}{4}\,(1-\xi) \\ \frac{\partial L_2}{\partial \xi} &= \frac{1}{4}\,(1-\eta) \qquad & \frac{\partial L_2}{\partial \eta} &= -\frac{1}{4}\,(1+\xi) \\ \frac{\partial L_3}{\partial \xi} &= \frac{1}{4}\,(1+\eta) \qquad & \frac{\partial L_3}{\partial \eta} &= \frac{1}{4}\,(1+\xi) \\ \frac{\partial L_4}{\partial \xi} &= -\frac{1}{4}\,(1+\eta) \qquad & \frac{\partial L_4}{\partial \eta} &= \frac{1}{4}\,(1-\xi) \end{aligned} \tag{7.14}$$

The approximation to $u(x, y)$ over the complete domain A can be expressed equivalently in terms of six two-dimensional roof functions $R_i(x, y)$, $i = 1, 2, \ldots, 6$, one for each node. We assume that each element is transformed by (7.12b) into the standard square (see Figure 6.12(a)) so that the global/local node number relations are as in Table 7.1.

global nodes		local nodes
$[e] = 1$	$[e] = 2$	
1	3	1
2	4	2
4	6	3
3	5	4

Table 7.1

We can then define the roof functions as

$$R_1 = \begin{cases} L_1 & \text{in } [1] \\ 0 & \text{in } [2] \end{cases} \qquad R_2 = \begin{cases} L_2 & \text{in } [1] \\ 0 & \text{in } [2] \end{cases}$$

$$R_3 = \begin{cases} L_4 & \text{in } [1] \\ L_1 & \text{in } [2] \end{cases} \qquad R_4 = \begin{cases} L_3 & \text{in } [1] \\ L_2 & \text{in } [2] \end{cases} \tag{7.15}$$

$$R_5 = \begin{cases} 0 & \text{in } [1] \\ L_4 & \text{in } [2] \end{cases} \qquad R_6 = \begin{cases} 0 & \text{in } [1] \\ L_3 & \text{in } [2] \end{cases}$$

Hence, throughout A, $u(x, y)$ is approximated by

$$\tilde{u}(x, y) = \sum_{i=1}^{6} \tilde{u}_i \, R_i\,(x, y) \tag{7.16}$$

Note that because of the essential boundary conditions

$$\tilde{u}_1 = \tilde{u}_2 = \tilde{u}_3 = \tilde{u}_5 = \tilde{u}_6 = k^*$$

so that we can re-express (7.16) as

$$\tilde{u}(x, y) = k^*(R_1 + R_2 + R_3 + R_5 + R_6) + \tilde{u}_4 \, R_4 \tag{7.17}$$

where $\tilde{u}_4$ is the only unknown. Comparing (7.17) with (7.2) we see that

$$\theta_0(x, y) = k^*(R_1 + R_2 + R_3 + R_5 + R_6)$$

Also

$$\theta_1(x, y) = R_4$$

is the single coordinate function for this approximation.

Owing to the simple mesh being used and because of the essential boundary conditions the numerical approximation we are considering is reduced to a single-parameter problem. We now (finally) begin using the modified Galerkin method to determine the value of this parameter.

• Finite element calculations

We split the double integral and line integral in (7.10) according to the mesh we have chosen:

$$-\iint_{[1]}\left\{\frac{\partial\theta_i}{\partial x}\frac{\partial\tilde{u}}{\partial x}+\frac{\partial\theta_i}{\partial y}\frac{\partial\tilde{u}}{\partial y}\right\}dx\,dy-\iint_{[2]}\left\{\frac{\partial\theta_i}{\partial x}\frac{\partial\tilde{u}}{\partial x}+\frac{\partial\theta_i}{\partial y}\frac{\partial\tilde{u}}{\partial y}\right\}dx\,dy$$

$$+\int_{C^{[1]}}k\,\theta_i\,dy+\int_{C^{[2]}}k\,\theta_i\,dy=0\qquad i=1\qquad\textbf{(7.18)}$$

See Figure 7.2 for the significance of $C^{[1]}$, $C^{[2]}$. Note that the mesh we have used for A naturally induces a mesh (of two-noded linear elements in this case) onto the boundary of A.

Even for this relatively simple mesh the hand calculations are heavy so we will simplify the problem by considering the same special case as in Example 7.1 and put $k^* = 0$, so that in the notation of (7.2)

$$\tilde{u}(x,y)=0+a_1\,\theta_1(x,y)\qquad\textbf{(7.19)}$$

in which (directly from (7.17)):

$$a_1\longleftrightarrow\tilde{u}_4\qquad\theta_1\longleftrightarrow R_4$$

The remainder of the calculation is mostly straightforward numerical work. We perform the calculation of the double integrals element by element, each element being transformed to the standard square via the coordinate transformations (7.12b). We make use of Section 6.4, with particular reference to the effect of transforming a double integral from one region to another and also note the following easily obtained integrals (involving products of derivatives of the shape functions) taken over the standard square:

$$\int_{-1}^{1}\int_{-1}^{1}(1-\eta)^2d\xi\,d\eta=\frac{16}{3}\qquad\int_{-1}^{1}\int_{-1}^{1}(1-\eta^2)d\xi\,d\eta=\frac{8}{3}$$

$$\int_{-1}^{1}\int_{-1}^{1}(1-\eta)(1-\xi)d\xi\,d\eta=4\qquad\textbf{(7.20)}$$

• Element [1]

Here the transformations (7.12b) are

$$x = \sum_{i=1}^{4} x_i^{[1]} L_i(\xi,\eta) = x_2 L_2 + x_4 L_3$$
$$= 2(L_2 + L_3) = 1 + \xi \tag{7.21}$$

and

$$y = \sum_{i=1}^{4} y_i^{[1]} L_i(\xi,\eta) = y_1 L_1 + y_2 L_2$$
$$= (-2)(L_1 + L_2) = \eta - 1 \tag{7.22}$$

Thus, for this element the Jacobian of the transformation is:

$$J = \begin{vmatrix} \dfrac{\partial x}{\partial \xi} & \dfrac{\partial x}{\partial \eta} \\ \dfrac{\partial y}{\partial \xi} & \dfrac{\partial y}{\partial \eta} \end{vmatrix} = \begin{vmatrix} 1 & 0 \\ 0 & 1 \end{vmatrix} = 1$$

(as we might expect since element [1] is simply a square which is being transformed into the standard square without change of shape or dimension).

The integrand in the double integrals of (7.18) contain derivative terms with respect to x and y and these must be changed to derivatives with respect to ξ and η before we can proceed.

Now if $\alpha(\xi,\eta)$ is any function then, using the chain rule for partial derivatives:

$$\frac{\partial \alpha}{\partial x} = \frac{\partial \alpha}{\partial \xi}\frac{\partial \xi}{\partial x} + \frac{\partial \alpha}{\partial \eta}\frac{\partial \eta}{\partial x} = \frac{\partial \alpha}{\partial \xi}$$

using (7.21).

Similarly for the y-derivative

$$\frac{\partial \alpha}{\partial y} = \frac{\partial \alpha}{\partial \xi}\frac{\partial \xi}{\partial y} + \frac{\partial \alpha}{\partial \eta}\frac{\partial \eta}{\partial y} = \frac{\partial \alpha}{\partial \eta}$$

using (7.22).

Hence denoting the first integral in (7.18) (over element [1]) by $I^{[1]}$ we have:

$$I^{[1]} = -\int_{-1}^{1}\int_{-1}^{1}\left\{\frac{\partial \theta_1}{\partial \xi}\frac{\partial \tilde{u}}{\partial \xi} + \frac{\partial \theta_1}{\partial \eta}\frac{\partial \tilde{u}}{\partial \eta}\right\} d\xi\, d\eta \tag{7.23}$$

But $\tilde{u} = \tilde{u}_4\, R_4 = \tilde{u}_4\, L_3(\xi, \eta)$ within element [1]. Hence

$$\frac{\partial \tilde{u}}{\partial \xi} = \tilde{u}_4 \left[\frac{1}{4}(1+\eta)\right] \quad \text{and} \quad \frac{\partial \tilde{u}}{\partial \eta} = \tilde{u}_4 \left[\frac{1}{4}(1+\xi)\right]$$

so that

$$\begin{aligned} I^{[1]} &= -\int_{-1}^{1}\int_{-1}^{1} \left\{ \frac{\partial L_3}{\partial \xi}\frac{\partial \tilde{u}}{\partial \xi} + \frac{\partial L_3}{\partial \eta}\frac{\partial \tilde{u}}{\partial \eta} \right\} d\xi\, d\eta \\ &= -\tilde{u}_4 \int_{-1}^{1}\int_{-1}^{1} \left\{ \frac{1}{16}(1+\eta)^2 + \frac{1}{16}(1+\xi)^2 \right\} d\xi\, d\eta \\ &= -\frac{2}{3}\,\tilde{u}_4 \end{aligned}$$

by straightforward calculation using the first of (7.20).

- **Element [2]**

Proceeding as for element [1] the transformation to the standard square is accomplished using

$$x = 1 + \xi \qquad y = \eta + 1$$

so that for element [2] the Jacobian is again equal to 1 and the derivatives are again simply related:

$$\frac{\partial}{\partial x} \equiv \frac{\partial}{\partial \xi} \qquad \frac{\partial}{\partial y} \equiv \frac{\partial}{\partial \eta}$$

Denoting the second double integral in (7.18) by $I^{[2]}$ we have

$$I^{[2]} = -\iint_{[2]} \left\{ \frac{\partial L_2}{\partial \xi}\frac{\partial \tilde{u}}{\partial \xi} + \frac{\partial L_2}{\partial \eta}\frac{\partial \tilde{u}}{\partial \eta} \right\} d\xi\, d\eta$$

in which $\tilde{u} = \tilde{u}_4\, R_4 = \tilde{u}_4\, L_2\,(\xi, \eta)$ within element [2].
Hence

$$\frac{\partial \tilde{u}}{\partial \xi} = \tilde{u}_4 \left[\frac{1}{4}(1-\eta)\right] \qquad \frac{\partial \tilde{u}}{\partial \eta} = \tilde{u}_4 \left[-\frac{1}{4}(1+\xi)\right]$$

and so

$$\begin{aligned} I^{[2]} &= -\tilde{u}_4 \int_{-1}^{1}\int_{-1}^{1} \left\{ \frac{1}{16}(1-\eta)^2 + \frac{1}{16}(1+\xi)^2 \right\} d\xi\, d\eta \\ &= -\frac{2}{3}\,\tilde{u}_4 \qquad \text{using (7.20)} \end{aligned}$$

- **Line integrals**

We complete the calculation by evaluating the line integrals in (7.18). We define

$$B^{[1]} = \int_{C^{[1]}} k\,\theta_1\,dy \qquad \text{and} \qquad B^{[2]} = \int_{C^{[2]}} k\,\theta_1\,dy$$

Using (7.22) we easily find the following results:

$$\begin{aligned} B^{[1]} &= \int_{-1}^{1} k\,\theta_1\,d\eta = \int_{-1}^{1} k\,L_3\,(+1,\eta)\,d\eta \\ &= \int_{-1}^{1} k\,\frac{1}{2}\,(1+\eta)\,d\eta = k \end{aligned}$$

and

$$\begin{aligned} B^{[2]} &= \int_{-1}^{1} k\,\theta_1\,d\eta = \int_{-1}^{1} k\,L_2\,(+1,\eta)\,d\eta \\ &= \int_{-1}^{1} k\,\frac{1}{2}\,(1-\eta)\,d\eta = k \end{aligned}$$

Thus, the single Galerkin equation (7.18)

$$I^{[1]} + I^{[2]} + B^{[1]} + B^{[2]} = 0$$

becomes as a result of these calculations

$$-\frac{4}{3}\,\tilde{u}_4 + 2k = 0$$

giving

$$\tilde{u}_4 = 1.5k$$

Considering that only two four-noded elements were used, this approximate value is in reasonable agreement with the exact value at this node, which to two decimal places is $1.35k$. A more accurate finite element calculation was carried out by computer on this problem using sixteen eight-noded elements. The corresponding value at the above node was found to be

$$\tilde{u}_4 = 1.344k$$

Of course the exact solution is not normally available, and then the only indication of the accuracy of a finite element calculation is given by the size of the residual function ((7.5) for this problem) whose value, of course, will vary from point to point. The determination of this function involves a long numerical calculation and, as such, will itself be subject to computational error, particularly since it involves the calculation of second derivatives. It may be shown that the calculation of second derivatives is most accurate at the centre of an element. For this specific problem we have calculated $\epsilon(x, y)$ at element centres for the two-element and sixteen-element meshes. The results are shown in Figures 7.3 (a) and (b) (which are not to scale of course).

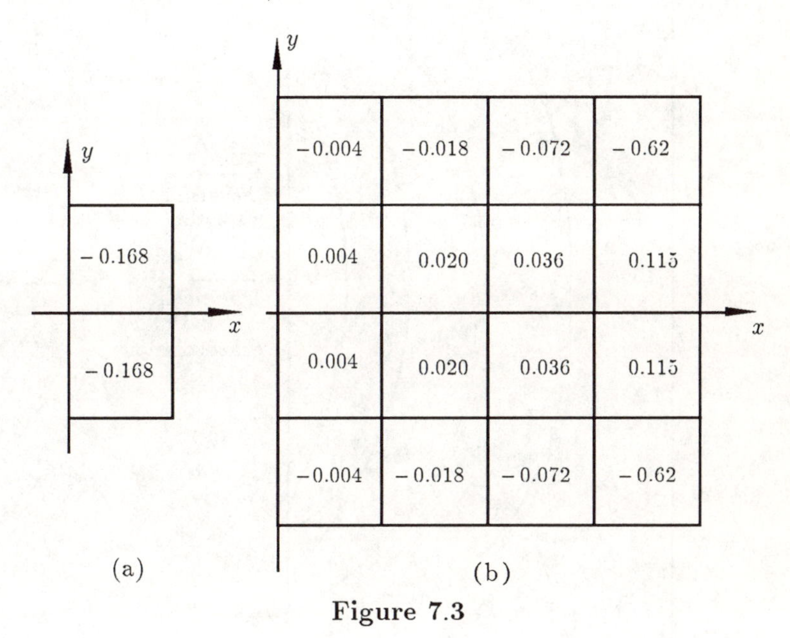

Figure 7.3

We have also drawn, for comparison purposes, contour maps for the two-element, sixteen-element and exact solutions. These are displayed in Figure 7.4. The accuracy obtained for the sixteen-element solution is quite impressive, as is indicated by the fact that the contour curves for this approximation and for the exact solution are almost indistinguishable.

7.4 Matrix formulation for two-dimensional finite elements

In the above example we have used the finite element procedure to obtain an approximate solution to a boundary value problem. We chose to work with only two elements since the computations were carried through without the aid of a computer.

In practical problems involving many tens (if not hundreds) of elements this 'direct' approach is unworkable. Fortunately, as with one-dimensional problems, a matrix formulation for the finite element equations arising from two-dimensional boundary value problems can be developed. Indeed it is unlikely that the finite element method would have attained its pre-eminent position amongst numerical techniques for solving boundary value problems had such a formulation not been possible.

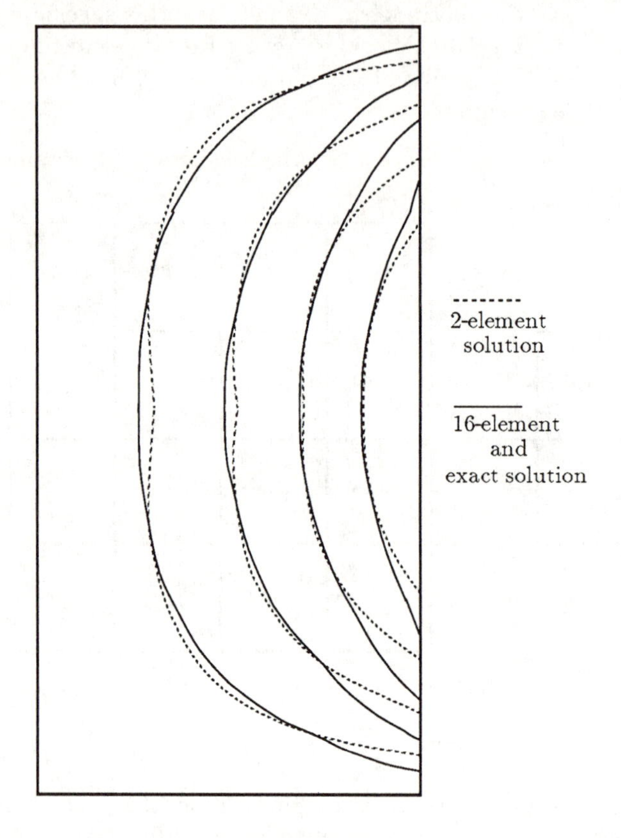

Figure 7.4

Much of the material of this section will generalize the procedures and techniques used in the matrix formulation of one-dimensional problems. To keep matters relatively straightforward we shall examine the solution of Poisson's equation over a general region A of the xy plane. Specifically we consider the boundary value problem:

$$\frac{\partial^2 u}{\partial x^2} + \frac{\partial^2 u}{\partial y^2} + h = 0 \qquad \text{within } A \tag{7.24a}$$

$$u = k^* \text{ on } C^* \qquad \frac{\partial u}{\partial n} = k \text{ on } C \tag{7.24b}$$

In other words the boundary of A is composed of two parts C and C^*. On C^* an essential boundary condition is imposed and there is a suppressible boundary condition on C.

For this problem the Galerkin formulation is (7.2) which we restate here:

$$\tilde{u}(x,y) = \theta_0(x,y) + \sum_{i=1}^{n} a_i\,\theta_i(x,y) \qquad \textbf{(7.25)}$$

Performing a similar analysis to that which led to (7.10) we find, in this case, the modified Galerkin formula:

$$-\iint_A \left\{ \frac{\partial\theta_i}{\partial x}\frac{\partial\tilde{u}}{\partial x} + \frac{\partial\theta_i}{\partial y}\frac{\partial\tilde{u}}{\partial y} \right\} dx\,dy + \iint_A h\theta_i\,dx\,dy$$
$$+ \int_C k\,\theta_i\,ds = 0 \qquad i = 1, 2, \ldots, n \qquad \textbf{(7.26)}$$

in which, we repeat, θ_0 (which is implicitly present within $\tilde{u}$) satisfies the essential boundary conditions of the problem and θ_i satisfies the corresponding homogeneous form of these essential boundary conditions. (Here the line integral over C is written with respect to the arc-length parameter s.)

In setting up the finite element equations for treatment on a computer we proceed in a different manner from that used in hand calculations.

Firstly we mesh the region with elements of our choice. Let *nde* denote the total number of nodes in the mesh. Our finite element approximation is **always** of the form

$$\tilde{u}(x,y) = \sum_{i=1}^{nde} \tilde{u}_i\,R_i\,(x,y) \qquad \textbf{(7.27a)}$$

in which $\tilde{u}_i$ are the nodal values of the unknown function $\tilde{u}(x,y)$ and $R_i(x,y)$ are roof functions. There is a similarity between (7.25) and (7.27a) but the two are **not** identical. We note that, depending on the problem under consideration, some of the nodal values $\tilde{u}_i$ may not be genuine unknowns as they might be fixed by essential boundary conditions. (In (7.24) the nodal values on C^* are known.) However, we formulate the finite element equations using (7.27a) taking $\theta_i \equiv R_i$ in (7.26) and **agree to rectify any inconsistencies** (due to the boundary conditions) at a later stage. As in the corresponding one-dimensional case this is done in order to maintain a certain amount of structure in the finite element equations which is advantageous for computer implementation.

Now, from (7.27a)

$$\frac{\partial \tilde{u}}{\partial x} = \sum_{j=1}^{nde} \tilde{u}_j \frac{\partial R_j}{\partial x} \qquad \frac{\partial \tilde{u}}{\partial y} = \sum_{j=1}^{nde} \tilde{u}_j \frac{\partial R_j}{\partial y} \tag{7.27b}$$

(where we have changed the index label from i to j to avoid later confusion). Hence (7.26) becomes

$$-\sum_{j=1}^{nde} \tilde{u}_j \iint_A \left\{ \frac{\partial R_i}{\partial x}\frac{\partial R_j}{\partial x} + \frac{\partial R_i}{\partial y}\frac{\partial R_j}{\partial y} \right\} dx\, dy + \iint_A hR_i \, dx dy$$
$$+ \int_C k\, R_i \, ds = 0 \qquad i = 1, 2, \ldots, n \tag{7.28}$$

This system of equations, (7.28), can be written in matrix form:

$$[K]\{\tilde{u}\} = \{F\} + \{B\} \tag{7.29}$$

in which the components of $[K]$ (a square matrix of order $nde \times nde$) are:

$$K_{ij} = \iint_A \left\{ \frac{\partial R_i}{\partial x}\frac{\partial R_j}{\partial x} + \frac{\partial R_i}{\partial y}\frac{\partial R_j}{\partial y} \right\} dx\, dy$$
$$i, j = 1, 2, \ldots, nde \tag{7.30}$$

The components of $\{F\}$ and $\{B\}$ (both $nde \times 1$ column vectors) are

$$F_i = \iint_A h\, R_i \, dx\, dy \qquad i = 1, 2, \ldots, nde \tag{7.31a}$$

$$B_i = \int_C k\, R_i \, ds \qquad i = 1, 2, \ldots, nde \tag{7.31b}$$

and

$$\{\tilde{u}\} = \{\tilde{u}_1, \tilde{u}_2, \ldots, \tilde{u}_{nde}\}^T \tag{7.32}$$

Using the chosen mesh we can write

$$K_{ij} = \sum_{e=1}^{nel} \iint_{[e]} \left\{ \frac{\partial R_i}{\partial x}\frac{\partial R_j}{\partial x} + \frac{\partial R_i}{\partial y}\frac{\partial R_j}{\partial y} \right\} dx\, dy \tag{7.33}$$

$$F_i = \sum_{e=1}^{nel} \iint_{[e]} hR_i \, dx\, dy \qquad B_i = \sum_{b=1}^{nbe} \int_{[b]} k\, R_i \, ds \tag{7.34}$$

in which *nel* is the total number of (interior) elements in A and *nbe* is the total number of (boundary) elements on that part of the boundary C on which suppressible boundary conditions are imposed. The term 'boundary elements' refers to the edges of those interior elements which form part of the boundary of the region. Figure 7.5 shows part of a general problem domain (segmented into eight-noded quadrilateral elements) near the boundary and highlights the two types of elements that naturally arise in two-dimensional problems.

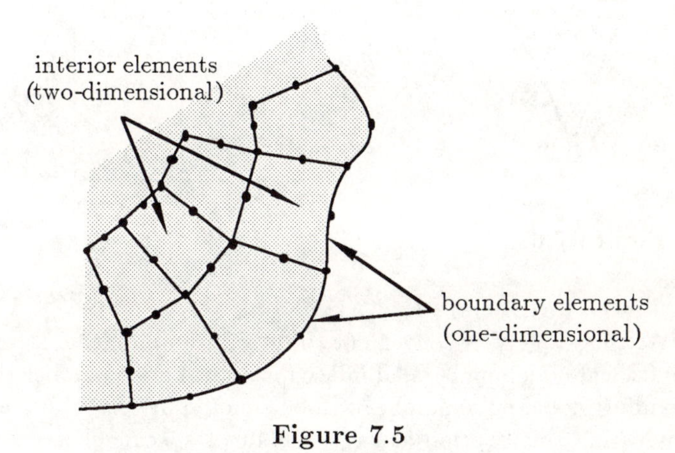

Figure 7.5

7.4.1 Element matrices and assembly

Although roof functions are useful in the **formal** process of developing the finite element equations we always express these functions in terms of the element shape functions when evaluating the integrals which arise. To do this we need to know, for each element, the relation between the global node numbers $i = 1, 2 \ldots, nde$, and the local node numbers attached to the standard element.

Suppose, for example, that in a three-noded triangular element $[e]$ (Figure 7.6) the relation between global node numbers, denoted by a, b and c, and local node numbers is as shown in Table 7.2. (Of course the choice as to the ordering of the local nodes is ours.) Then, **within this element**, the roof functions R_a, R_b, R_c are

$$R_a \equiv T_1 \quad , \quad R_b \equiv T_2 \quad , \quad R_c \equiv T_3 \tag{7.35}$$

in which T_i, $i = 1, 2, 3$, are the shape functions (6.19) for a three-noded triangular element.

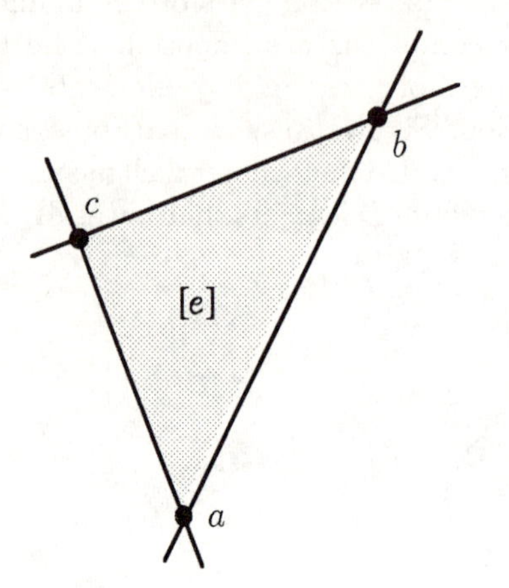

Figure 7.6

element [e]	
global	local
a	1
b	2
c	3

Table 7.2

At this stage it is convenient to introduce a notation for relating global and local node numbers. If index i is a local node number then the corresponding global node number will be denoted by $< i >$. It should be emphasised that this relation **always** refers to a specific element.

For the example shown in Figure 7.6 then, clearly,

$$\underset{\text{global}}{\underset{\uparrow}{a}} = <\underset{\text{local}}{\underset{\uparrow}{1}}> \qquad b = < 2 > \qquad c = < 3 >$$

This notation applies to any kind of element. Consider for example the eight-noded element of Figure 7.7 with global nodes as shown. If the global/local node relations are as in Table 7.3 then we can write

$$101 = < 1 > \quad , \quad 122 = < 2 > \quad \ldots \quad 100 = < 8 >$$

Within this element the roof functions are related to the bi-quadratic shape functions $Q_i(\xi, \eta)$, $i = 1, 2, \ldots, 8$, appropriate to an eight-noded quadrilateral element (see 6.25) by

$$R_{101} = R_{<1>} = Q_1 \quad , \quad R_{122} = R_{<2>} = Q_2 \quad \ldots \quad R_{100} = R_{<8>} = Q_8$$

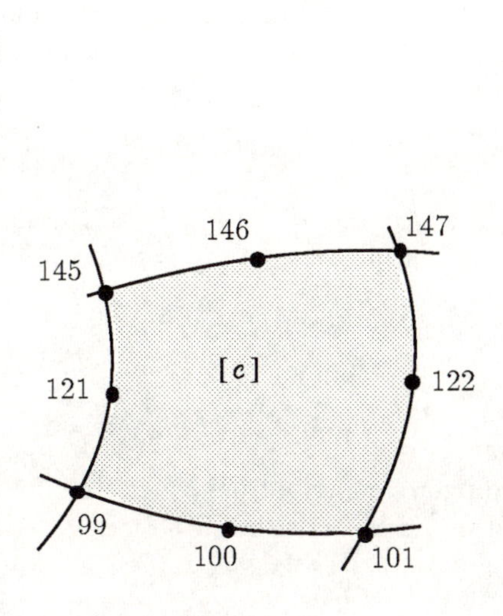

Figure 7.7

element [e]	
global	local
101	1
122	2
147	3
146	4
145	5
121	6
99	7
100	8

Table 7.3

- **Element construction of $[\mathrm{K}], \{\mathrm{F}\}$ and $\{\mathrm{B}\}$**

In the above formulation the components of $[K], \{F\}$ and $\{B\}$ can be determined using (7.33) and (7.34). As a specific example consider a typical node 'a' within a domain 'meshed' with triangular elements (see Figure 7.8(a)).

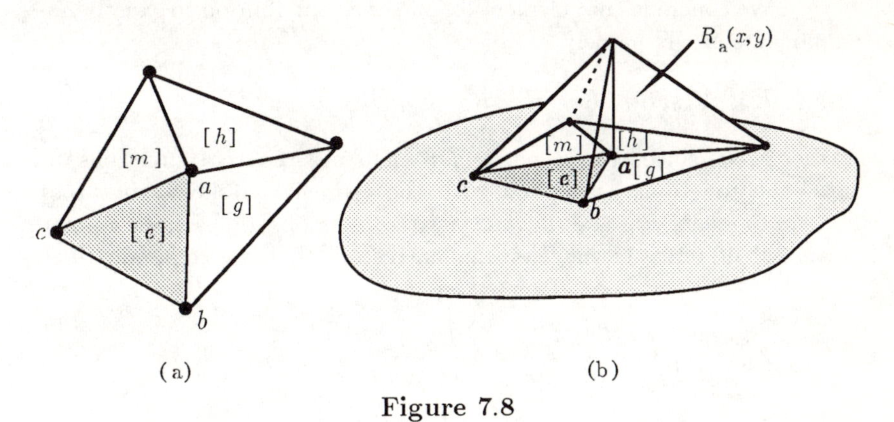

Figure 7.8

The node 'a' is attached to four elements $[e], [g], [m]$ and $[h]$. Now, by definition, the roof function $R_a(x, y)$ has the value 1 at node a, is zero at all other nodes, varies linearly with x and y (for triangular elements) within all elements attached to node 'a' and is zero in all other elements (see Figure 7.8(b)). It follows from these properties that the only non-zero contributions to the term K_{aj} in (7.33) come from those elements attached to node 'a'. Hence as far as this specific component is concerned we can write

$$K_{aj} = \iint_{[e]} D_{aj} dx dy + \iint_{[g]} D_{aj} dx dy + \iint_{[h]} D_{aj} dx dy + \iint_{[m]} D_{aj} dx dy$$

where, for convenience, we have defined, from (7.33),

$$D_{ij} = \frac{\partial R_i}{\partial x}\frac{\partial R_j}{\partial x} + \frac{\partial R_i}{\partial y}\frac{\partial R_j}{\partial y}$$

It follows further that the only non-zero contributions from the first integral (over element $[e]$) will be if the index j is either a, b or c.

Hence

$$K_{aa} = \iint_{[e]} D_{aa} dx dy \quad + \text{(possibly) terms from other elements}$$

$$K_{ab} = \iint_{[e]} D_{ab} dx dy \quad + \text{(possibly) terms from other elements}$$

$$K_{ac} = \iint_{[e]} D_{ac} dx dy \quad + \text{(possibly) terms from other elements}$$

Similar considerations apply to the other two nodes b and c attached to element $[e]$.

We conclude that element $[e]$ makes a contribution to exactly nine components of $[K]$ namely:

$$K_{aa}, K_{ab}, K_{ac}, K_{ba}, K_{bb}, K_{bc}, K_{ca}, K_{cb}, K_{cc}$$

Furthermore, since, for a particular element, the roof functions are the shape functions associated with the element it follows that these nine components are (taking due account of global/local node number relations) the terms of the 'element matrix' $[K]^{[e]}$ whose components are

$$K_{\alpha\beta}^{[e]} = \iint_{[e]} \left(\frac{\partial T_\alpha}{\partial x}\frac{\partial T_\beta}{\partial x} + \frac{\partial T_\alpha}{\partial y}\frac{\partial T_\beta}{\partial y}\right) dx dy$$

This analysis suggests the following alternative formulation for the construction of $[K]$ and also of $\{F\}$ and $\{B\}$:

For each element we construct 'local' element matrices $[K]^{[e]}$, $\{F\}^{[e]}$ and $\{B\}^{[b]}$ with components

$$K_{\alpha\beta}^{[e]} = \iint_{[e]} \left(\frac{\partial S_\alpha}{\partial x}\frac{\partial S_\beta}{\partial x} + \frac{\partial S_\alpha}{\partial y}\frac{\partial S_\beta}{\partial y}\right) dx dy \tag{7.36}$$

$$F_\alpha^{[e]} = \iint_{[e]} h\, S_\alpha \, dx\, dy \qquad B_\alpha^{[b]} = \int_{[b]} k S_\alpha \, ds \tag{7.37}$$

Here we have considered a general situation in which an n-noded element is used with associated shape functions S_α. Note that the integrands in (7.36) and (7.37) are precisely those of (7.33) and (7.34) except that the roof function label 'R' is **replaced** by the shape function label 'S' under consideration. Each of these local components makes a contribution to the components $K_{<\alpha><\beta>}, F_{<\alpha>}, B_{<\alpha>}$ of the global matrices $[K]$, $\{F\}$ $\{B\}$ respectively.

In a certain sense the global matrices $[K]$, $\{F\}$ and $\{B\}$ can be regarded as a 'sum' of the local matrices and, using an accepted abuse of notation, we write

$$[K] = \sum_{e=1}^{nel} [K]^{[e]} \quad \{F\} = \sum_{e=1}^{nel} \{F\}^{[e]} \quad \{B\} = \sum_{b=1}^{nbe} \{B\}^{[b]} \tag{7.38}$$

The procedure for the practical implementation of the finite element method is then as follows:

(i) Evaluate the local element matrix $[K]^{[e]}$ for each element.

(ii) Evaluate the local element column vector $\{F\}^{[e]}$ for each element. (If $[e]$ is a three-noded triangular element the range α (and β) is $1, 2, 3$ whilst if it is an eight-noded quadrilateral the range would be $1, 2, \ldots, 8$.)

(iii) Evaluate the local element column vector $\{B\}^{[b]}$ for each boundary element on which suppressible boundary conditions have been imposed. (Note however, that the range of the index α in $B_\alpha^{[b]}$ will only be 1,2 for triangular elements as such an element degenerates to a two-noded linear element along the domain boundary. For similar reasons the range of α would only be 1,2,3 if eight-noded elements are used.)

(iv) Superpose local element contributions into global counterparts. This superposition process is simplified if, for each 'local' matrix, global node numbers are attached to each row of the matrix. For the example shown in Figure 7.6 (and Table 7.3) we would write

$$\begin{bmatrix} K_{11}^{[e]} & K_{12}^{[e]} & K_{13}^{[e]} \\ K_{21}^{[e]} & K_{22}^{[e]} & K_{23}^{[e]} \\ K_{31}^{[e]} & K_{32}^{[e]} & K_{33}^{[e]} \end{bmatrix} \begin{matrix} a \\ b \\ c \end{matrix}$$

so that the 1-1 term of $[K]^{[e]}$ would be superimposed into K_{aa} whilst the 2-3 term of $[K]^{[e]}$ would be superimposed into K_{bc} and so on.

Of course, once the integrals are written in element forms (7.36) or (7.37) they can be transformed into integrals over standard elements and then evaluated as outlined earlier in this chapter.

- **Solution of** $[\mathbf{K}]\{\tilde{\mathbf{u}}\} = \{\mathbf{F}\} + \{\mathbf{B}\}$

Having constructed the equations (7.29) we can now consider their solution. If all the boundary conditions of the problem are suppressible then (7.29) can be solved immediately using any convenient numerical technique for solving large systems of linear equations. Two such techniques are briefly described in Chapter 9.

However, if some of the boundary conditions are essential then, as we have already noted, some of the equations of this system will be invalid and need amending before a solution can be obtained. The method of doing this is exactly the same as that used in one-dimensional problems (see Chapter 4).

Example 7.2

Using the matrix formulation described above solve the boundary value problem

$$\frac{\partial^2 u}{\partial x^2} + \frac{\partial^2 u}{\partial y^2} = 0 \qquad (x, y) \text{ within A}$$

$$u = 0 \quad \text{on} \quad C^* : x = 0, \quad y = -2, \quad y = +2$$

$$\frac{\partial u}{\partial x} = k \quad \text{on} \quad C : x = 2$$

using the mesh of four-noded quadrilateral elements shown in Figure 7.9.

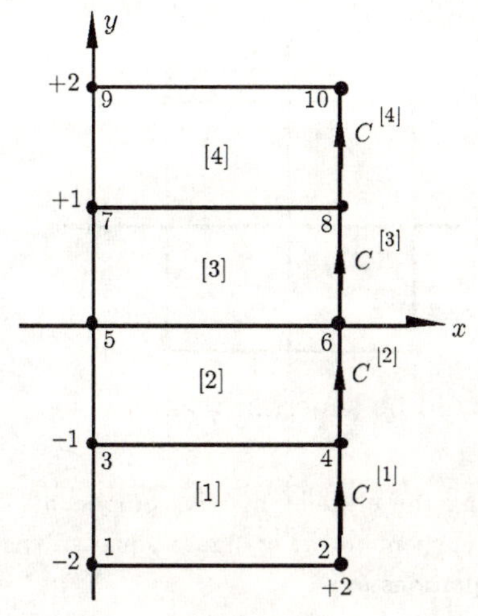

Figure 7.9

Solution

This problem is identical to that described in (7.1) with $k^* = 0$ and we have in Section 7.3 obtained an approximate solution using two four-noded quadrilateral elements. The matrix formulation relating to the use of a general element has been described above. When this formulation is amended (in an obvious way) for use with four-noded quadrilaterals we obtain the finite element equations in matrix form

$$[K]\{\tilde{u}\} = \{B\} \tag{7.39}$$

where

$$[K] = \sum_{e=1}^{4}[K]^{[e]} \qquad \{B\} = \sum_{b=1}^{4}\{B\}^{[b]} \tag{7.40}$$

(By (7.31a) $\{F\} \equiv 0$ here.)
The components of $[K]^{[e]}$ and $\{B\}^{[b]}$ are, using (7.36) and (7.37),

$$K^{[e]}_{\alpha\beta} = \iint_{[e]} \left\{ \frac{\partial L_\alpha}{\partial x}\frac{\partial L_\beta}{\partial x} + \frac{\partial L_\alpha}{\partial y}\frac{\partial L_\beta}{\partial y} \right\} dx\, dy \tag{7.41}$$

and

$$B^{[b]}_\alpha = \int_{C^{[b]}} k\, L_\alpha dy \tag{7.42}$$

Here L_α, $\alpha = 1, 2, 3, 4$, are the bi-linear shape functions (7.13) associated with four-noded quadrilaterals. (See Figure 7.10 for the local node numbering convention being employed.)

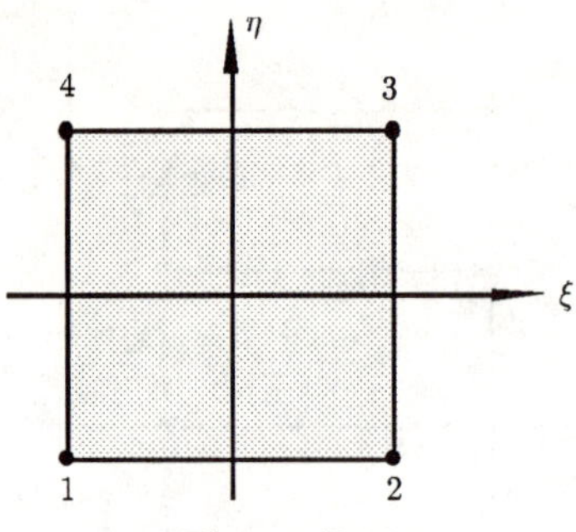

Figure 7.10

In calculating the element matrix components ((7.41) and (7.42)) we transform each element to the standard square. It is easy to show that the transforming equations are

$$\text{element [1]} \quad x = 1 + \xi, \quad y = \frac{1}{2}\eta - \frac{3}{2}$$

$$\text{element [2]} \quad x = 1 + \xi, \quad y = \frac{1}{2}\eta - \frac{1}{2}$$

$$\text{element [3]} \quad x = 1 + \xi, \quad y = \frac{1}{2}\eta + \frac{1}{2}$$

$$\text{element [4]} \quad x = 1 + \xi, \quad y = \frac{1}{2}\eta + \frac{3}{2}$$

Hence for each element

$$\frac{\partial}{\partial x} \equiv \frac{\partial}{\partial \xi} \qquad \frac{\partial}{\partial y} \equiv 2\frac{\partial}{\partial \eta} \qquad J = \frac{1}{2}$$

and we can write from (7.41)

$$K_{\alpha\beta}^{[e]} = \int_{-1}^{1}\int_{-1}^{1}\left\{\frac{\partial L_\alpha}{\partial \xi}\frac{\partial L_\beta}{\partial \xi} + 4\frac{\partial L_\alpha}{\partial \eta}\frac{\partial L_\beta}{\partial \eta}\right\}\frac{1}{2}\,d\xi\,d\eta \tag{7.43}$$

and from (7.42)

$$B_\alpha^{[b]} = \int_{-1}^{1} k\,L_\alpha \frac{1}{2}\,d\eta \tag{7.44}$$

To evaluate the integrals use is made of the results (7.20). (All double integrals that arise in this problem can be written in terms of these due to the fact that integrals of 'odd' powers of ξ or of η over $(-1, +1)$ vanish.)

Performing the straightforward calculations of (7.43) which the reader is encouraged to carry through we find

$$[K]^{[1]} = \begin{bmatrix} 5/6 & 1/6 & -5/12 & -7/12 \\ 1/6 & 5/6 & -7/12 & -5/12 \\ -5/12 & -7/12 & 5/6 & 1/6 \\ -7/12 & -5/12 & 1/6 & 5/6 \end{bmatrix} \begin{matrix} 1 \\ 2 \\ 4 \\ 3 \end{matrix} \qquad \textbf{(7.45a)}$$

Note that the numbers 'attached' to the local matrix are used to indicate, for element [1], precisely where each term will be located in the global matrix $[K]$ after superposition according to (7.40). As mentioned earlier in this section the summations implied in (7.40) are an accepted abuse of notation. Of course the local matrices cannot be directly added together (even though each is 4×4) as each component need not refer to the same node. Due account must be taken of the global node numbers attached to each element. The 'global nodes' 1,2,4,3 are those attached to element [1] taken in the order chosen for the standard element (see Figure 7.10).

In an exactly similar manner we obtain

$$[K]^{[2]} = \begin{bmatrix} 5/6 & 1/6 & -5/12 & -7/12 \\ 1/6 & 5/6 & -7/12 & -5/12 \\ -5/12 & -7/12 & 5/6 & 1/6 \\ -7/12 & -5/12 & 1/6 & 5/6 \end{bmatrix} \begin{matrix} 3 \\ 4 \\ 6 \\ 5 \end{matrix} \qquad \textbf{(7.45b)}$$

$$[K]^{[3]} = \begin{bmatrix} 5/6 & 1/6 & -5/12 & -7/12 \\ 1/6 & 5/6 & -7/12 & -5/12 \\ -5/12 & -7/12 & 5/6 & 1/6 \\ -7/12 & -5/12 & 1/6 & 5/6 \end{bmatrix} \begin{matrix} 5 \\ 6 \\ 8 \\ 7 \end{matrix} \qquad \textbf{(7.45c)}$$

$$[K]^{[4]} = \begin{bmatrix} 5/6 & 1/6 & -5/12 & -7/12 \\ 1/6 & 5/6 & -7/12 & -5/12 \\ -5/12 & -7/12 & 5/6 & 1/6 \\ -7/12 & -5/12 & 1/6 & 5/6 \end{bmatrix} \begin{matrix} 7 \\ 8 \\ 10 \\ 9 \end{matrix} \qquad \textbf{(7.45d)}$$

We should not be too surprised at the similarity of each of these local element matrices. This is entirely due to the similarity of the elements to which they refer (each one has exactly the same orientation and dimensions) and to the nature of the underlying boundary value problem.

Also, on each of the boundary elements $C_{[i]}$, $i = 1, 2, 3, 4$, we have

$$L_1 = 0, \quad L_2 = \tfrac{1}{2}(1-\eta), \quad L_3 = \tfrac{1}{2}(1+\eta), \quad L_4 = 0$$

Hence from (7.44) we obtain

$$\{B\}^{[1]} = \begin{Bmatrix} 0 \\ k/2 \\ k/2 \\ 0 \end{Bmatrix} \begin{matrix} 1 \\ 2 \\ 4 \\ 3 \end{matrix} \qquad \{B\}^{[2]} = \begin{Bmatrix} 0 \\ k/2 \\ k/2 \\ 0 \end{Bmatrix} \begin{matrix} 3 \\ 4 \\ 6 \\ 5 \end{matrix}$$

$$\{B\}^{[3]} = \begin{Bmatrix} 0 \\ k/2 \\ k/2 \\ 0 \end{Bmatrix} \begin{matrix} 5 \\ 6 \\ 8 \\ 7 \end{matrix} \qquad \{B\}^{[4]} = \begin{Bmatrix} 0 \\ k/2 \\ k/2 \\ 0 \end{Bmatrix} \begin{matrix} 7 \\ 8 \\ 10 \\ 9 \end{matrix}$$

where again we have attached global node numbers so that each term of $\{B\}^{[b]}$ can be located in $\{B\}$ in the correct position. When the superposition is carried out we obtain for (7.39) (after multiplying throughout by 12):

$$\begin{bmatrix} 10 & 2 & -7 & -5 & 0 & 0 & 0 & 0 & 0 & 0 \\ 2 & 10 & -5 & -7 & 0 & 0 & 0 & 0 & 0 & 0 \\ -7 & -5 & 20 & 4 & -7 & -5 & 0 & 0 & 0 & 0 \\ -5 & -7 & 4 & 20 & -5 & -7 & 0 & 0 & 0 & 0 \\ 0 & 0 & -7 & -5 & 20 & 4 & -7 & -5 & 0 & 0 \\ 0 & 0 & -5 & -7 & 4 & 20 & -5 & -7 & 0 & 0 \\ 0 & 0 & 0 & 0 & -7 & -5 & 20 & 4 & -7 & -5 \\ 0 & 0 & 0 & 0 & -5 & -7 & 4 & 20 & -5 & -7 \\ 0 & 0 & 0 & 0 & 0 & 0 & -7 & -5 & 10 & 2 \\ 0 & 0 & 0 & 0 & 0 & 0 & -5 & -7 & 2 & 10 \end{bmatrix} \begin{Bmatrix} \tilde{u}_1 \\ \tilde{u}_2 \\ \tilde{u}_3 \\ \tilde{u}_4 \\ \tilde{u}_5 \\ \tilde{u}_6 \\ \tilde{u}_7 \\ \tilde{u}_8 \\ \tilde{u}_9 \\ \tilde{u}_{10} \end{Bmatrix} = \begin{Bmatrix} 0 \\ 6k \\ 0 \\ 12k \\ 0 \\ 12k \\ 0 \\ 12k \\ 0 \\ 6k \end{Bmatrix}$$

Of this system of ten equations in ten unknowns only three (specifically equations 4, 6 and 8) are valid. The other equations are meaningless and have only been carried through in order to maintain 'structure' in the equations. As mentioned earlier, the invalid equations are those that correspond to nodes at which essential boundary conditions are imposed.

Selecting only the valid equations and imposing the essential boundary conditions:

$$\tilde{u}_1 = \tilde{u}_2 = \tilde{u}_3 = \tilde{u}_5 = \tilde{u}_7 = \tilde{u}_9 = \tilde{u}_{10} = 0$$

we find

$$\begin{aligned} 20\,\tilde{u}_4 \; - \; 7\,\tilde{u}_6 \qquad\qquad &= 12k \\ -7\,\tilde{u}_4 + 20\,\tilde{u}_6 - 7\,\tilde{u}_8 &= 12k \\ - \; 7\,\tilde{u}_6 + 20\,\tilde{u}_8 &= 12k \end{aligned}$$

with solution

$$\tilde{u}_4 = \tilde{u}_8 = 1.0728k \qquad \text{and} \qquad \tilde{u}_6 = 1.3509k$$

These compare well with values of the exact solution at corresponding points:

$$u(2,1) = u(2,-1) = 1.255k \qquad \text{and} \qquad u(2,0) = 1.35063k$$

7.5 Stages in the finite element method

As we did for one-dimensional problems in Section 4.6 it is useful to summarize the steps carried out in obtaining the finite element solution of two-dimensional boundary value problems. The steps that we outline below are those to be followed when working by 'hand'. Some of the steps would be altered slightly if we were to automate the process for implementation on a computer.

- 1 Reformulate the boundary value problem using the modified Galerkin method.
- 2 Mesh the domain of the problem using suitably chosen elements, numbering the nodes, the interior elements and the boundary elements.
- 3 Approximate the unknown function (surface) according to the chosen elements (see (7.12) for the example of four-noded quadrilaterals).
- 4 For each element note the relation between global node numbers and local node numbers on the standard element (cf. Table 7.1).
- 5 Note the relation between roof functions and shape functions (cf. (7.15)).
- 6 Formulate the approximation over the complete domain in terms of roof functions (cf. (7.16)).
- 7 Form the overall matrix components K_{ij}, F_i, B_i (cf. (7.30),(7.31)) and define corresponding element matrix components (cf. (7.36),(7.37)).
- 8 Evaluate each integral. In practical problems this is normally done numerically.
- 9 Impose essential boundary conditions as in Example 7.2.
- 10 Solve the resulting system of (linear) equations for the unknown nodal values.

Although this seems (and indeed is) a somewhat laborious procedure when working by hand we should always keep in mind that all stages can, with modifications, be fully automated (including the initial 'meshing') for implementation on a computer.

Our aim in this text is to help the reader to understand how the finite element method works as well as to be able to solve some problems by hand using the method. We are fully aware that, with the present plethora of finite element computer packages available, the average engineer is unlikely to solve many problems of the type considered in this text by hand. Nevertheless any user of a finite element package should have some understanding of the technique and be aware of its limitations. We feel that only by following through the solution of some problems by hand can the technique be fully appreciated.

7.6 Three-noded triangular elements

If the differential equation in the boundary value problem under consideration is a simple constant coefficient type (such as Poisson's equation (7.24a)) and if particularly simple elements are used, for example three-noded triangles, the local matrix formulation (7.36) and (7.37) can be taken a stage further. That is, for an arbitrarily placed element, each component of $[K]^{[e]}, \{F\}^{[e]}, \{B\}^{[b]}$ can be determined explicitly. Similar calculations may be performed for arbitrary rectangular elements whose sides are parallel to the coordinate axes. See Exercise Q7.7, the results of which could have been used to obtain the element matrices (7.45) in Example 7.2.

Consider then a mesh of three-noded triangular elements, as in Figure 7.11(a), in which a single element has been highlighted.

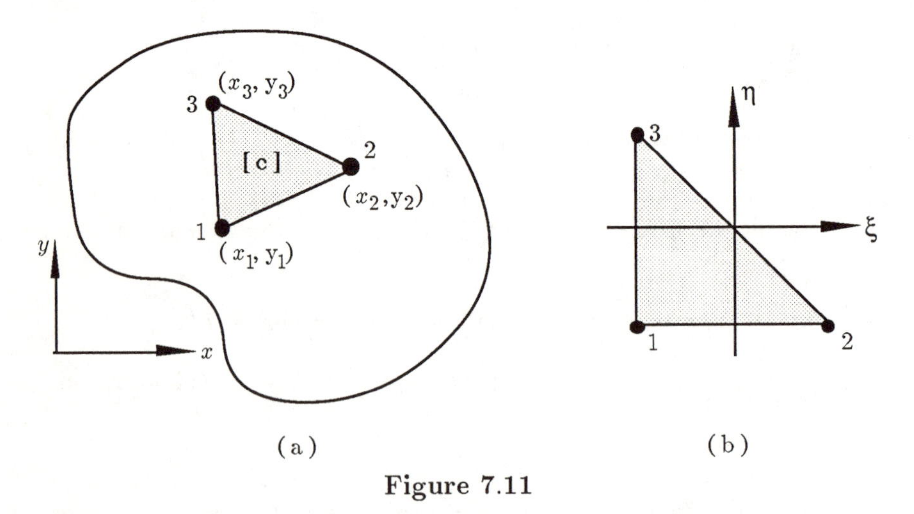

Figure 7.11

For such triangular elements the local matrices $[K]^{[e]}, \{F\}^{[e]}$ appropriate to Poisson's equation, have, according to (7.36) and (7.37), components:

$$K_{\alpha\beta}^{[e]} = \iint_{[e]} \left[\frac{\partial T_\alpha}{\partial x}\frac{\partial T_\beta}{\partial x} + \frac{\partial T_\alpha}{\partial y}\frac{\partial T_\beta}{\partial y} \right] dxdy \quad \alpha, \beta = 1, 2, 3 \qquad \textbf{(7.46a)}$$

$$F_\alpha^{[e]} = \iint_{[e]} hT_\alpha \; dxdy \qquad \alpha = 1, 2, 3 \qquad \textbf{(7.46b)}$$

where the T_α are the shape functions appropriate to a three-noded triangle viz.

$$T_1 = -\frac{1}{2}(\xi + \eta), \qquad T_2 = \frac{1}{2}(1 + \xi), \qquad T_3 = \frac{1}{2}(1 + \eta) \quad \textbf{(7.47a)}$$

To evaluate the double integrals we transform to the standard triangular element (see Figure 7.11(b)) using the transformation (first used in (6.18)):

$$x = \sum_{i=1}^{3} x_i T_i(\xi, \eta) \qquad y = \sum_{i=1}^{3} y_i T_i(\xi, \eta) \qquad \textbf{(7.47b)}$$

It is convenient to introduce the following quantities which involve various combinations of the nodal coordinates of the triangle $[e]$:

$$\begin{aligned}
&b_1 = y_2 - y_3 \qquad c_1 = x_3 - x_2 \\
&b_2 = y_3 - y_1 \qquad c_2 = x_1 - x_3 \\
&b_3 = y_1 - y_2 \qquad c_3 = x_2 - x_1 \\
&D = b_2 c_3 - b_3 c_2 = 2 \times \text{area of } [e]
\end{aligned}$$

With this notation we find, for the relations between derivatives:

$$\begin{aligned}
\frac{\partial}{\partial x} &= \frac{2b_2}{D}\frac{\partial}{\partial \xi} + \frac{2b_3}{D}\frac{\partial}{\partial \eta} \\
\frac{\partial}{\partial y} &= \frac{2c_2}{D}\frac{\partial}{\partial \xi} + \frac{2c_3}{D}\frac{\partial}{\partial \eta}
\end{aligned}$$

The Jacobian of the transformation (7.47b) is found to be

$$J = \begin{vmatrix} \dfrac{\partial x}{\partial \xi} & \dfrac{\partial x}{\partial \eta} \\[2ex] \dfrac{\partial y}{\partial \xi} & \dfrac{\partial y}{\partial \eta} \end{vmatrix} = \frac{1}{4}D = \frac{1}{2}A^{[e]}$$

as we would expect because we are transforming a triangle of area $A^{[e]}$

into a standard triangle of area 2.

Substituting these relations into (7.46a) we find, for example,

$$K_{11}^{[e]} = \iint_{[e]} \left[\left(\frac{\partial T_1}{\partial x} \right)^2 + \left(\frac{\partial T_1}{\partial y} \right)^2 \right] dx dy$$

in which

$$\begin{aligned} \left(\frac{\partial T_1}{\partial x} \right)^2 &= \left(\frac{2b_2}{D} \frac{\partial T_1}{\partial \xi} + \frac{2b_3}{D} \frac{\partial T_1}{\partial \eta} \right)^2 \\ &= \frac{(b_2 + b_3)^2}{D^2} \\ &= \frac{b_1^2}{D^2} \end{aligned}$$

Similarly

$$\left(\frac{\partial T_1}{\partial y} \right)^2 = \frac{c_1^2}{D^2}$$

and so

$$\begin{aligned} K_{11}^{[e]} &= \iint_{[e]} \left[\frac{b_1^2 + c_1^2}{D^2} \right] \frac{D}{4} \, d\xi \, d\eta \\ &= \frac{b_1^2 + c_1^2}{2D} \quad \text{using } (6.20a) \end{aligned}$$

Continuing in this way for the other components we find:

$$K_{ij}^{[e]} = \frac{1}{2D} \begin{bmatrix} b_1^2 + c_1^2 & b_1 b_2 + c_1 c_2 & b_1 b_3 + c_1 c_3 \\ b_1 b_2 + c_1 c_2 & b_2^2 + c_2^2 & b_2 b_3 + c_2 c_3 \\ b_1 b_3 + c_1 c_3 & b_2 b_3 + c_2 c_3 & b_3^2 + c_3^2 \end{bmatrix} \tag{7.48}$$

Similarly from (7.46b) we obtain

$$F_i^{[e]} = \frac{hD}{6} \begin{Bmatrix} 1 \\ 1 \\ 1 \end{Bmatrix} \tag{7.49}$$

We can also find an explicit form for the components of $\{B\}^{[b]}$ arising from boundary elements. To obtain this we consider the particular boundary element highlighted in Figure 7.12

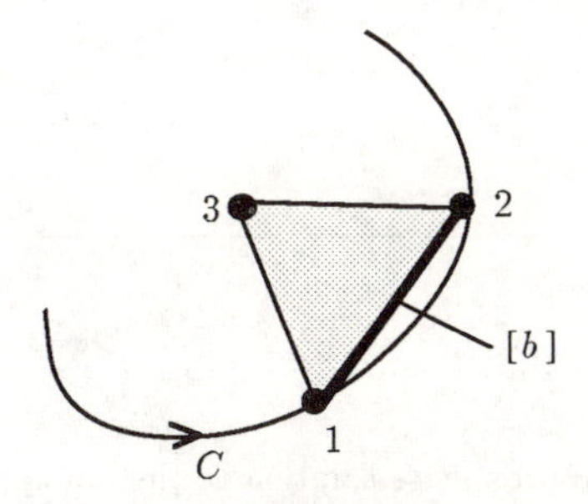

Figure 7.12

Of course the actual boundary C is assumed to be replaced by the edges of the triangular elements. On the particular element shown only T_1 and T_2 need be considered as $T_3 = 0$ on this edge $[b]$. Now, from standard properties of shape functions:

$$T_1 = \begin{cases} 1 & \text{at node 1} \\ 0 & \text{at node 2} \end{cases} \qquad T_2 = \begin{cases} 0 & \text{at node 1} \\ 1 & \text{at node 2} \end{cases}$$

Both of these functions vary linearly between nodes 1 and 2 and hence we can conveniently express the shape functions in terms of arc-length s along the boundary of the triangle.

If $s = s_1$ at node 1 and $s = s_2$ at node 2 then

$$T_1 = \frac{s_2 - s}{s_2 - s_1} \qquad T_2 = \frac{s - s_1}{s_2 - s_1}$$

(compare (3.3) and (3.4)) so that, from the second equation of (7.37),

$$\begin{aligned} B_1^{[b]} &= \int_{[b]} kT_1 \; ds = k \int_{s_1}^{s_2} \left(\frac{s - s_2}{s_1 - s_2} \right) ds \\ &= k\frac{\ell}{2} \end{aligned}$$

where ℓ is the length of the element.
Similarly

$$B_2^{[b]} = \int_{[b]} kT_2 \; ds = k\frac{\ell}{2}$$

leading to

$$B_\alpha^{[b]} = \frac{k\ell}{2} \begin{Bmatrix} 1 \\ 1 \end{Bmatrix} \qquad \textbf{(7.50)}$$

We utilize the results (7.48), (7.49) and (7.50) in the following example which involves triangular elements. In our solution we shall adhere to the finite element stages as outlined in Section 7.5.

Example 7.3

Determine an approximate solution to the boundary value problem:

$$\frac{\partial^2 u}{\partial x^2} + \frac{\partial^2 u}{\partial y^2} + 2 = 0 \qquad \text{within } D$$
$$u = 0 \qquad \text{on the boundary of } D$$

where D is the interior of the ellipse $x^2/36 + y^2/4 = 1$ (see Figure 7.13a).

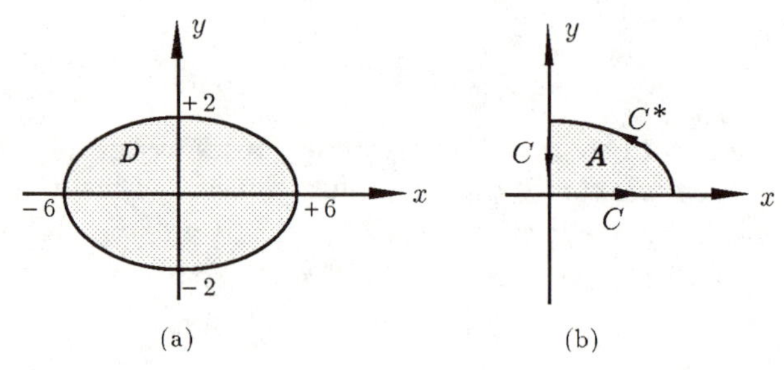

Figure 7.13

This problem arises in the analysis of beam-twisting but again, as far as its numerical solution is concerned, the physical interpretation has little significance. The differential equation is of course a two-dimensional Poisson equation which, as we have seen, arises in other contexts.

Solution

By symmetry, we need only consider the region in the positive quadrant. To obtain a unique solution to the modified problem we must impose appropriate boundary conditions on the 'cut'-edges $x = 0$ and $y = 0$. These may also be deduced by considerations of symmetry.

By inspection of the boundary value problem, it is clear that the solution we seek must be an 'even' function of both the x and y variables separately. We can deduce therefore that $\partial u/\partial x$ must be an odd function of x and $\partial u/\partial y$ an odd function of y. If we assume that both $\partial u/\partial x$ and $\partial u/\partial y$ are continuous functions of x and y respectively then we can conclude that

$$\frac{\partial u}{\partial x} = 0 \quad \text{on 'cut' edge} \quad x = 0$$

$$\frac{\partial u}{\partial y} = 0 \quad \text{on 'cut' edge} \quad y = 0$$

These two conditions can be combined by use of the n-coordinate (which points in the negative x-direction on $x = 0$ and in the negative y-direction on $y = 0$):

$$\frac{\partial u}{\partial n} = 0 \quad \text{on 'cut' edges} \quad x = 0 \text{ and } y = 0$$

We have thus arrived at a new boundary value problem over a reduced domain A (see Figure 7.13b):

$$\frac{\partial^2 u}{\partial x^2} + \frac{\partial^2 u}{\partial y^2} + 2 = 0 \qquad \text{within } A \tag{7.51a}$$

$$u = 0 \text{ on } C^* : \text{ the quarter ellipse } \frac{x^2}{36} + \frac{y^2}{4} = 1 \tag{7.51b}$$

$$\frac{\partial u}{\partial n} = 0 \text{ on } C : x = 0, \ y = 0 \tag{7.51c}$$

Stage 1 Galerkin formulation

Assuming an approximate solution $\tilde{u}(x, y)$ of the form (7.2) the Galerkin equations for this example are

$$\iint_A \left\{ \frac{\partial^2 \tilde{u}}{\partial x^2} + \frac{\partial^2 \tilde{u}}{\partial y^2} + 2 \right\} \theta_i(x, y) dx\, dy = 0 \quad i = 1, 2 \ldots, n \tag{7.52}$$

in which, since all the boundary conditions are homogeneous, $\theta_0(x, y) \equiv 0$ and

$$\theta_i(x, y) = 0 \quad \text{on} \quad C^*$$

$$\frac{\partial \theta_i}{\partial n} = 0 \text{ on } x = 0 \text{ and } y = 0 \quad i = 1, 2, \ldots, n$$

($\tilde{u}$ will then satisfy **all** the boundary conditions). Applying Green's Theorem in the Plane to the second derivative terms in (7.52) gives the following (see (7.8) and (7.26)):

$$\iint_A \left\{ -\frac{\partial \theta_i}{\partial x}\frac{\partial \tilde{u}}{\partial x} - \frac{\partial \theta_i}{\partial y}\frac{\partial \tilde{u}}{\partial y} + 2\,\theta_i \right\} dx\,dy$$

$$+ \oint_\Gamma \theta_i \frac{\partial \tilde{u}}{\partial n}\,ds = 0 \qquad i = 1, 2 \ldots, n \qquad \textbf{(7.53)}$$

where Γ is the boundary of A. To obtain the line integral in (7.53) we have used the relation:

$$-\frac{\partial \tilde{u}}{\partial y}dx + \frac{\partial \tilde{u}}{\partial x}dy = \frac{\partial \tilde{u}}{\partial n}ds$$

Now on the portion C^* of the boundary, all the coordinate functions θ_i are zero. Hence the only contribution to the line integral is along C. The suppressible boundary condition (7.51c) tells us that this integral is zero so that (7.53) becomes

$$\iint_A \left\{ -\frac{\partial \theta_i}{\partial x}\frac{\partial \tilde{u}}{\partial x} - \frac{\partial \theta_i}{\partial y}\frac{\partial \tilde{u}}{\partial y} + 2\,\theta_i \right\} dx\,dy = 0 \qquad i = 1, 2, \ldots, n \quad \textbf{(7.54)}$$

in which $\displaystyle \tilde{u} = \sum_{i=1}^{n} a_i\,\theta_i\,(x, y)$

with θ_i needing to satisfy only the **essential** boundary conditions $\theta_i(x, y) = 0$ on C^*. This completes the formulation stage.

Stage 2 Meshing the domain A

We choose the rather crude mesh of four three-noded triangular elements shown in Figure 7.14. The choice of mesh is purely for illustrative purposes. We should not expect the numerical solution obtained to be particularly accurate as the curved part of the boundary is being approximated by the straight edges of element [1] and element [4]. Since there are no line integrals to deal with in (7.54) we need not concern ourselves with boundary elements. This completes the meshing stage.

Stage 3 Element approximation

Within each element the true solution $u(x, y)$ will be approximated by

$$\tilde{u}(x, y) = \sum_{i=1}^{3} \tilde{u}_i^{[e]}\,T_i\,(\xi, \eta) \qquad \textbf{(7.55)}$$

in which

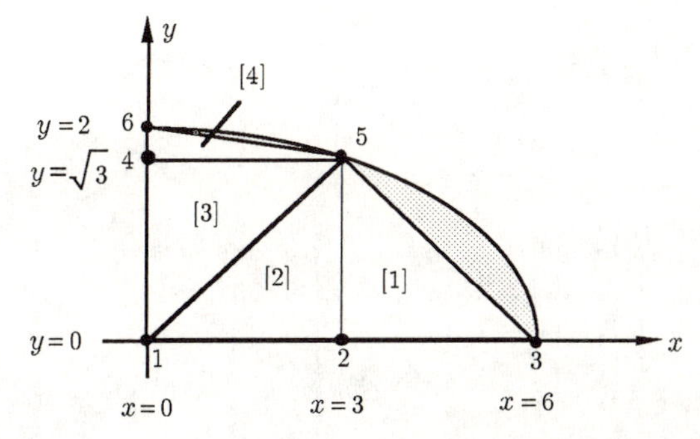

Figure 7.14

$$x = \sum_{i=1}^{3} x_i^{[e]} T_i(\xi, \eta) \quad , \quad y = \sum_{i=1}^{3} y_i^{[e]} T_i(\xi, \eta) \tag{7.56}$$

where $T_i(\xi, \eta)$ $i = 1, 2, 3$ are the linear shape functions (7.47a) for a standard triangular element.

Stage 4 Global/local node number relations

A global node number is assigned to each node as shown in Figure 7.14. For each element we also assign a local number to each node, and of many possible choices, we use the numbering given in Table 7.4.

global nodes				local nodes
$[e] = 1$	$[e] = 2$	$[e] = 3$	$[e] = 4$	
2	1	5	4	1
3	2	4	5	2
5	5	1	6	3

Table 7.4

Stage 5 Roof function/shape function relations

The roof function/shape function relations follow from the global/local node number relations chosen at Stage 4.

$$R_1 = \begin{cases} 0 & [e] = 1 \\ T_1 & [e] = 2 \\ T_3 & [e] = 3 \\ 0 & [e] = 4 \end{cases} \qquad R_2 = \begin{cases} T_1 & [e] = 1 \\ T_2 & [e] = 2 \\ 0 & [e] = 3 \\ 0 & [e] = 4 \end{cases}$$

$$R_3 = \begin{cases} T_2 & [e]=1 \\ 0 & [e]=2 \\ 0 & [e]=3 \\ 0 & [e]=4 \end{cases} \qquad R_4 = \begin{cases} 0 & [e]=1 \\ 0 & [e]=2 \\ T_2 & [e]=3 \\ T_1 & [e]=4 \end{cases}$$

$$R_5 = \begin{cases} T_3 & [e]=1 \\ T_3 & [e]=2 \\ T_1 & [e]=3 \\ T_2 & [e]=4 \end{cases} \qquad R_6 = \begin{cases} 0 & [e]=1 \\ 0 & [e]=2 \\ 0 & [e]=3 \\ T_3 & [e]=4 \end{cases} \tag{7.57}$$

Stage 6 Formulate overall approximation

Over the complete domain the approximation (7.55) is equivalent to

$$\tilde{u}(x,y) = \sum_{i=1}^{6} \tilde{u}_i \, R_i\,(x,y) \tag{7.58}$$

Stages 7,8 Construct local and global matrices

Here we use immediately the results (7.48), (7.49) for the local matrices.

- **Element [1]**

$x_1 = 3 \qquad x_2 = 6 \qquad x_3 = 3$

$y_1 = 0 \qquad y_2 = 0 \qquad y_3 = \sqrt{3}$

$b_1 = -\sqrt{3} \qquad b_2 = \sqrt{3} \qquad b_3 = 0$

$c_1 = -3 \qquad c_2 = 0 \qquad c_3 = 3$

$D = 3\sqrt{3}$

Substitution into (7.48) gives

$$K_{ij}^{[1]} = \frac{1}{6\sqrt{3}} \begin{bmatrix} 12 & -3 & -9 \\ -3 & 3 & 0 \\ -9 & 0 & 9 \end{bmatrix} \begin{matrix} 2 \\ 3 \\ 5 \end{matrix} \tag{7.59a}$$

and from (7.49)

$$F_i^{[1]} = \sqrt{3}\begin{Bmatrix} 1 \\ 1 \\ 1 \end{Bmatrix} \begin{matrix} 2 \\ 3 \\ 5 \end{matrix} \tag{7.59b}$$

- **Element [2]**

$$\begin{array}{lll} x_1 = 0 & x_2 = 3 & x_3 = 3 \\ y_1 = 0 & y_2 = 0 & y_3 = \sqrt{3} \\ b_1 = -\sqrt{3} & b_2 = \sqrt{3} & b_3 = 0 \\ c_1 = 0 & c_2 = -3 & c_3 = 3 \\ D = 3\sqrt{3} & & \end{array}$$

Hence

$$K_{ij}^{[2]} = \frac{1}{6\sqrt{3}}\begin{bmatrix} 3 & -3 & 0 \\ -3 & 12 & -9 \\ 0 & -9 & 9 \end{bmatrix} \begin{matrix} 1 \\ 2 \\ 5 \end{matrix} \tag{7.60a}$$

$$F_i^{[2]} = \sqrt{3}\begin{Bmatrix} 1 \\ 1 \\ 1 \end{Bmatrix} \begin{matrix} 1 \\ 2 \\ 5 \end{matrix} \tag{7.60b}$$

- **Element [3]**

We find

$$K_{ij}^{[3]} = \frac{1}{6\sqrt{3}}\begin{bmatrix} 3 & -3 & 0 \\ -3 & 12 & -9 \\ 0 & -9 & 9 \end{bmatrix} \begin{matrix} 5 \\ 4 \\ 1 \end{matrix} \tag{7.61a}$$

$$F_i^{[3]} = \sqrt{3}\begin{Bmatrix} 1 \\ 1 \\ 1 \end{Bmatrix} \begin{matrix} 5 \\ 4 \\ 1 \end{matrix} \tag{7.61b}$$

- **Element [4]**

We find

$$K_{ij}^{[4]} = \frac{1}{6(2-\sqrt{3})}\begin{bmatrix} 16-4\sqrt{3} & -7+4\sqrt{3} & -9 \\ -7+4\sqrt{3} & 7-4\sqrt{3} & 0 \\ -9 & 0 & 9 \end{bmatrix} \begin{matrix} 4 \\ 5 \\ 6 \end{matrix} \tag{7.62a}$$

$$F_i^{[4]} = (2-\sqrt{3})\begin{Bmatrix} 1 \\ 1 \\ 1 \end{Bmatrix} \begin{matrix} 4 \\ 5 \\ 6 \end{matrix} \tag{7.62b}$$

The local contributions in (7.59)-(7.62) are now superimposed into the overall system matrices as in (7.38). We find, after also inserting the essential boundary conditions,

$$\begin{bmatrix} 12 & -3 & 0 & -9 & 0 & 0 \\ -3 & 24 & -3 & 0 & -18 & 0 \\ 0 & -3 & 3 & 0 & 0 & 0 \\ -9 & 0 & 0 & 70.64 & -3.46 & -58.17 \\ 0 & -18 & 0 & -3.46 & 21.46 & 0 \\ 0 & 0 & 0 & -58.17 & 0 & 9 \end{bmatrix} \begin{Bmatrix} \tilde{u}_1 \\ \tilde{u}_2 \\ 0 \\ \tilde{u}_4 \\ 0 \\ 0 \end{Bmatrix} = \begin{Bmatrix} 36 \\ 36 \\ 18 \\ 20.78 \\ 56.78 \\ 2.78 \end{Bmatrix}$$

Of these six equations only three are valid – the first, second and fourth. When these three equations are extracted we find

$$\begin{bmatrix} 12 & -3 & -9 \\ -3 & 24 & 0 \\ -9 & 0 & 70.64 \end{bmatrix} \begin{Bmatrix} \tilde{u}_1 \\ \tilde{u}_2 \\ \tilde{u}_4 \end{Bmatrix} = \begin{Bmatrix} 36 \\ 36 \\ 20.78 \end{Bmatrix}$$

with solution

$$\tilde{u}_1 = 4.117 \qquad \tilde{u}_2 = 2.0147 \qquad \tilde{u}_4 = 0.818$$

These nodal approximations can be compared with the exact solution to this boundary value problem which is

$$u(x, y) = -3.6(x^2/36 + y^2/4 - 1)$$

which gives at the nodes:

$$u(0,0) = 3.6 \qquad u(3,0) = 2.7 \qquad u(0, \sqrt{3}) = 0.9$$

The approximate solutions are only moderately accurate. This is almost entirely due to two factors. Firstly we have used only four elements. A realistic application would use many more elements for a problem of this type. Secondly, the four elements used have not accurately approximated the curved boundary of the region. Much better approximations using only a few elements would be obtained by using six-noded triangular elements which can accommodate curved boundaries. However, we cannot hope to obtain highly accurate approximations without the use of a computer.

7.7 Matrix formulation of finite elements in plane strain

In the final section of Chapter 5 we examined briefly the application of triangular elements to a simple problem in plane strain. In this section we derive a matrix formulation of finite elements in plane strain valid for any type of element. Before studying this rather taxing section, which necessarily involves substantial algebra, the reader is urged to re-read Sections 5.5 and 5.10.

Suppose a region A is segmented into 'nel' finite elements each element possessing 'num' nodes with shape functions $S_i(x,y)$, $i = 1, 2, \ldots, num$. Within each element it is assumed

$$\tilde{u}(x,y) = \sum_{i=1}^{num} \tilde{u}_{<i>} S_i(x,y) \qquad \tilde{v}(x,y) = \sum_{i=1}^{num} \tilde{v}_{<i>} S_i(x,y) \tag{7.63}$$

(Note that, by using the $<i>$ notation, introduced in Section 7.4.1, we have distinguished between global and local node numbers.) The element formulation (7.63) can be re-expressed in terms of roof functions $R_i(x,y)\, i = 1, 2, \ldots, nde$, ($nde$ being the total number of nodes used):

$$\tilde{u}(x,y) = \sum_{i=1}^{nde} \tilde{u}_i R_i(x,y) \qquad \tilde{v}(x,y) = \sum_{i=1}^{nde} \tilde{v}_i R_i(x,y) \tag{7.64}$$

in which

$$R_i(x,y) = \begin{cases} S_{<i>}(x,y) & \text{any } [e] \text{ attached to node } i \\ 0 & \text{elsewhere} \end{cases} \tag{7.65}$$

Now, using the matrix formulation of the plane strain equations (5.24) we have, as the approximation to the displacement vector,

$$\{\tilde{d}\} = \begin{Bmatrix} \tilde{u}(x,y) \\ \tilde{v}(x,y) \end{Bmatrix} = \sum_{i=1}^{nde} \{\tilde{d}\}_i R_i(x,y) \tag{7.66}$$

where

$$\{\tilde{d}\}_i = \begin{Bmatrix} \tilde{u}_i \\ \tilde{v}_i \end{Bmatrix}$$

is the nodal displacement for node i.

The expression for the strain components is, from (5.26),

$$\{\tilde{s}\} = [\partial]\{\tilde{d}\} = \sum_{i=1}^{nde} [\partial] R_i \{\tilde{d}\}_i$$
$$= \sum_{i=1}^{nde} [\partial R_i]\{\tilde{d}\}_i \tag{7.67}$$

in which, from (5.22),

$$[\partial R_i] = \begin{bmatrix} \frac{\partial R_i}{\partial x} & 0 \\ 0 & \frac{\partial R_i}{\partial y} \\ \frac{\partial R_i}{\partial y} & \frac{\partial R_i}{\partial x} \end{bmatrix} \tag{7.68}$$

The stress components are then obtained:

$$\{\tilde{\Omega}\} = [C]\{\tilde{s}\}$$
$$= \sum_{j=1}^{nde} [C][\partial R_j]\{\tilde{d}\}_j \tag{7.69}$$

Hence the strain energy in plane strain is

$$\frac{h}{2} \iint_A \{\tilde{s}\}^T \{\tilde{\Omega}\}\ dxdy$$
$$= \frac{h}{2} \iint_A \{ \sum_{i=1}^{nde} \{\tilde{d}\}_i^T [\partial R_i]^T \sum_{j=1}^{nde} [C][\partial R_j]\{\tilde{d}\}_j \}\ dxdy$$
$$= \frac{h}{2} \sum_{i=1}^{nde} \sum_{j=1}^{nde} \iint_A \{\tilde{d}\}_i^T [\partial R_i]^T [C][\partial R_j]\{\tilde{d}\}_j\ dxdy \tag{7.70}$$

Thus, in the absence of body forces, we obtain an expression for the potential energy (from (5.31)):

$$U = \frac{h}{2} \sum_{i=1}^{nde} \sum_{j=1}^{nde} \iint_A \{\tilde{d}\}_i^T [\partial R_i]^T [C][\partial R_i]\{\tilde{d}\}_j\ dxdy$$
$$- h \sum_{i=1}^{nde} \int_{\Gamma_0} \{\tilde{d}\}_i^T R_i \{T\}\ d\ell \tag{7.71}$$

The nodal displacement components $\{\tilde{d}_i\}$ are now chosen so that U is minimised. In this calculation use is made of results to be obtained in the supplement to Chapter 8 (amended slightly to encompass the formalism of this chapter):

$$\frac{\partial U}{\partial\{\tilde{d}_k\}} = h\sum_{j=1}^{nde}\iint_A [\partial R_k]^T[C][\partial R_j]\{\tilde{d}_j\}\,dx\,dy$$

$$-h\int_{\Gamma_0} R_k\{T\}\,d\ell = 0 \qquad k = 1,2,\ldots,nde \quad \textbf{(7.72)}$$

This is a matrix system of $2(nde)$ equations in the $2(nde)$ unknowns $\{\tilde{u}_i, \tilde{v}_i\}^T$ $i = 1,2,\ldots,nde$, and it may be re-expressed in the form

$$[G]\{\tilde{D}\} = \{H\} \qquad \textbf{(7.73)}$$

in which $[G]$ is a symmetric matrix

$$[G] = \begin{bmatrix} [a]_{11} & [a]_{12} & \cdots & [a]_{1\,nde} \\ [a]_{21} & [a]_{22} & \cdots & [a]_{2\,nde} \\ & & & \\ [a]_{nde\,1} & [a]_{nde\,2} & \cdots & [a]_{nde\,nde} \end{bmatrix} \qquad \textbf{(7.74a)}$$

and $\{\tilde{D}\}, \{H\}$ are column vectors:

$$\{\tilde{D}\} = \begin{Bmatrix} \tilde{u}_1 \\ \tilde{v}_1 \\ \cdot \\ \cdot \\ \tilde{u}_{nde} \\ \tilde{v}_{nde} \end{Bmatrix} \qquad \{H\} = \begin{Bmatrix} \{F\}_1 \\ \{F\}_2 \\ \cdot \\ \cdot \\ \{F\}_{nde} \end{Bmatrix} \qquad \textbf{(7.74b)}$$

where

$$[a]_{ij} = \begin{bmatrix} \alpha_{ij} & \beta_{ij} \\ \gamma_{ij} & \delta_{ij} \end{bmatrix} \qquad \{F\}_i = \begin{Bmatrix} F_{1i} \\ F_{2i} \end{Bmatrix} \qquad \textbf{(7.75)}$$

in which

$$\alpha_{ij} = \alpha_{ji} = \iint_A \left\{(2\mu+\lambda)\frac{\partial R_i}{\partial x}\frac{\partial R_j}{\partial x} + \mu\frac{\partial R_i}{\partial y}\frac{\partial R_j}{\partial y}\right\} dx\,dy$$

$$\gamma_{ji} = \beta_{ij} = \iint_A \left\{ \lambda \frac{\partial R_i}{\partial x} \frac{\partial R_j}{\partial y} + \mu \frac{\partial R_i}{\partial y} \frac{\partial R_j}{\partial x} \right\} dx dy$$

$$\delta_{ij} = \delta_{ji} = \iint_A \left\{ (2\mu + \lambda) \frac{\partial R_i}{\partial y} \frac{\partial R_j}{\partial y} + \mu \frac{\partial R_i}{\partial x} \frac{\partial R_j}{\partial x} \right\} dx dy$$

and

$$F_{1i} = \int_{\Gamma_0} T_1 R_i \, d\ell \qquad F_{2i} = \int_{\Gamma_0} T_2 R_i \, d\ell$$

(The details of this reformulation should be verified by the reader.)

The evaluation of each term of $[G]$ and $\{H\}$ follows exactly as in other finite element problems. Each double integral is written as a sum of integrals taken over each element. Within each element the roof function is replaced by the appropriate shape function. Each of the 'element' integrals can be evaluated by transforming to a standard region and employing Gaussian quadrature techniques.

The displacement boundary conditions are then imposed, again in the normal manner, to amend the system of equations $[G]\{\tilde{D}\} = \{H\}$ into a valid set. From (5.26), specific values of $\{\tilde{d}\}_i$ can be found for those nodes lying on Γ_1. (Those equations, obtained in the minimization process, which correspond to nodes on Γ_1 will thus be invalid.)

7.7.1 Unstressed circular hole in a plate under tension

As a specific example of the application of the finite element technique to plane elasticity we consider the stress analysis of a plate with a small circular hole of radius a. The plate is stretched by tensile forces along the y-direction as shown in Figure 7.15.

We choose to analyse this problem not only because it is an important one in the field of elasticity (with significant applications to the understanding of plate weakening due to small imperfections) but also because the finite element solution can be compared with a closely related exact solution of the plane strain equations. The exact solution applies to a hole in an **infinite** plate so that the finite element solution will be expected to approximate the exact solution only if the diameter of the hole is 'small' compared with the dimensions of the plate.

In this problem the stress around the hole is of particular interest, and it is convenient to express the stress components in terms of polar coordinates. It is found that the exact solution for these components is

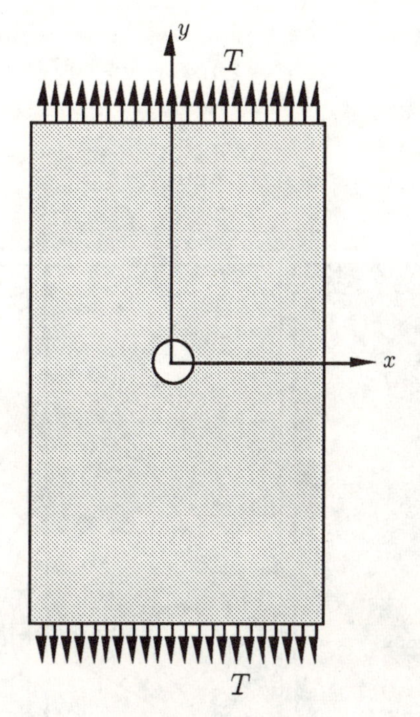

Figure 7.15

$$\sigma_{rr} = \frac{T}{2}\left[\left(1 - \frac{a^2}{r^2}\right) + \left(1 - \frac{4a^2}{r^2} + \frac{3a^4}{r^4}\right)\cos 2\theta\right]$$

$$\sigma_{\theta\theta} = \frac{T}{2}\left[\left(1 + \frac{a^2}{r^2}\right) - \left(1 + \frac{3a^4}{r^4}\right)\cos 2\theta\right]$$

$$\sigma_{r\theta} = -\frac{T}{2}\left[1 + \frac{2a^2}{r^2} - \frac{3a^4}{r^4}\right]\sin 2\theta$$

Of particular interest, from a physical point of view, is the stress at the hole boundary ($r = a$) at which only the so-called 'hoop stress' $\sigma_{\theta\theta}$ is non-zero, this having a value

$$\sigma_{\theta\theta}\big|_{r=a} = T(1 - 2\cos 2\theta)$$

which assumes a maximum value of $3T$ at $\theta = \pi/2,\ 3\pi/2$. Thus the presence of the hole, no matter how small, leads to a threefold increase in stress levels at these particular points. Under large loading the plate will begin to tear apart at these points.

Particular aspects of this exact solution can be compared with experiment. The photoelastic fringe pattern of a stretched plate with a small hole is shown in Figure 7.16(a)*. The dark areas locate the lines along which the maximum shear stress takes constant values. These are the so-called isochromatic curves:

$$(\sigma_{rr} - \sigma_{\theta\theta})^2 + 4\ \sigma_{r\theta}^2 = \text{constant}$$

photoelastic results (a) exact solution (b) finite element solution (c)

Figure 7.16

If the isochromatic curves are plotted from the exact solution we obtain the contour plot shown in Figure 7.16(b). The gross features of the two sets of curves correspond closely. A detailed calculation using the photoelastic data confirms the maximum hoop stress levels predicted from the exact solution.

In the finite element modelling of this problem we exploit the inherent symmetry and only consider a quarter of the plate (see Figure 7.17).

Along the 'cut-edges' appropriate boundary conditions, consistent with the original configuration, must be imposed. These can be deduced using symmetry arguments.

For the full plate the displacement component u is an odd function of x and an even function of y whilst the component v is an even function of x and an odd function of y.

---

* Kindly supplied by K. Sharples of Sharples Engineering, Bamber Bridge, Preston England.

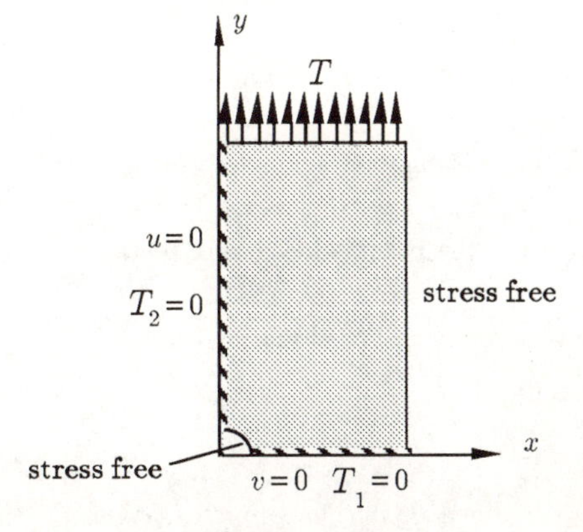

Figure 7.17

$$\text{Hence} \quad \sigma_{12} = 2\mu\varepsilon_{12} = \frac{1}{2}\left(\frac{\partial u}{\partial y} + \frac{\partial v}{\partial x}\right) \quad \text{is odd in } x \text{ and } y.$$

It follows that along the left cut-edge ($x = 0$) both u and $T_2(= -\sigma_{21})$ are zero on this part of the boundary. Similarly both v and T_1 are zero along $y = 0$. Figure 7.17 specifies in full the boundary conditions to be applied on this reduced problem. Using 64 eight-noded elements graded in size (with small elements near the hole, and larger ones further away) we can obtain a numerical solution to the problem. From this solution for the displacement components the stresses and strains can be determined. If we then plot the isochromatic curves we obtain the contour plot shown in Figure 7.16(c). In this plot we have displayed the region near the hole. The gross features of the curves are in excellent agreement with the exact solution to the problem.

The exact solution is relatively easy to obtain here due to the high degree of symmetry present both in the region A and also in the loading conditions. If no such symmetry is present an exact solution would be very difficult, if not impossible, to obtain. In such cases finite element analysis is often an excellent alternative. In fact, in most practical plane strain problems the finite element method, together with the experimental photoelastic technique, offers the only realistic method of analysis.

EXERCISES

7.1 Using a simple one-term multinomial determine, by the Galerkin method, an approximation to the so-called 'torsion problem':

$$\frac{\partial^2 u}{\partial x^2} + \frac{\partial^2 u}{\partial y^2} = -2 \qquad \text{in } A$$

$$u = 0 \quad \text{on } \Gamma \text{ the boundary of } A$$

in which A is the rectangular region $-a \leq x \leq a, \; -b \leq y \leq b$.

7.2 Rework Q7.1 using the modified Galerkin method.

7.3 Using the two-parameter approximation

$$\tilde{u} = a_1 xy + a_2 xy(x + y)$$

determine a modified Galerkin approximation to the boundary value problem

$$-\frac{\partial^2 u}{\partial x^2} - \frac{\partial^2 u}{\partial y^2} = 1 \qquad \text{within } D$$

$$u = 0 \quad \text{on } x = 0 \text{ and on } y = 0$$

$$\frac{\partial u}{\partial n} = 0 \quad \text{on } x + y - 1 = 0$$

Here D is the triangular region bounded by $x = 0$, $y = 0$ and $y = 1 - x$.

7.4 Deduce the four element matrices in (7.45).

7.5 Deduce fully the element matrix $[K^{[e]}]$ in (7.48) and the vector $\{F^{[e]}\}$ in (7.49) for the three noded triangle.

7.6 Consider the boundary value problem:

$$\frac{\partial^2 u}{\partial x^2} + \frac{\partial^2 u}{\partial y^2} = 0 \qquad 0 \leq x, y \leq 1$$

with all boundary conditions essential.

Using the four element model shown in Figure Q7.6 deduce that

$$\tilde{u}_3 = \frac{1}{4}[\tilde{u}_1 + \tilde{u}_2 + \tilde{u}_4 + \tilde{u}_5]$$

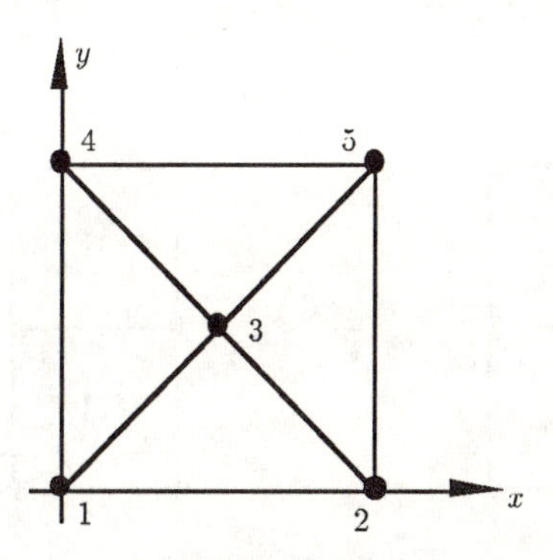

Figure Q7.6

7.7 Show that for a rectangular element

$$x_1 \le x \le x_1 + 2b \qquad y_1 \le y \le y_1 + 2a$$

Poisson's equation

$$\frac{\partial^2 u}{\partial x^2} + \frac{\partial^2 u}{\partial y^2} + h = 0$$

gives rise to an element matrix

$$[K^{[e]}] = \frac{a}{6b}\begin{bmatrix} 2 & -2 & -1 & 1 \\ -2 & 2 & 1 & -1 \\ -1 & 1 & -1 & -2 \\ 1 & -1 & -2 & 2 \end{bmatrix} + \frac{b}{6a}\begin{bmatrix} 2 & 1 & -1 & -2 \\ 1 & 2 & -2 & -1 \\ -1 & 2 & 2 & 1 \\ -2 & 1 & 1 & 2 \end{bmatrix}$$

and, if h is constant, to a vector

$$\{F^{[e]}\} = \frac{hA}{4}\begin{Bmatrix} 1 \\ 1 \\ 1 \\ 1 \end{Bmatrix} \qquad \text{where } A = 4ab.$$

7.8 Consider the boundary value problem

$$\frac{\partial^2 u}{\partial x^2} + \frac{\partial^2 u}{\partial y^2} = -1 \qquad 0 \le x, y \le 1$$

$$u(1, y) = u(x, 1) = 0 \qquad \frac{\partial u}{\partial x}(0, y) = \frac{\partial u}{\partial y}(x, 0) = 0$$

Solve the problem for each of the finite element models shown in Figure Q7.8.

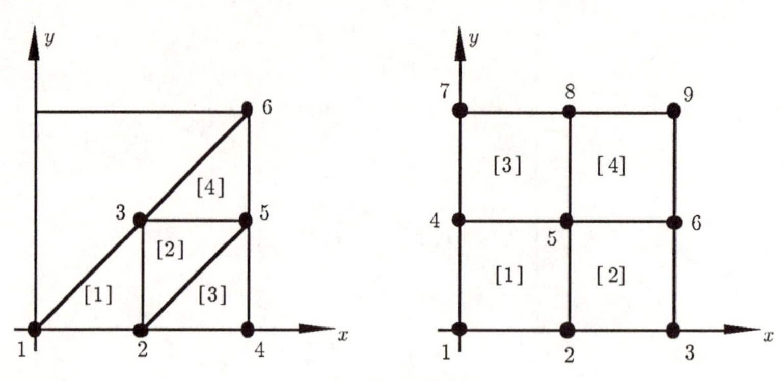

Figure Q7.8

(Note the use of symmetry in connection with the triangular elements. The boundary condition on the line of symmetry $y = x$ is $\dfrac{\partial u}{\partial n} = 0$.)

7.9 The Poisson equation:

$$\frac{\partial^2 u}{\partial x^2} + \frac{\partial^2 u}{\partial y^2} + h = 0$$

with $h = 2790$ and all boundary conditions (on u) zero arises in the twisting of the square shaft shown in Figure Q7.9(a). Using symmetry and the four triangular element mesh shown determine the three unknown nodal values.

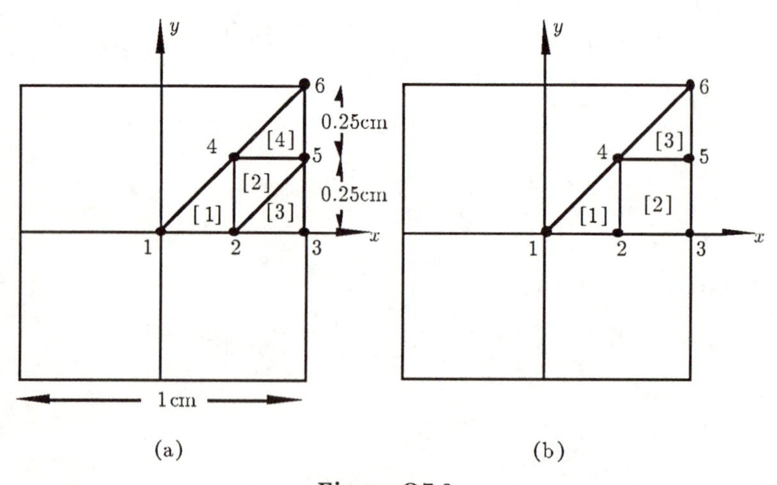

Figure Q7.9

7.10 Repeat Q7.9 for the mesh shown in in Figure Q7.9(b). (Note now the use of distinct element types. This is perfectly acceptable as the elements are compatible in the sense that the unknown function $\tilde{u}(x, y)$ is continuous across common element boundaries due to the fact that both elements used, triangular and quadrilateral, have the same number of nodes 'per edge'. This point is discussed more fully in the Appendix.)

7.11 Repeat Q7.9 for the four rectangular elements shown in Figure Q7.11.

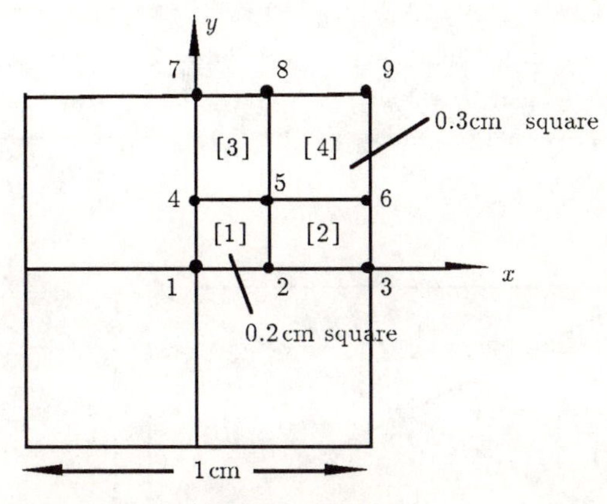

Figure Q7.11

7.12 Consider the two-dimensional Helmholtz equation

$$\frac{\partial^2 u}{\partial x^2} + \frac{\partial^2 u}{\partial y^2} + \beta^2 u = 0$$

(i) Show that for a three noded triangular element of area A, a contribution, say $[K_\beta^{[e]}]$, is made to the element matrix $[K^{[e]}]$ by the final term in the equation where

$$[K_\beta^{[e]}] = \frac{\beta^2 A}{12}\begin{bmatrix} 2 & 1 & 1 \\ 1 & 2 & 1 \\ 1 & 1 & 2 \end{bmatrix}$$

(ii) Repeat for a rectangular element

$$x_1 \leq x \leq x_1 + 2b \qquad y_1 \leq y \leq y_1 + 2a$$

to show

$$[K_\beta^{[e]}] = \frac{\beta^2 ab}{9}\begin{bmatrix} 4 & 2 & 1 & 2 \\ 2 & 4 & 2 & 1 \\ 1 & 2 & 4 & 2 \\ 2 & 1 & 2 & 4 \end{bmatrix}$$

7.13 Two-dimensional acoustic vibrations are governed by the boundary value problem:

$$\frac{\partial^2 u}{\partial x^2} + \frac{\partial^2 u}{\partial y^2} + \frac{\omega^2}{c^2}u = 0 \qquad (x, y) \text{ within } A$$

$$\frac{\partial u}{\partial n} = 0 \text{ on } \Gamma \text{ the boundary of } A.$$

Formulate, for the rectangular region shown in Figure Q7.13 and the four triangular element model shown, a system of equations $[K]\{\tilde{u}\} = \{0\}$. Do not solve the equations unless you have a computer program available.

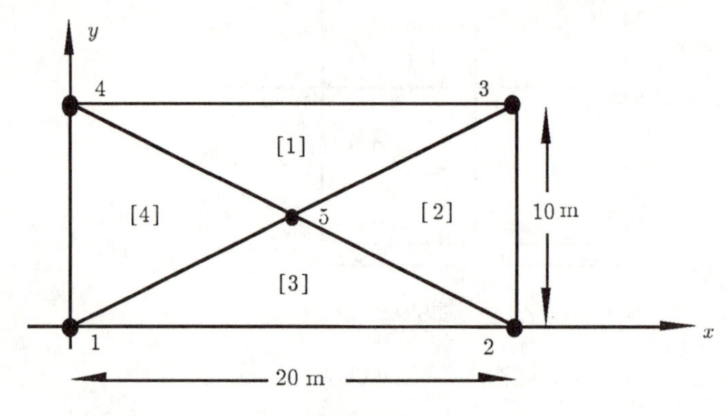

Figure Q7.13

7.14 Rework Example 7.3 by

(a) replacing element [1] by two smaller elements as shown in Figure Q7.14(a)

(b) replacing elements [2], [3] by a single four-noded quadrilateral as in Figure Q7.14(b)

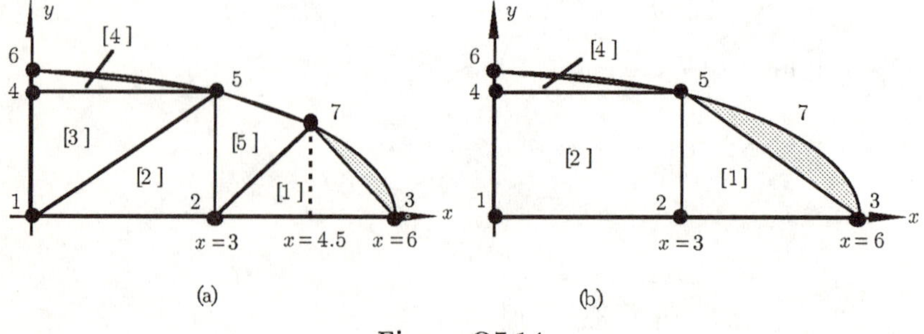

Figure Q7.14

7.15 Consider the six-noded standard triangular element shown in Figure Q7.15.

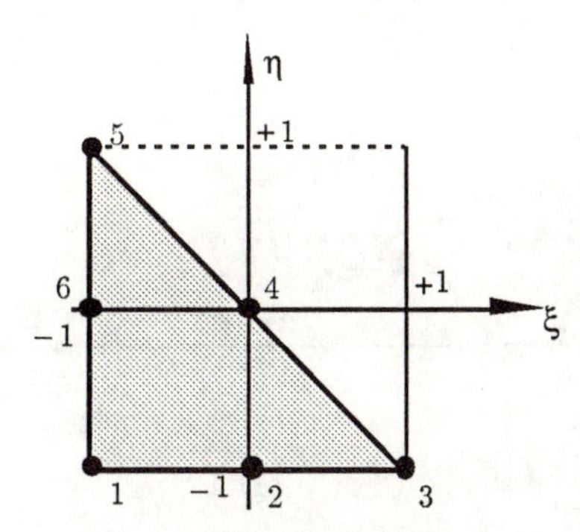

Figure Q7.15

The shape functions for this element are (see Appendix)

$$Q_1(\xi,\eta) = \frac{1}{2}(\xi+\eta)(\xi+\eta+1) \qquad Q_2(\xi,\eta) = -(\xi+1)(\xi+\eta)$$

$$Q_3(\xi,\eta) = \frac{1}{2}\xi(\xi+1) \qquad Q_4(\xi,\eta) = -(\xi+1)(\eta+1)$$

$$Q_5(\xi,\eta) = \frac{1}{2}\eta(\eta+1) \qquad Q_7(\xi,\eta) = -(\eta+1)(\xi+\eta)$$

(i) If (ξ_i, η_i) are the nodal coordinates verify the following properties:

(a) $\sum_{i=1}^{6} Q_i(\xi,\eta) = 1$

(b) $\sum_{i=1}^{6} \xi_i Q_i(\xi,\eta) = \xi \qquad \sum_{i=1}^{6} \eta_i Q_i(\xi,\eta) = \eta$

(c) $\sum_{i=1}^{6} \xi_i^2 Q_i(\xi,\eta) = \xi^2 \qquad \sum_{i=1}^{6} \eta_i \xi_i Q_i(\xi,\eta) = \xi\eta \qquad \sum_{i=1}^{6} \eta_i^2 Q_i(\xi,\eta) = \eta^2$

(d) $Q_i(\xi,\eta) = \begin{cases} 1 & \text{at node } i \\ 0 & \text{at every other node} \end{cases}$

(ii) Examine the effect of the transformation:

$$x = \sum_{i=1}^{6} x_i Q_i(\xi,\eta) \qquad y = \sum_{i=1}^{6} y_i Q_i(\xi,\eta)$$

on the standard triangle

7.16 Rework Example 7.3 using a single six-noded triangular element (cf Q7.15) as in Figure Q7.16.

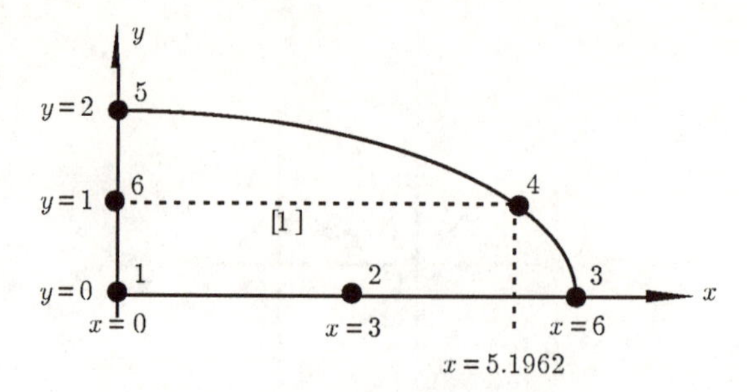

Figure Q7.16

Show that the Jacobian of the transformation for this element is

$$J = 2.1962\eta + 5.1962$$

(Does this vanish within the standard element?) Also show that the relations between the derivatives in the xy and $\xi\eta$ variables are:

$$\frac{\partial}{\partial x} = \frac{1}{J}\frac{\partial}{\partial \xi}, \qquad \frac{\partial}{\partial y} = -\frac{2.1962(1+\xi)}{J}\frac{\partial}{\partial \xi} + \frac{\partial}{\partial \eta}$$

Convince yourself that with a complete reworking of Example 7.3 using this single element the integrals which arise could be determined analytically but only after much tedious algebra.

8

Variational Formulation of Boundary Value Problems

PREVIEW

In this chapter we study an alternative integral formulation of boundary value problems which is widely used in connection with the finite element method. This new formulation is based upon variational calculus. An approximate variationally based technique due to Ritz is discussed and used to solve one- and two-dimensional linear boundary value problems. The equivalence of the Ritz method to the modified Galerkin technique for such problems is demonstrated.

8.1 Introduction

This text deals primarily with the finite element method as an approximate numerical technique for solving boundary value problems. The approach that we have adopted, both for one-dimensional and two-dimensional problems, has been based on the idea of weighted residuals and specifically the method of Galerkin. We have chosen to emphasize this approach as it offers the most straightforward method of obtaining the final system of equations. The Galerkin method is essentially an integral formulation but it is not the only such formulation. Another popular integral re-formulation of a boundary value problem is based on a topic known as the **variational calculus.**

In this chapter we will discuss variational formulation of boundary value problems and a widely used approximate technique – the Ritz method – for solving variational problems. We will show that the method can be used in conjunction with finite elements and discuss in detail the case of linear second-order differential equations. We shall show that the Ritz and Galerkin formulations are essentially equivalent for this case in that each method gives rise to the same system of linear equations for the same approximate trial solution.

8.2 Functionals and variational calculus

As is well-known, a **function** is essentially a rule which transforms a number into another number. For example the function $\sin x$ transforms any number x into a number y, where $y = \sin x$. The set of values of y (here $-1 \leq y \leq +1$) is called the range of the function. In many applications it is important to know the maximum and minimum values of y and the values of x at which they occur. To determine local maxima and minima the techniques of differential calculus are often used and the equation

$$\frac{dy}{dx} = 0$$

is solved to determine possible extreme values of y.

A **functional** is a generalization of a function in that it is a rule which transforms a function into a number. Perhaps the most common example is a definite integral:

$$J[\phi] = \int_a^b \phi(x)\,dx$$

where the limits a and b are given. (The notation $J[\phi]$ is common for a functional.) Clearly $\phi(x)$ is any function which we can integrate between

a and b.

If, for example, $a = 0$ and $b = 1$ so that the functional is

$$J[\phi] = \int_0^1 \phi(x)\, dx$$

then $\phi(x) = \sin x$ gives $J[\phi] = 1 - \cos 1 = 0.4597$ whilst choosing $\phi(x) = x^2$ gives $J[\phi] = 1/3$ and so on. The transformation from a function to a number is clear. The question of particular interest to us is which particular function $\phi(x)$ (for which $J[\phi]$ is defined) gives the minimum (or maximum) value of $J[\phi]$.

An important type of functional that we need to study is

$$J[\phi] = \int_a^b F(x, \phi, \phi')\, dx \qquad \textbf{(8.1)}$$

where F is a **given** function of its arguments $x, \phi(x)$ and $\phi'(x) \equiv d\phi/dx$. The precise set of functions $\phi(x)$ for which $J[\phi]$ in (8.1) is defined will depend on the given function F but will normally include only functions which are differentiable between a and b. We shall refer to such functions as **admissible**.

A number of functionals of the form (8.1) can be quoted:

(i) $$J[\phi] = \int_a^b \sqrt{1 + (\phi')^2}\, dx$$

which is the length of the curve $y = \phi(x)$ between $x = a$ and $x = b$

(ii) $$J[\phi] = 2\pi \int_a^b \phi\sqrt{1 + (\phi')^2}\, dx$$

The interpretation of $J[\phi]$ here is that it is the area, between the planes $x = a$ and $x = b$, of the surface obtained when the curve $y = \phi(x)$ is rotated about the x-axis.

(iii) $$J[\phi] = \frac{1}{\sqrt{2g}} \int_a^b \frac{\sqrt{1 + (\phi')^2}}{\sqrt{\phi}}\, dx$$

The interpretation of $J[\phi]$ in this case comes from a problem in elementary mechanics: given a frictionless wire whose equation is $y = \phi(x)$, calculate the time taken for a particle to slide between the points $(x = a, y = A)$ and $(x = b, y = B)$. This time clearly depends on the particular path $y = \phi(x)$ involved and the basic problem is to determine which path minimizes the sliding time. (A child's slide made using this curve would give the fastest ride.) This problem was one of the first studied in the subject of variational calculus which is concerned with minimizing (or maximizing) functionals. The reason for studying this topic in a finite element text is that, as we shall demonstrate shortly, finding the extreme values (minima or maxima) of a functional is equivalent to solving a boundary value problem.

Example 8.1

Show that minimizing the functional

$$J[\phi] = \int_0^1 (\phi^2 + (\phi')^2)\, dx$$

with $\phi(0) = u_0$ and $\phi(1) = u_1$ is equivalent to solving a boundary value problem.

Solution

The problem is to find the specific function say $u(x)$ (from the set of admissible functions for which $J[\phi]$ is defined) which minimizes the value of J. That is, u must be such that

$$J[u] < J[\phi], \quad \text{all} \quad \phi \neq u$$

We proceed by writing

$$\phi(x) = u(x) + \varepsilon\, \eta\,(x) \tag{8.2}$$

where ε is a parameter independent of x. The function $\eta(x)$ is an almost arbitrary function but it must satisfy

$$\eta(0) = 0 \quad \text{and} \quad \eta(1) = 0 \tag{8.3}$$

since $\phi(x) = u(x)$ (for all ϕ) at $x = 0$ and $x = 1$. Geometrically, we are considering a range of curves $\phi(x)$ in the 'close neighbourhood' of the function we seek. Each 'comparison curve' must, of course, pass through the points $(0, u_0); (1, u_1)$ as this is a constraint imposed on the functional (see Figure 8.1).

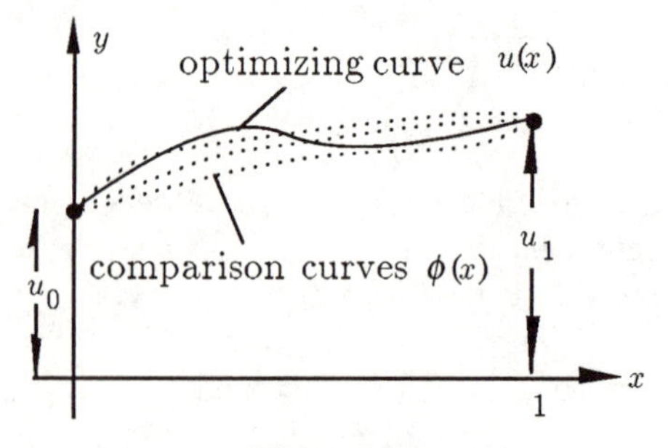

Figure 8.1

We substitute the assumed form (8.2) into the given functional producing

$$J[u+\varepsilon\,\eta]=\int_0^1\left\{(u+\varepsilon\eta)^2+(u'+\varepsilon\eta')^2\right\}dx \tag{8.4}$$

which, when we carry out the integration, is a function only of the parameter ε and will be denoted by $\Phi(\varepsilon)$.

We find

$$\Phi(\varepsilon)=\int_0^1\left\{u^2+(u')^2+(\varepsilon\eta)^2+(\varepsilon\eta')^2+2\varepsilon(u\eta+u'\eta')\right\}\,dx$$

$$=J[u]+\varepsilon^2J[\eta]+2\varepsilon\int_0^1(u\eta+u'\eta')\,dx$$

so that

$$\frac{d\Phi}{d\varepsilon}=2\varepsilon J[\eta]+2\int_0^1(u\eta+u'\eta')\,dx$$

If this is to be a minimum when $\varepsilon=0$ we have from elementary calculus

$$\int_0^1(u\eta+u'\eta')\,dx=0$$

Integrating the second term by parts and using (8.3) we obtain

$$\int_0^1(u\eta-u''\eta)dx=0$$

or

$$\int_0^1\eta(u-u'')dx=0$$

from which, since $\eta(x)$ is almost arbitrary, it seems plausible to assume

$$u''-u=0 \qquad 0\le x\le 1$$

(In fact, as shown in the first supplement to Chapter 2, a rigorous deduction of this result is based on the fundamental lemma of the variational calculus.) This differential equation, together with the given conditions $u(0)=u_0$ and $u(1)=u_1$, constitutes a standard boundary value problem which is readily solved.
We now do another, but more difficult, example showing the equivalence of minimizing a functional to solving a boundary value problem.

Example 8.2

Deduce the boundary value problem equivalent to minimizing the functional

$$J[\phi] = \int_1^3 \{-(\phi')^2 - 4\phi\} dx + 12\,\phi(3)$$

where

$$\phi(1) = 1$$

Solution

Clearly the admissible functions $\phi(x)$ must be differentiable and must each satisfy $\phi(1) = 1$.

Proceeding in a similar manner to the previous example, the function $u(x)$ that we seek is obtained by writing any other admissible function $\phi(x)$ as

$$\phi(x) = u(x) + \varepsilon\,\eta\,(x)$$

where $\eta(1) = 0$ but otherwise $\eta(x)$ is arbitrary.

Then

$$\begin{aligned}
\Phi(\varepsilon) &= J[u + \varepsilon\,\eta] \\
&= \int_1^3 \{-(u' + \varepsilon\,\eta')^2 - 4(u + \varepsilon\,\eta)\}\, dx + 12(u(3) + \varepsilon\,\eta(3)) \\
&= \int_1^3 \{-(u')^2 - 4u - 2\varepsilon u'\eta' - (\varepsilon\eta')^2 - 4\varepsilon\eta\}\ dx \\
&\qquad + 12(u(3) + \varepsilon\eta(3)) \\
&= J[u] - \varepsilon\left\{ \int_1^3 \{2u'\eta' + 4\eta\}\ dx - 12\eta(3) \right\} \\
&\qquad - \varepsilon^2 \int_1^3 (\eta')^2\ dx
\end{aligned}$$

Hence

$$\frac{d\Phi}{d\varepsilon} = -\left\{ \int_1^3 \{2u'\eta' + 4\eta\}\ dx - 12\eta(3) \right\} - 2\varepsilon \int_1^3 (\eta')^2\ dx$$

so to obtain $\dfrac{d\Phi}{d\varepsilon} = 0$ when $\varepsilon = 0$ we must have

$$\int_1^3 \{-2u'\eta' - 4\eta\} dx + 12\,\eta\,(3) = 0 \qquad (8.5)$$

The first term in (8.5) can be transformed using integration by parts:

$$\int_1^3 u'\eta' dx = [u'\eta]_1^3 - \int_1^3 u''\eta\, dx$$

$$= u'(3)\eta(3) - \int_1^3 u''\eta\, dx$$

Hence (8.5) can be written

$$\int_1^3 (2u'' - 4)\eta\, dx + (12 - 2u'(3))\,\eta\,(3) = 0 \tag{8.6}$$

We now seek to extract $u(x)$ from (8.6). It is vital to realize that this is not a single equation but must be satisfied for **all** functions $\eta(x)$ such that $\eta(1) = 0$ (and there are infinitely many such functions). Suppose that from the set of all such functions we choose a subset satisfying both $\eta(1) = 0$ **and** $\eta(3) = 0$ (see Figure 8.2).

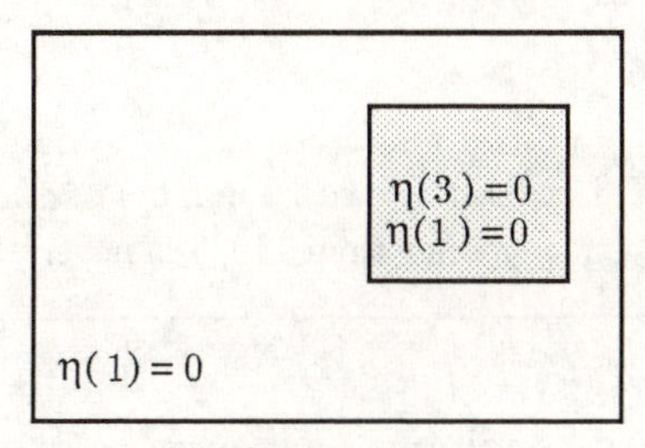

Figure 8.2

For this 'smaller' set of functions, which still contains infinitely many members, (8.6) becomes

$$\int_1^3 (2u'' - 4)\eta\, dx = 0$$

Since $\eta(x)$ here is arbitrary apart from satisfying $\eta(1) = 0$ and $\eta(3) = 0$ it is reasonable to conclude that

$$2u'' - 4 = 0 \qquad 1 \le x \le 3 \tag{8.7}$$

We now revert to (8.6) and to functions for which only $\eta(1) = 0$ is demanded. Using (8.7) this becomes

$$(12 - 2u'(3))\,\eta\,(3) = 0$$

This equation must hold for **all** functions $\eta(x)$ in the set of functions satisfying $\eta(1) = 0$ and not just for those in the smaller set satisfying $\eta(3) = 0$. Consequently

$$u'(3) = 6 \tag{8.8}$$

In other words we have shown that the particular function $u(x)$ (taken from the general class of functions satisfying $\phi(1) = 1$) which minimizes the given functional is the solution to the boundary value problem

$$u'' = 2 \qquad 1 \le x \le 3 \tag{8.9a}$$

with (from the original constraint $\phi(1) = 1$)

$$u(1) = 1 \tag{8.9b}$$

and

$$u'(3) = 6 \tag{8.9c}$$

Clearly the exact solution to this problem is

$$u(x) = x^2$$

which yields a minimum value for the given functional of

$$J[u] = \int_1^3 (-4x^2 - 4x^2)dx + (12)(9)$$

$$= \frac{116}{3}$$

The reader should experiment with a number of other admissible functions to show that a value smaller than 116/3 is never obtained.

The boundary conditions that arise in the above example are noteworthy. The first condition (8.9b) was specified at the outset and all the admissible functions $\phi(x)$ had to satisfy this condition. We have already referred to such a boundary condition by the name **essential condition**. On the other hand the condition (8.9c) was **not** known at the outset and not all the admissible functions $\phi(x)$ need satisfy it. Again we have already given a name to a boundary condition of this type viz. **suppressible.**

8.3 Approximate solution to variational problems

We have now demonstrated a correspondence between minimizing a functional and finding the solution to a boundary value problem. It is not surprising therefore that we can find an approximate solution to a boundary value problem by working with the corresponding functional.

In a one-dimensional finite element problem this involves segmenting the domain into elements, using low order polynomial approximations to the unknown function $u(x)$ within each element and choosing the parameters in the polynomials so as to minimize the functional. This is essentially the **Ritz method** which we will discuss more formally later.

Example 8.3

Find an approximate solution, using two linear elements, to the boundary value problem (8.9) using the corresponding functional.

Solution

(Note that we have already, in Sections 4.1 and 4.2, used a two-element approximation and a weighted residual approach to this problem.)

The corresponding variational problem has been shown to be that of minimizing the functional

$$J[\phi] = \int_1^3 \{-(\phi')^2 - 4\phi\}\, dx + 12\,\phi\,(3) \tag{8.10}$$

with $\phi(1) = 1$.

We use the two-element model shown in Figure 4.1. As usual we let $\tilde{u}$ be the approximation to $u(x)$ which is the function that optimizes $J[\phi]$.

The approximate solutions are, in terms of the nodal values $\tilde{u}_1, \tilde{u}_2, \tilde{u}_3$,

$$\tilde{u}^{[1]} = (2 - x)\tilde{u}_1 + (x - 1)\tilde{u}_2$$

$$\tilde{u}^{[2]} = (3 - x)\tilde{u}_2 + (x - 2)\tilde{u}_3$$

which have already been obtained as (4.3) and (4.4) respectively.

The segmentation of the domain into elements means that the integral in the functional (8.10) also has to be segmented. The contribution from element [1] where $\tilde{u}' = -\tilde{u}_1 + \tilde{u}_2$ is

$$J_1[\tilde{u}] = \int_1^2 [-(-\tilde{u}_1 + \tilde{u}_2)^2 - 4\{(2 - x)\tilde{u}_1 + (x - 1)\tilde{u}_2\}]dx$$

while due to element [2] where $\tilde{u}' = -\tilde{u}_2 + \tilde{u}_3$ we have

$$J_2[\tilde{u}] = \int_2^3 [-(-\tilde{u}_2 + \tilde{u}_3)^2 - 4\{(3 - x)\tilde{u}_2 + (x - 2)\tilde{u}_3\}]dx$$

Substituting into (8.10), performing the integrations and noting that $u(3)$ is the nodal value $\tilde{u}_3$, we find that

$$\begin{aligned} J[\tilde{u}] = -\tilde{u}_1^2 - 2\,\tilde{u}_2^2 - \tilde{u}_3^2 + 2\,\tilde{u}_1\tilde{u}_2 + 2\,\tilde{u}_2\tilde{u}_3 - 2\,\tilde{u}_1 \\ - 4\,\tilde{u}_2 + 10\,\tilde{u}_3 \end{aligned} \tag{8.11}$$

Inserting the essential boundary condition

$$\tilde{u}(1) \equiv \tilde{u}_1 = 1$$

we have

$$J[\tilde{u}] = -2\,\tilde{u}_2^2 - \tilde{u}_3^2 + 2\,\tilde{u}_2\tilde{u}_3 - 2\,\tilde{u}_2 + 10\,\tilde{u}_3 - 3$$

which depends on the two parameters (nodal values) $\tilde{u}_2, \tilde{u}_3$. These values are determined so that (for this approximation) J has an extreme value. Thus we must solve

$$\frac{\partial J}{\partial \tilde{u}_2} = -4\,\tilde{u}_2 + 2\,\tilde{u}_3 - 2 = 0$$

and

$$\frac{\partial J}{\partial \tilde{u}_3} = 2\tilde{u}_2 - 2\tilde{u}_3 + 10 = 0$$

These equations are the same as (4.9a) and (4.9b) which were obtained by the Galerkin method. They give rise to nodal solutions $\tilde{u}_2 = 4$ and $\tilde{u}_3 = 9$, which are exact.

8.4 Construction of functionals I

In Examples 8.1 and 8.2 we derived the boundary value problems equivalent to two particular variational problems. A number of techniques are available to enable us to do this for more general problems.

Perhaps the most straightforward case is when

$$J[\phi] = \int_a^b F(x, \phi, \phi')dx \qquad \textbf{(8.12)}$$

and ϕ is the set of functions each satisfying $\phi(a) = A$ and $\phi(b) = B$ where A and B are given constants.

For this case, the particular function $u(x)$ which makes $J(\phi)$ an extremum (maximum or minimum) is a solution of the **Euler–Lagrange equation**

$$\frac{\partial F}{\partial u} - \frac{d}{dx}\left(\frac{\partial F}{\partial u'}\right) = 0 \qquad a \leq x \leq b \qquad \textbf{(8.13)}$$

with

$$u(a) = A \quad \text{and} \quad u(b) = B \qquad \textbf{(8.14)}$$

The proof of this result may be found in any book on variational calculus. For our purposes it is sufficient to note that (8.13) and (8.14) constitute the boundary value problem equivalent to the given variational problem. We note that (8.13) is a **second**-order differential equation. The reader is invited to re-work Example 8.1 using (8.13) and (8.14); a straightforward exercise.

The Euler–Lagrange equation can be generalized beyond (8.13). If the integrand in (8.12) is $F(x, \phi, \phi', \phi'', \phi''', \ldots)$ then the corresponding Euler–Lagrange equation can be shown to be

$$\frac{\partial F}{\partial u} - \frac{d}{dx}\left(\frac{\partial F}{\partial u'}\right) + \frac{d^2}{dx^2}\left(\frac{\partial F}{\partial u''}\right) - \ \ldots \ = 0 \quad a \leq x \leq b \qquad \textbf{(8.15)}$$

In general if the highest derivative of u appearing as an argument in F is of nth order then (8.15) is an ordinary differential equation of order $2n$. This is typical of a differential equation equivalent to a variational process – such an equation is always of **even** order.

We now consider the reverse question to the above viz. finding the variational problem equivalent to a given even-ordered boundary value problem. This is, in general, a considerably more difficult process but fortunately is one which has been extensively studied for certain classes of problems. In particular for linear second-order boundary value problems we can give a complete prescription for finding the equivalent variational problem. We omit all proofs due to their technical nature.

Theorem 8.1

The second-order boundary value problem:

$$-\frac{d}{dx}\left(p(x)\frac{du}{dx}\right) + q(x)\,u = r(x) \qquad a \le x \le b \tag{8.16a}$$

$$\begin{aligned} \alpha_0\,u(a) + \alpha_1\,u'(a) &= \gamma_1 \\ \beta_0\,u(b) + \beta_1\,u'(b) &= \gamma_2 \end{aligned} \tag{8.16b}$$

may be written as the necessary conditions to be satisfied by the solution of the variational problem:

$$J[\phi] = \text{extremum}$$

for each of the following cases.

- case 1 $\quad \alpha_1 \ne 0,\ \beta_1 \ne 0$

that is, with suppressible boundary conditions at $x = a$ and $x = b$. In this case the corresponding functional is

$$J[\phi] = \int_a^b \{p(\phi')^2 + q\phi^2 - 2r\phi\}\,dx + B_a + B_b \tag{8.17a}$$

where

$$\begin{aligned} B_a &= \frac{p(a)}{\alpha_1}\left(-\alpha_0\phi^2(a) + 2\,\gamma_1\,\phi\,(a)\right) \\ B_b &= \frac{p(b)}{\beta_1}\left(\beta_0\phi^2(b) - 2\,\gamma_2\,\phi\,(b)\right) \end{aligned} \tag{8.17b}$$

and $\phi(x)$ need not satisfy either boundary condition.

• case 2 $\alpha_1 = 0,\ \beta_1 \neq 0$

that is, with an essential boundary condition at $x = a$ and a suppressible condition at $x = b$. Here $J[\phi]$ is identical to (8.17a) except that B_a is omitted and the functions $\phi(x)$ **must** satisfy the essential boundary condition

$$\alpha_0\, \phi(a) = \gamma_1$$

at $x = a$.

• case 3 $\alpha_1 \neq 0,\ \beta_1 = 0$

that is, with a suppressible boundary condition at $x = a$ and an essential condition at $x = b$. Here $J[\phi]$ is identical to (8.17a) except that B_b is omitted and the functions $\phi(x)$ **must** satisfy the essential boundary condition

$$\beta_0\, \phi(b) = \gamma_2$$

at $x = b$.

• case 4 $\alpha_1 = 0,\ \beta_1 = 0$

such that both boundary conditions are essential. In this case $J[\phi]$ is identical to (8.17a) except that both B_a and B_b are omitted and the functions $\phi(x)$ **must** satisfy both the essential boundary conditions

$$\alpha_0\, \phi(a) = \gamma_1 \qquad \beta_0\, \phi(b) = \gamma_2$$

It is perhaps not immediately obvious that all second-order, linear ordinary differential equations can be expressed in the form (8.16a) which is called the **Sturm–Liouville** form.

To show that this is the case, we consider any second-order linear ordinary differential equation of the form

$$a_2(x)u'' + a_1(x)u' + a_0(x)u = c(x) \tag{8.18}$$

We note firstly that the form of (8.16a) is arbitrary up to a multiplicative function, say $f(x)$. Thus we can write:

$$-f(x)\,(p(x)u')' + f(x)\,q(x)u = f(x)\,r(x)$$

or, on expanding,

$$-f(x)\,p(x)u'' - f(x)\,p'(x)\,u' + f(x)\,q(x)\,u = f(x)r(x) \tag{8.19}$$

For (8.19) to be equivalent to (8.18) we must have

$$\begin{aligned} a_2(x) &= -f(x)\,p(x) \\ a_1(x) &= -f(x)\,p'(x) \\ a_0(x) &= f(x)\,q(x) \\ c(x) &= f(x)\,r(x) \end{aligned}$$

from which we deduce that

$$\frac{p'(x)}{p(x)} = \frac{a_1(x)}{a_2(x)}$$

Hence

$$\ell n\ p(x) = \int \frac{a_1(x)}{a_2(x)}\ dx$$

or

$$p(x) = \exp\left\{\int \frac{a_1(x)}{a_2(x)}\ dx\right\} \tag{8.20}$$

giving

$$f(x) = -\frac{a_2(x)}{p(x)} \quad q(x) = \frac{a_0(x)}{f(x)} \quad \text{and} \quad r(x) = \frac{c(x)}{f(x)} \tag{8.21}$$

Example 8.4

Determine the variational problem corresponding to the boundary value problem:

$$x^2\, u'' + x\, u' + (x^2 - 1)u = 0 \qquad 0 \le x \le 1$$

$$u(0) = 1 \qquad u'(1) = 6$$

Solution

We first re-write the given differential equation in Sturm–Liouville form. The coefficients of u and its derivatives are, by comparison with (8.18),

$$a_2(x) = x^2 \qquad a_1(x) = x \qquad a_0(x) = x^2 - 1$$

Hence from (8.20) and (8.21) we find

$$p(x) = \exp\left\{\int \frac{1}{x}\ dx\right\} = \exp(\ell n\ x) = x$$

$$f(x) = -\frac{x^2}{x} = -x \qquad q(x) = -\frac{(x^2-1)}{x} \qquad r(x) = 0$$

Thus the given differential equation may be expressed in the form (8.16a) as

$$-\frac{d}{dx}\left(x\,\frac{du}{dx}\right) - \left(x - \frac{1}{x}\right) u = 0$$

which is the form required for the application of Theorem 8.1. We find using (8.17) (case 2)

$$J[\phi] = \int_0^1 \left\{x(\phi')^2 - \left(x - \frac{1}{x}\right)\phi^2\right\} dx - 12\,\phi\,(1)$$

and the functions $\phi(x)$ must all satisfy $\phi(0) = 1$.

The reader is invited to work Examples 8.1 and 8.2 in reverse i.e. to use Theorem 8.1 (case 4 and case 2 respectively) to re-obtain $J[\phi]$ in each case from the derived boundary value problem.

8.5 The Ritz method

This is one method used for finding approximate solutions to variational problems of the form

$$J[\phi] = \text{extremum}$$

The Ritz technique, like the Galerkin weighted residual method, pre-dates the finite element method but its implementation in a **piecewise** manner i.e. after segmenting the domain of the problem, has proved a most useful tool in the solution of boundary value problems. We have in fact already (in Example 8.3) studied a simple application of the piecewise Ritz method by using a two element approximation to Poisson's equation.

We consider firstly the Ritz method applied to one-dimensional boundary value problems where the whole domain is considered as one element i.e. no segmentation is carried out.

In order to use the Ritz method to solve

$$J[\phi] = \text{extremum}$$

over the set of admissible functions ϕ which satisfy the given essential boundary conditions the following procedure is carried out.

- Select a **sequence of admissible functions**

$$\tilde{u}_1(x, a_1), \quad \tilde{u}_2(x, a_1, a_2), \ldots, \tilde{u}_k(x, a_1, a_2, \ldots, a_k), \ldots$$

where the a_i are parameters to be determined. The functions are chosen so that $\tilde{u}_k$ contains $\tilde{u}_{k-1}$, in other words the $(k-1)$th function is **embedded** in the kth function.

For example, for the variational problem

$$J[\phi] = \int_0^1 \left\{ \left(\frac{d\phi}{dx} \right)^2 + \phi^2 \right\} dx = \text{extremum}$$

with

$$\phi(0) = 1$$

a possible sequence of functions would be

$$\begin{aligned}\tilde{u}_1 &= 1 + a_1\, x \\ \tilde{u}_2 &= 1 + x(a_1 + a_2\, x) = \tilde{u}_1 + a_2\, x^2 \\ \tilde{u}_3 &= 1 + x(a_1 + a_2\, x + a_3\, x^2) = \tilde{u}_2 + a_3\, x^2\end{aligned}$$

and so on. The embedding property (that $\tilde{u}_k$ becomes $\tilde{u}_{k-1}$ for a particular value of the parameter a_k) can be clearly seen here.

- Substitute each function in the sequence into the functional and optimize the functional with respect to the parameters. Thus for the functional $J[\tilde{u}_k]$ which is a function of $a_1, a_2, \ldots, a_k$ we optimize by determining the parameters from the k equations

$$\frac{\partial J}{\partial a_1} = 0 \quad \frac{\partial J}{\partial a_2} = 0 \quad \ldots \quad \frac{\partial J}{\partial a_k} = 0 \qquad \textbf{(8.22)}$$

- If the solutions to (8.22) are denoted by $\tilde{a}_1, \tilde{a}_2, \ldots, \tilde{a}_k$ and the corresponding functions by $\tilde{u}_1(x, \tilde{a}_1)$, $\tilde{u}_2(x, \tilde{a}_1, \tilde{a}_2)$, $\tilde{u}_k(x, \tilde{a}_1, \tilde{a}_2, \ldots, \tilde{a}_k)$, then, because of the embedding property, it can be shown that if u is the actual function that minimizes $J[\phi]$ then

$$J[\tilde{u}_1] \geq J[\tilde{u}_2] \geq \ldots \geq J[\tilde{u}_k] \geq \ldots \geq J[u]$$

that is, we obtain a sequence of values for the functional bounded below by the minimum value which is obtained from the true solution. (The inequalities are reversed if u maximizes $J[\phi]$).

Informally, each function $\tilde{u}_k$ is an approximation to u, the approximation improving as k increases i.e. as more parameters are used.

- In the Ritz method, the same form is used for the kth function in the sequence $\tilde{u}_k$ as has already been used in the Galerkin method viz.

$$\begin{aligned}\tilde{u}_k &= \theta_0(x) + a_1\, \theta_1(x) + a_2\, \theta_2(x) + \ldots + a_k\, \theta_k(x) \\ &= \theta_0(x) + \sum_{i=1}^{k} a_i\, \theta_i(x)\end{aligned} \qquad \textbf{(8.23)}$$

As in the modified Galerkin approach the function $\theta_0(x)$ is chosen to satisfy the essential boundary conditions of the problem and the coordinate functions $\theta_i(x)$ must satisfy the corresponding homogeneous form of these essential boundary conditions.

Example 8.5

Use the Ritz method to obtain (i) a one-parameter approximation (ii) a two-parameter approximation to the boundary value problem:

$$\frac{d^2u}{dx^2} + u = -x \qquad 0 < x < 1$$

$$u(0) = u(1) = 0$$

Solution

We must firstly recast the given problem in variational form. This is readily done using Theorem 8.1 (case 4). For our problem (8.17a) becomes

$$J[\phi] = \int_0^1 \left\{ -\left(\frac{d\phi}{dx}\right)^2 + \phi^2 + 2\,x\,\phi \right\} dx \tag{8.24}$$

so the variational problem to solve is

$$J[\phi] = \text{extremum}$$

$$\phi(0) = \phi(1) = 0$$

(i) We choose a one-parameter trial solution of the form (8.23). Since both (essential) boundary conditions are homogeneous we may choose $\theta_0(x) \equiv 0$ and, as a simple one-term approximation satisfying both the (essential) boundary conditions, we use

$$\theta_1(x) = x(1-x)$$

so that $\tilde{u}_1(x) = a_1 x(1-x)$.

Substituting into (8.24) we obtain

$$J[\tilde{u}_1] = \int_0^1 \{-a_1^2(1-2x)^2 + a_1^2\,x^2(1-x)^2 + 2\,a_1\,x^2(1-x)\}\,dx$$
$$= -\frac{3}{10}\,a_1^2 + \frac{a_1}{6}$$

We find a_1 such that J is optimized by putting

$$\frac{dJ}{da_1} = -\frac{6}{10}a_1 + \frac{1}{6} = 0$$

i.e.

$$a_1 = \frac{5}{18}$$

so our one-parameter approximation is

$$\tilde{u}_1(x) = \frac{5}{18}\,x(1-x)$$

(ii) A possible two parameter approximation is

$$\tilde{u}_2(x) = x(1-x)(a_1 + a_2\,x)$$

(Note that $\tilde{u}_2(x) = \tilde{u}_1(x)$ when $a_2 = 0$ verifying that $\tilde{u}_1(x)$ is indeed embedded in $\tilde{u}_2(x)$.) Using this approximation in (8.24) gives after integrating

$$J[\tilde{u}_2] = -\left[\frac{9}{30}a_1^2 + \frac{13}{105}a_2^2 + \frac{9}{30}a_1 a_2 - \frac{a_1}{6} - \frac{a_2}{10}\right]$$

Again we optimize $J[\tilde{u}_2]$, now with respect to a_1 and a_2. We obtain

$$\frac{\partial J}{\partial a_1} = \frac{18}{30}a_1 + \frac{9}{30}a_2 - \frac{1}{6} = 0$$

and

$$\frac{\partial J}{\partial a_2} = \frac{9}{30}a_1 + \frac{26}{105}a_2 - \frac{1}{10} = 0$$

with solutions

$$a_1 = 71/369 \qquad a_2 = 7/41$$

Hence a two parameter approximation to the solution of the given boundary value problem is

$$\tilde{u}_2(x) = x(1-x)\left[\frac{71}{369} + \frac{7}{41}x\right]$$

For this particular boundary value problem the exact solution is easily obtained as

$$u = \frac{\sin x}{\sin 1} - x$$

As shown in Figure 8.3 the agreement between the approximation $\tilde{u}_2(x)$ and $u(x)$ is quite impressive.

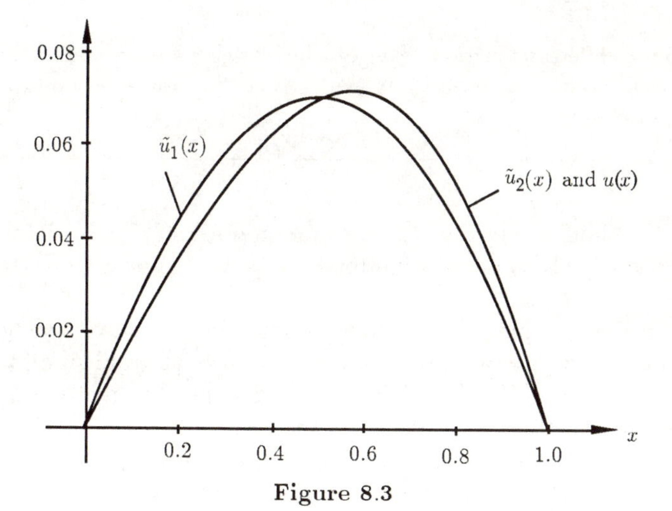

Figure 8.3

A further indication of the accuracy is to calculate the functional $J[\phi]$ for each approximation and (in this case) compare with the value of $J[\phi]$ for the exact solution. The results are

$$J[\tilde{u}_1] = 0.02314$$
$$J[\tilde{u}_2] = 0.02456$$
$$J[u] = 0.02457$$

It is interesting to note that the extremum in $J[\phi]$ in this case is in fact neither a maximum nor a minimum but a point of inflexion. To show this we evaluate $J[\phi]$ for a collection of functions of the form

$$\phi(x) = u(x) + \varepsilon\, \eta\,(x) \tag{8.25}$$

where $\eta(0) = \eta(1) = 0$

Substituting (8.25) into (8.24)

$$\begin{aligned} J[\phi] &= \int_0^1 \{-(u' + \varepsilon\,\eta')^2 + (u + \varepsilon\,\eta)^2 + 2\,x(u + \varepsilon\,\eta)\}\, dx \\ &= \int_0^1 (-u'^2 + u^2 + 2\,x\,u)dx \\ &\qquad + \varepsilon \int_0^1 \{-2\,u'\,\eta' + 2\,u\,\eta + 2\,x\,\eta\}\, dx + \varepsilon^2 \int_0^1 \{-\eta'^2 + \eta^2\} \\ &= J[u] + 2\,\varepsilon \int_0^1 \{u'' + u + x\}\,\eta\, dx + \varepsilon^2 \int_0^1 \{-\eta'^2 + \eta^2\}\, dx \end{aligned}$$

where we have used integration by parts in the second term. Using the original differential equation the first integral clearly vanishes and we obtain

$$J[\phi] = J[u] + \varepsilon^2 \int_0^1 \{-\eta'^2 + \eta^2\}\, dx$$

Since different functions $\eta(x)$ can be chosen so as to make the integral either positive or negative, it is clear that $J[u]$ is neither a maximum nor a minimum.

We should note that the Ritz method for solving boundary value problems is not always as straightforward as may appear from the above example.

In the first place the method depends on the very existence of an equivalent variational problem. No such variational problem exists if the boundary value problem is of **odd** order and the Ritz method is clearly not applicable in this case. Even an even-order problem may be difficult to recast in variational form if the differential equation is non-linear.

Difficulties can also arise with the choice of coordinate functions. We illustrate briefly by two examples.

The boundary value problem

$$\frac{d^2u}{dx^2} = -\cos \pi x \qquad 0 \leq x \leq 1$$

$$u(0) = u(1) = 0$$

may be recast as the variational problem

$$J[\phi] = \int_0^1 \{(\phi')^2 - 2\cos \pi x \; \phi\}\, dx = \text{extremum} \tag{8.26a}$$

$$\phi(0) = \phi(1) = 0 \tag{8.26b}$$

A possible sequence for use with the Ritz method is

$$\tilde{u}_k(x) = a_1 \sin \pi x + a_2 \sin 2\pi x + \ldots + a_k \sin k\pi x \tag{8.27}$$

(which clearly satisfies the essential boundary conditions (8.26b)). It turns out that as k increases $\tilde{u}_k(x)$ does indeed give better and better approximations to the true solution to this problem.

Now consider the similar boundary value problem:

$$\frac{d^2u}{dx^2} = -\cos \pi x \qquad 0 \leq x \leq 1$$

$$u(0) = 0 \quad u'(1) = 0$$

(that is, one essential and one suppressible condition). The corresponding functional is again (8.26a) but the admissible functions need only satisfy

$$\phi(0) = 0 \tag{8.28}$$

In this case use of (8.27), which of course does satisfy (8.28), in the Ritz method does not produce an approximation which tends to the correct solution as $k \to \infty$. The problem here is that the choice (8.27) is over-restrictive in that

$$\tilde{u}_k(1) = 0 \qquad \text{for all } k$$

which is not a constraint in the variational problem. To overcome this difficulty we choose a sequence of coordinate functions which allow for the possibility of $\tilde{u}_k(1)$ being **non-zero**. For example

$$\tilde{u}_k(x) = a_1 x + a_2 \sin \pi x + a_3 \sin 2\pi x + \ldots + a_k \sin(k-1)\pi x$$

It turns out that this choice does yield a good approximation to the solution of the second boundary value problem.

8.6 The Ritz method and finite elements

In the previous section we used the Ritz technique in conjunction with approximating functions $\tilde{u}_k(x)$ defined over the complete domain of the boundary value problem. The essence of the finite element method, as we know, is to segment the domain and use, as approximating functions, low order polynomials (usually linear or quadratic) defined separately in each element but satisfying continuity conditions at element boundaries.

Consequently we discuss in this section how to adapt the Ritz technique to use simple piecewise approximating functions. The discussion will be relatively brief and will draw on many of the ideas from earlier chapters.

For segmentation of a domain such that *nde* nodes are created, we have seen in Chapter 4 that the artifice of roof functions, one for each node, is a convenient way of expressing an approximate solution:

$$\tilde{u}(x) = \sum_{i=1}^{nde} R_i(x)\,\tilde{u}_i \tag{8.29}$$

where the $\tilde{u}_i$ denote nodal values. For each node, i, $R_i(x)$ is obtained from the shape functions associated with that node. We have also discussed the equivalence of (8.29) to the normal form (2.2) of the Galerkin approximation which we have pointed out is also used in the Ritz method. It follows therefore that (8.29), or its equivalent in terms of shape functions, can be used in conjunction with the Ritz method. Indeed we have already done this, using linear shape functions, in Example 8.3. Consequently we will discuss here the use of quadratic shape functions in conjunction with a piecewise application of the Ritz method.

Example 8.6

Use the Ritz method to obtain a two-element quadratic approximation to the boundary value problem

$$\frac{d^2u}{dx^2} + u = -x \qquad 0 < x < 1$$

with $u(0) = u(1) = 0$

Solution

The boundary value problem is the same as in Example 8.5 and has the equivalent variational form

$$J[\phi] = \int_0^1 \left\{ -\left(\frac{d\phi}{dx}\right)^2 + \phi^2 + 2x\phi \right\} dx = \text{extremum} \tag{8.30a}$$

$$\phi(0) = \phi(1) = 0 \tag{8.30b}$$

We segment the domain into two quadratic elements, each with a central node (see Figure 8.4).

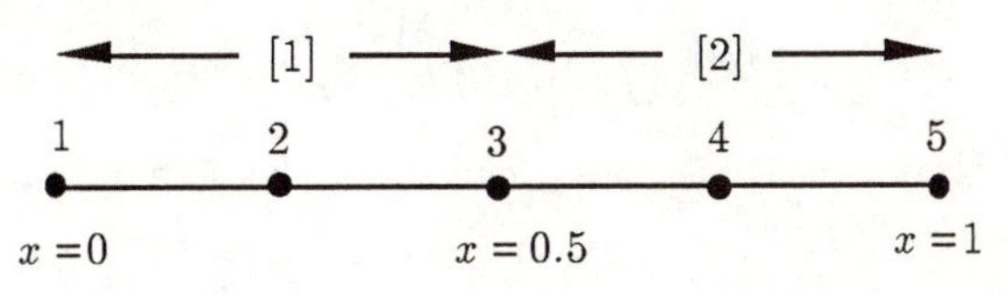

Figure 8.4

The relation between an approximate trial solution in terms of five roof functions and in terms of quadratic shape functions has already been given in (4.63) and (4.64) and we use the latter equations in evaluating $J(\tilde{u})$.

In element [1]:

$$\begin{aligned}\tilde{u} &= \sum_{i=1}^{3} Q_i^{[1]}(x)\tilde{u}_i \\ &= Q_2(x)\,\tilde{u}_2 + Q_3(x)\,\tilde{u}_3\end{aligned}$$

after inserting the essential boundary condition $u(0) = 0$. (The superscript [1] is omitted for clarity in the following calculations.) Hence, for the element [1] the three contributions to the functional (8.30a) are

$$\begin{aligned}\text{(i)}\quad -\int_{[1]}\left(\frac{d\tilde{u}}{dx}\right)^2 dx &= -\tilde{u}_2^2\int_{[1]}\left(\frac{dQ_2}{dx}\right)^2 dx - \tilde{u}_3^2\int_{[1]}\left(\frac{dQ_3}{dx}\right)^2 dx \\ &\qquad - 2\,\tilde{u}_2\,\tilde{u}_3\int_{[1]}\frac{dQ_2}{dx}\frac{dQ_3}{dx}\,dx \\ &= -\tilde{u}_2^2\left(\frac{32}{3}\right) - \tilde{u}_3^2\left(\frac{14}{3}\right) + 2\,\tilde{u}_2\,\tilde{u}_3\left(\frac{16}{3}\right)\end{aligned}$$

where we have used the standard results (3.18) with $\ell = 1/2$.

$$\begin{aligned}\text{(ii)}\quad \int_{[1]}\tilde{u}^2\,dx &= \tilde{u}_2^2\int_{[1]}Q_2^2\,dx + \tilde{u}_3^2\int_{[1]}Q_3^2\,dx \\ &\qquad + 2\,\tilde{u}_2\,\tilde{u}_3\int_{[1]}Q_2\,Q_3\,dx \\ &= \tilde{u}_2^2\left(\frac{4}{15}\right) + \tilde{u}_3^2\left(\frac{1}{15}\right) + 2\,\tilde{u}_2\,\tilde{u}_3\left(\frac{1}{30}\right)\end{aligned}$$

using the standard results (3.20)

$$\text{(iii)}\quad \int_{[1]} 2\,x\,\tilde{u}\,dx$$

This integral is readily evaluated if we transform element [1] to the standard element using the transformation $x = (1+\xi)/4$. Hence

$$\int_{[1]} 2\,x\,\tilde{u}\,dx = \int_{-1}^{1} \frac{1}{2}(1+\xi)\,\tilde{u}\,\frac{d\xi}{4}$$
$$= \frac{1}{8}\int_{-1}^{1}(1+\xi)(\tilde{u}_2\,Q_2(\xi) + \tilde{u}_3\,Q_3(\xi))d\xi$$
$$= \tilde{u}_2\left(\frac{1}{6}\right) + \tilde{u}_3\left(\frac{1}{12}\right)$$

by straightforward integration and using the standard quadratic shape functions (3.16).

Collecting terms we have

$$J_1[\tilde{u}] = \tilde{u}_2^2\left(-\frac{52}{5}\right) + \tilde{u}_3^2\left(-\frac{23}{5}\right) + \tilde{u}_2\,\tilde{u}_3\left(\frac{161}{15}\right) + \tilde{u}_2\left(\frac{1}{6}\right) + \tilde{u}_3\left(\frac{1}{12}\right)$$

In element [2]:

$$\tilde{u} = \sum_{i=1}^{3} Q_i^{[2]}(x)\tilde{u}_i$$
$$= Q_1^{[2]}(x)\tilde{u}_3 + Q_2^{[2]}(x)\tilde{u}_4$$

on inserting the essential boundary condition $\tilde{u}(1) = 0$.

Calculations similar to those performed for element [1] show that the contribution of this element to the functional (8.30a) is

$$J_2[\tilde{u}] = \tilde{u}_3^2\left(-\frac{23}{5}\right) + \tilde{u}_4^2\left(-\frac{52}{5}\right) + \tilde{u}_3\,\tilde{u}_4\left(\frac{161}{15}\right) + \tilde{u}_3\left(\frac{1}{12}\right) + \tilde{u}_4\left(\frac{1}{2}\right)$$

Adding $J_1[\tilde{u}]$ and $J_2[\tilde{u}]$ gives us the required functional $J[\tilde{u}]$. The required nodal values are found by solving the three equations

$$\frac{\partial J}{\partial \tilde{u}_2} = -\frac{104}{5}\tilde{u}_2 + \frac{165}{5}\tilde{u}_3 + \frac{1}{6} = 0$$

$$\frac{\partial J}{\partial \tilde{u}_3} = \frac{161}{15}\,\tilde{u}_2 - \frac{92}{5}\,\tilde{u}_3 + \frac{161}{15}\,\tilde{u}_4 + \frac{1}{6} = 0 \qquad \textbf{(8.31)}$$

$$\frac{\partial J}{\partial \tilde{u}_4} = \frac{161}{15}\,\tilde{u}_3 - \frac{104}{5}\,\tilde{u}_4 + \frac{1}{2} = 0$$

The solution of these equations is

$$\tilde{u}_2 = 0.04400 \qquad \tilde{u}_3 = 0.0697 \qquad \tilde{u}_4 = 0.0600$$

These compare remarkably well with the exact solution to this problem quoted in Example 8.5 and which give

$$u(0.25) = 0.04401 \qquad u(0.5) = 0.06975 \qquad u(0.75) = 0.06006$$

The full approximate solution can therefore be written:

$$\text{element}[1] : \tilde{u}(x) = 0.0440\ (1 - \xi^2) + \frac{0.0697}{2}\ \xi(\xi + 1)$$

$$\text{where } \xi = 4x - 1$$

$$\text{element}[2] : \tilde{u}(x) = \frac{0.0697}{2}\ \xi(\xi - 1) + 0.0600\ (1 - \xi^2)$$

$$\text{where } \xi = 4x - 3$$

Graphs of this approximate solution and the exact solution are shown in Figure 8.5. Clearly the approximate solution is a good one as can also be shown by evaluating the functionals. For the approximate solution we find

$$J[\tilde{u}] = 0.024485$$

whereas, using the exact solution,

$$J[u] = 0.02457$$

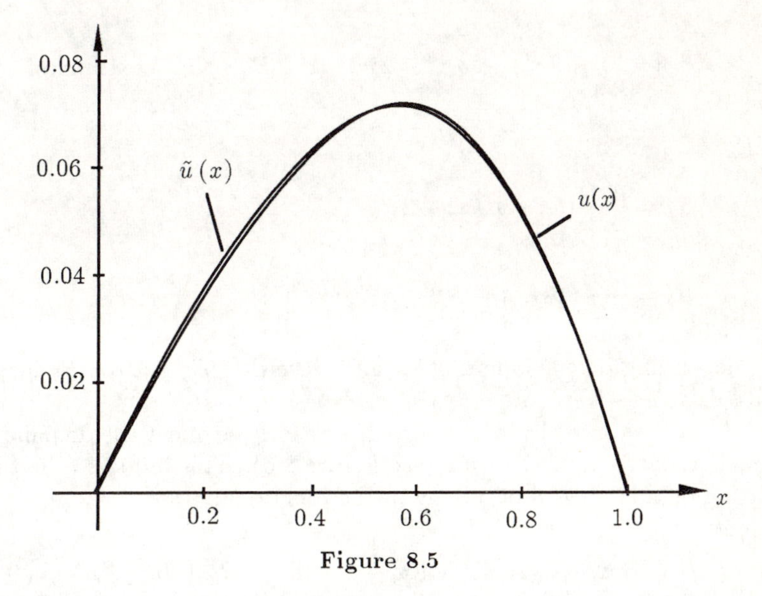

Figure 8.5

The reader might find it a useful revision exercise to repeat the above calculation using weighted residuals (Galerkin method). The same set of linear equations for the nodal values is found as the set (8.31) obtained by the Ritz method. Hence, in this example and, as we shall see later, more generally, the two techniques are equivalent.

8.7 Matrix formulation of the Ritz procedure

We are now in a position to describe a general approach, utilizing the Ritz method and piecewise polynomial coordinate functions, for obtaining a finite element approximation to the solution of one-dimensional second-order boundary value problems.

As we have already noted in Theorem 8.1 the boundary value problem:

$$-\frac{d}{dx}(p(x)\frac{du}{dx}) + q(x)u = r(x) \qquad a \leq x \leq b \tag{8.32a}$$

$$\begin{aligned} \alpha_0\, u(a) + \alpha_1\, u'(a) &= \gamma_1 \\ \beta_0\, u(b) + \beta_1\, u'(b) &= \gamma_2 \end{aligned} \tag{8.32b}$$

is equivalent to the variational problem:

$$J[\phi] = \int_a^b \{p(\phi')^2 + q\,\phi^2 - 2\,r\phi\}dx + B_a + B_b = \text{extremum}$$

where

$$B_a = \frac{p(a)}{\alpha_1}(-\alpha_0\,\phi^2\,(a) + 2\,\gamma_1\,\phi\,(a))$$

$$B_b = \frac{p(b)}{\beta_1}(\beta_0\,\phi^2\,(b) - 2\,\gamma_2\,\phi\,(b))$$

but, depending on the values of α_1 and β_1, either B_a and/or B_b may be omitted.

In the finite element approach we firstly segment the domain into a number, say *nel*, of elements (see Figure 8.6). This induces a 'natural' splitting of the integral occurring in the functional:

$$J[\phi] = \sum_{e=1}^{nel} \int_{[e]} \{p(\phi')^2 + q\phi^2 - 2r\phi\}dx + B_a + B_b \tag{8.33}$$

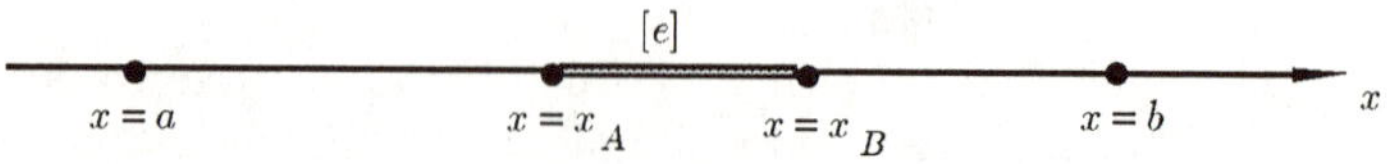

[e]

$x = a$ $\quad x = x_A$ $\quad x = x_B$ $\quad x = b$ $\quad x$

Figure 8.6

Unless the element [e] is linear it will contain internal nodes which are not shown.

We assume, over the complete domain, an approximate solution

$$\tilde{u}(x) = \sum_{i=1}^{nde} \tilde{u}_i \, R_i(x) \tag{8.34}$$

where *nde* is the total number of nodes. This implies that within each element with, say, *nint* nodes

$$\tilde{u}(x) = \sum_{j=1}^{nint} \tilde{u}_j^{[e]} \, S_j(\xi)$$

Here $S_j(\xi)$ are the shape functions appropriate to the type of element chosen. The x and ξ variables are related by

$$\begin{aligned} x = x_A \, L_1(\xi) + x_B \, L_2(\xi) &= \frac{(x_B - x_A)}{2} \xi + \frac{(x_B + x_A)}{2} \\ &= \left(x_B - \frac{\ell}{2} \right) + \frac{\ell}{2} \xi \end{aligned}$$

where $\ell = x_B - x_A$ is the length of the element.

As discussed in Section 4.3 if there is a single essential boundary condition (say at $x = a$) then $\tilde{u}_1$ is a given constant ($= \gamma_1/\alpha_0$ from (8.32b)) and not a proper parameter and so we can write (8.34) as

$$\tilde{u}(x) = \tilde{u}_1 R_1(x) + \sum_{i=2}^{nde} \tilde{u}_i R_i(x)$$

Due to the properties of the roof functions the first term satisfies the essential condition ($u = \tilde{u}_1$ at $x = a$) and so corresponds directly to $\theta_0(x)$ in the Ritz method. The remaining terms satisfy the homogeneous form of the essential boundary conditions. In such problems we can only optimize with respect to $\tilde{u}_2, \tilde{u}_3, \ldots, \tilde{u}_{nde}$. Other boundary conditions are treated similarly. For example, in the case in which both boundary conditions are essential we optimize with respect to all parameters except $\tilde{u}_1$ and $\tilde{u}_{nde}$.

In fact, however, we will leave the incorporation of any essential boundary conditions until a later stage of the analysis in order to keep the formulation of the finite element equations as simple as possible. Thus we shall proceed with the optimization of the functional (8.33).

We have, putting (8.34) and its derivative into (8.33),

$$\begin{aligned} J[\tilde{u}] = & \sum_{i=1}^{nde} \sum_{j=1}^{nde} \left[\tilde{u}_i \, \tilde{u}_j \int_a^b \left\{ p \frac{dR_i}{dx} \frac{dR_j}{dx} + q \, R_i \, R_j \right\} dx \right] \\ & - \sum_{i=1}^{nde} \left[\tilde{u}_i \int_a^b 2 \, r \, R_i \, dx \right] + (B_a + B_b) \end{aligned}$$

or

$$J[\tilde{u}] = \sum_{i=1}^{nde} \sum_{j=1}^{nde} K_{ij}\, \tilde{u}_i \tilde{u}_j - 2 \sum_{i=1}^{nde} F_i\, \tilde{u}_i + (B_a + B_b)$$

in which

$$B_a = \frac{p(a)}{\alpha_1}(-\alpha_0\, \tilde{u}_1^2 + 2\,\gamma_1\, \tilde{u}_1)$$

$$B_b = \frac{p(b)}{\beta_1}(\beta_0\, \tilde{u}_{nde}^2 - 2\,\gamma_2\, \tilde{u}_{nde})$$

In matrix form we can write

$$J[\tilde{u}] = \{\tilde{u}\}^T [K] \{\tilde{u}\} - 2\, \{\tilde{u}\}^T \{F\} + (B_a + B_b)$$

Optimising with respect to $\{\tilde{u}\}$ where $\{\tilde{u}\} = \{\tilde{u}_1, \tilde{u}_2, \ldots, \tilde{u}_{nde}\}^T$ we obtain

$$\frac{\partial J}{\partial \{\tilde{u}\}} = 2[K]\{\tilde{u}\} - 2\{F\} + 2\{B\} = \{0\} \qquad \textbf{(8.35)}$$

(See the supplement to this chapter.) Here $\{B\}$ is the $nde \times 1$ column vector:

$$\{B\} = \left\{ \frac{p(a)}{\alpha_1}(-\,\alpha_0\, \tilde{u}_1 +\, \gamma_1), 0, \ldots, 0, \frac{p(b)}{\beta_1}(\beta_0\, \tilde{u}_{nde} -\, \gamma_2) \right\}^T$$

and so (8.35) implies

$$[K]\{\tilde{u}\} = \{F\} + \{B\}$$

Here

$$K_{ij} = \int_a^b \left\{ p\, \frac{dR_i}{dx}\, \frac{dR_j}{dx} + q\, R_i\, R_j \right\} dx \qquad i, j = 1, \ldots, nde$$

$$F_i = \int_a^b r\, R_i\, dx \qquad i = 1, \ldots, nde$$

We have now formulated the problem in principle. What remains is simply to construct each term of the matrix $[K]$ and the column vector $\{F\}$. We note that each term generally involves the evaluation of an integral over the complete domain. Since the roof functions are piecewise polynomials these integrals are evaluated element by element. The techniques used to construct these matrix terms are identical to those used in the Galerkin reformulation of a boundary value problem and the reader is referred to Chapter 4.

8.8 Equivalence of Ritz and Galerkin procedures

It is a straightforward matter to demonstrate the equivalence of the Ritz and modified Galerkin procedures for linear, second-order, boundary value problems of the form (8.32a). We begin, as usual, with the trial solution

$$\tilde{u}(x) = \theta_0(x) + \sum_{i=1}^{n} a_i\,\theta_i(x) \tag{8.36}$$

so that the Galerkin weighted residual equations are:

$$\int_a^b \left\{ -\frac{d}{dx}\left(p(x)\,\frac{d\tilde{u}}{dx} \right) + q(x)\,\tilde{u} - r(x) \right\} \theta_i(x)\,dx = 0 \qquad i = 1, 2, \ldots, n$$

Integrating the first term by parts and using (8.32b) we obtain

$$\begin{aligned} \int_a^b \left\{ p(x)\frac{d\tilde{u}}{dx}\,\frac{d\theta_i}{dx} + q(x)\,\tilde{u}\,\theta_i - r\,(x)\,\theta_i \right\} dx \\ + \frac{p(b)}{\beta_1}(\beta_0\,\tilde{u}\,(b) - \gamma_2)\,\theta_i\,(b) \\ + \frac{p(a)}{\alpha_1}(\gamma_1 - \alpha_0\,\tilde{u}\,(a))\,\theta_i(a) = 0 \quad i = 1, 2, \ldots, n \end{aligned} \tag{8.37}$$

We now formally optimise the equivalent functional (8.17) and confirm that the equations obtained for the parameters a_i are identical to the modified Galerkin equations (8.37).

Differentiating (8.17a) under the integral sign and replacing ϕ by $\tilde{u}$ as the trial solution we obtain

$$\begin{aligned} \frac{\partial J}{\partial a_i} = \int_a^b \left\{ 2\,p\,\tilde{u}'\,\frac{\partial \tilde{u}'}{\partial a_i} + 2\,q\,\tilde{u}\,\frac{\partial \tilde{u}}{\partial a_i} - 2\,r\,\frac{\partial \tilde{u}}{\partial a_i} \right\} dx \\ + \frac{p(a)}{\alpha_1}\left(-2\,\alpha_0\,\tilde{u}\,(a)\frac{\partial \tilde{u}}{\partial a_i}(a) + 2\,\gamma_1\,\frac{\partial \tilde{u}}{\partial a_i}\,(a) \right) \\ + \frac{p(b)}{\beta_1}\left(2\,\beta_0\,\tilde{u}\,(b)\frac{\partial \tilde{u}}{\partial a_i}(b) - 2\,\gamma_2\,\frac{\partial \tilde{u}}{\partial a_i}\,(b) \right) = 0 \quad i = 1, 2, \ldots, n \end{aligned}$$

But from (8.36)

$$\frac{\partial \tilde{u}}{\partial a_i} = \theta_i \quad \text{and} \quad \frac{\partial \tilde{u}'}{\partial a_i} = \theta_i' \equiv \frac{d\theta_i}{dx}$$

so that

$$\frac{\partial J}{\partial a_i} = \int_a^b \left\{ 2\,p\,\tilde{u}'\frac{d\theta_i}{dx} + 2\,q\,\tilde{u}\,\theta_i - 2\,r\,\theta_i \right\} dx$$
$$+ \frac{p(a)}{\alpha_1}(-2\,\alpha_0\,\tilde{u}(a) + 2\,\gamma_1)\theta_i(a) \tag{8.38}$$
$$+ \frac{p(b)}{\beta_1}(2\,\beta_0\,\tilde{u}\,(b) - 2\,\gamma_2)\theta_i(b) = 0 \qquad i = 1, 2, \ldots, n$$

Apart from an irrelevant factor of 2, equations (8.38) are identical to equations (8.37) thus establishing the equivalence of the Ritz and Galerkin approaches.

The above analysis assumes that both boundary conditions are suppressible i.e. that $\alpha_1 \neq 0$ and $\beta_1 \neq 0$. The reader should confirm that similar conclusions can be reached for the other cases that can arise (cases 2, 3 and 4 in Theorem 8.1).

8.9 Two-dimensional problems

The functionals considered thus far in this chapter have involved one independent variable x and a function u of that variable. We have demonstrated that optimizing such a functional over an admissible class of functions ϕ is equivalent to solving a one-dimensional boundary value problem and have discusssed an approximate method based on functionals – the Ritz technique – for solving boundary value problems.

We now extend the theory to the case where the functional involves two independent variables. The analysis is a fairly straightforward extension of the one variable case but there are some additional complications. For example, corresponding to a domain $a \leq x \leq b$, a line segment, we will have a planar region bounded by a closed curve and ordinary integrals will of course become double integrals. Also boundary terms (which arose in the one-dimensional case from integration by parts) will be line integrals while partial, rather than ordinary, differential equations will be the governing equations in the boundary value problems that arise.

The first 'two-dimensional' functional that we study is

$$J[\phi] = \iint_D \left\{ \left(\frac{\partial \phi}{\partial x}\right)^2 + \left(\frac{\partial \phi}{\partial y}\right)^2 - 4\phi \right\} dx\,dy \tag{8.39a}$$

$$\phi = 0 \quad \text{on} \quad \Gamma \tag{8.39b}$$

a functional which arises for example in torsion problems in linear elasticity theory. Γ denotes the boundary curve of the region D which, for simplicity, we will take as the symmetrical rectangular region $-a \leq x \leq a$, $-b \leq y \leq b$ (see Figure 8.7).

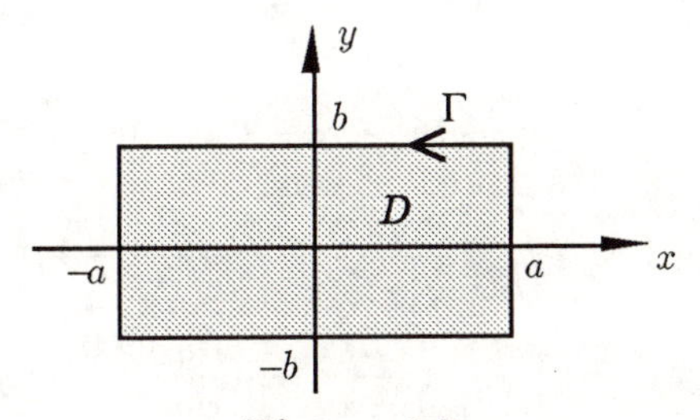

Figure 8.7

Clearly (8.39a) does indeed define a functional since a number is obtained when a specific admissible function ϕ (basically a function which vanishes on Γ and possesses continuous partial derivatives $\partial\phi/\partial x$ and $\partial\phi/\partial y$ over D) is substituted. Suppose $u(x,y)$ is the particular function from the admissible set that minimizes J i.e.

$$J[u] < J[\phi] \qquad \phi \neq u$$

To determine u, we adopt a similar approach to that used for functionals involving one independent variable and put

$$\phi(x,y) = u(x,y) + \varepsilon\,\eta\,(x,y) \tag{8.40}$$

where ε is a parameter independent of x and y and $\eta(x,y)$ is any arbitrary function satisfying

$$\eta(x,y) = 0 \quad \text{on} \quad \Gamma$$

This last condition follows from (8.39b) since u and ϕ must vanish on the boundary Γ.

Substituting (8.40) into (8.39a) and integrating produces a function $\Phi(\varepsilon)$. Since

$$\Phi(\varepsilon) = J[u + \varepsilon\,\eta] > J(u)$$

we have a minimum in Φ at $\varepsilon = 0$ so that

$$\frac{d\Phi}{d\varepsilon} = 0 \quad \text{when} \quad \varepsilon = 0 \tag{8.41}$$

which is, of course, exactly the same result as in the one-dimensional situation. For the functional (8.39a) we have

$$\Phi[\varepsilon] = \iint_D \left\{ (\frac{\partial u}{\partial x})^2 + (\frac{\partial u}{\partial y})^2 + 2\varepsilon \frac{\partial u}{\partial x}\frac{\partial \eta}{\partial x} + 2\varepsilon \frac{\partial u}{\partial y}\frac{\partial \eta}{\partial y} + \varepsilon^2 \left((\frac{\partial \eta}{\partial x})^2 + (\frac{\partial \eta}{\partial y})^2 \right) - 4(u + \varepsilon\eta) \right\} dx\, dy$$

$$= J[u] + \varepsilon \iint_D \left\{ 2\frac{\partial u}{\partial x}\frac{\partial \eta}{\partial x} + 2\frac{\partial u}{\partial y}\frac{\partial \eta}{\partial y} - 4\eta \right\} dx\, dy + \varepsilon^2 \iint_D \left\{ (\frac{\partial \eta}{\partial x})^2 + (\frac{\partial \eta}{\partial y})^2 \right\} dx\, dy$$

The condition

$$\frac{d\Phi}{d\varepsilon} = 0 \quad \text{at } \varepsilon = 0 \text{ then gives}$$

$$\iint_D \left\{ 2\frac{\partial u}{\partial x}\frac{\partial \eta}{\partial x} + 2\frac{\partial u}{\partial y}\frac{\partial \eta}{\partial y} - 4\,\eta \right\} dx\, dy = 0 \tag{8.42}$$

Now at the corresponding stage of a one-dimensional problem (Example 8.2) over the domain $1 \leq x \leq 3$ we had equation (8.5)

$$\int_1^3 \left\{ -2\,u' \frac{d\eta}{dx} - 4\,\eta \right\} dx + 12\,\eta\,(3) = 0$$

We then integrated the first term by parts so as to remove the term $d\eta/dx$. We need to perform an equivalent operation on (8.42) to eliminate the terms in $\partial\eta/\partial x$ and $\partial\eta/\partial y$. This can be done using Green's Theorem in the Plane (7.7) viz.

$$\iint_D \left(\frac{\partial G}{\partial x} - \frac{\partial F}{\partial y} \right) dx\, dy = \oint_\Gamma F\, dx + G\, dy$$

To use this theorem in (8.42) we put

$$F = -\eta\,\frac{\partial u}{\partial y} \quad \text{and} \quad G = +\eta\,\frac{\partial u}{\partial x}$$

so that

$$\iint_D \left\{ \frac{\partial}{\partial x}\left(\eta\,\frac{\partial u}{\partial x} \right) + \frac{\partial}{\partial y}\left(\eta\,\frac{\partial u}{\partial y} \right) \right\} dx\, dy = \oint_\Gamma \left(-\eta\,\frac{\partial u}{\partial y}\, dx + \eta\,\frac{\partial u}{\partial x}\, dy \right)$$

Expanding the derivatives in the double integral and re-arranging we obtain

$$\iint_D \left\{ \frac{\partial \eta}{\partial x}\frac{\partial u}{\partial x} + \frac{\partial \eta}{\partial y}\frac{\partial u}{\partial y} \right\} dx\,dy$$
$$= -\iint_D \eta \left\{ \frac{\partial^2 u}{\partial x^2} + \frac{\partial^2 u}{\partial y^2} \right\} dx\,dy + \oint_\Gamma \eta \left(-\frac{\partial u}{\partial y}\,dx + \frac{\partial u}{dx}\,dy \right) \tag{8.43}$$

Putting (8.43) in (8.42) and using the fact that $\eta = 0$ on Γ we obtain

$$-2 \iint_D \left\{ \frac{\partial^2 u}{\partial x^2} + \frac{\partial^2 u}{\partial y^2} + 2 \right\} \eta \, dx\,dy = 0 \tag{8.44}$$

Since (8.44) must be true for all functions $\eta(x, y)$, which vanish on the boundary Γ, we conclude that

$$\frac{\partial^2 u}{\partial x^2} + \frac{\partial^2 u}{\partial y^2} + 2 = 0 \qquad \text{all } (x, y) \text{ within } D \tag{8.45a}$$

(To deduce (8.45a) rigorously from (8.44) requires an extension, to two dimensions, of the fundamental lemma of the variational calculus described in the first supplement to Chapter 2.)

As well as (8.45a) we have the condition specified at the outset:

$$u = 0 \quad \text{on} \quad \Gamma \tag{8.45b}$$

Thus, as in one dimension, the problem of optimizing a functional does not lead to an explicit form for the optimizing function u but specifies u as the solution to a boundary value problem.

If we are to use variational techniques to solve given boundary value problems we must be able to obtain the equivalent functional. Various results from variational calculus are available to help us to do this. Perhaps the simplest case is a functional of the form

$$J[\phi] = \iint_D F(x, y, \phi, \phi_x, \phi_y)\,dx\,dy \tag{8.46}$$

where ϕ_x denotes $\partial\phi/\partial x$, ϕ_y denotes $\partial\phi/\partial y$ and where $\phi(x, y)$ is given on Γ, the boundary of D. (Compare (8.46) with (8.12) in the one-dimensional case.)

For a functional of the form (8.46), the particular function $u(x, y)$ which optimizes J is a solution of the second-order partial differential equation

$$\frac{\partial F}{\partial u} - \frac{\partial}{\partial x}\left(\frac{\partial F}{\partial u_x}\right) - \frac{\partial}{\partial y}\left(\frac{\partial F}{\partial u_y}\right) = 0 \quad \text{all points in } D \tag{8.47}$$

with u given on Γ.

This equation, (8.47), is a two-dimensional form of the Euler–Lagrange equation (8.13).

The problem of finding equivalent functionals for two-dimensional boundary value problems is considered in the next section. The reader who wishes to see the Ritz method used for two-dimensional problems may proceed directly to Section 8.11.

8.10 Construction of functionals II

In the previous section we obtained a boundary value problem, (8.45), defined over a rectangular region whose sides were parallel to the x and y coordinate axes. For other domains, other coordinate systems might be more appropriate – for example, cylindrical polar coordinates for a circular domain, elliptical polars for an elliptical region and so on. However, in a practical problem, the domain is not necessarily describable using one of the familiar coordinate systems (see Figure 8.8).

Consequently we will introduce a coordinate system which is 'natural' for an arbitrary planar region D. The two coordinates for this system are

(i) the arc length parameter s defined on the boundary Γ of the region and with an arbitrary origin $s = 0$

(ii) a parameter n which, for any point, is the distance from Γ measured along the normal to Γ.

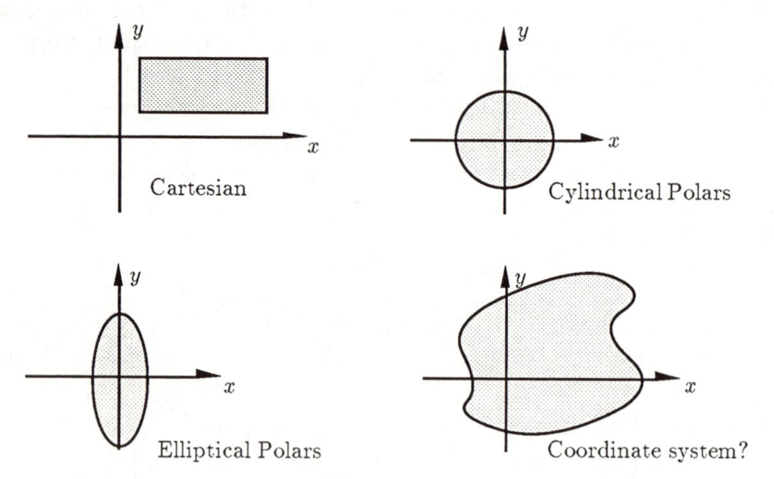

Figure 8.8

Given a point P we can determine its ns coordinates by drawing the shortest normal to Γ through P. The point of intersection P' of this normal with Γ determines the n coordinate of P as the length PP', this being taken as positive if P is exterior to D and negative if it is interior. The s coordinate of P is the arc length from the origin of s to the intersection point P' (see Figure 8.9).

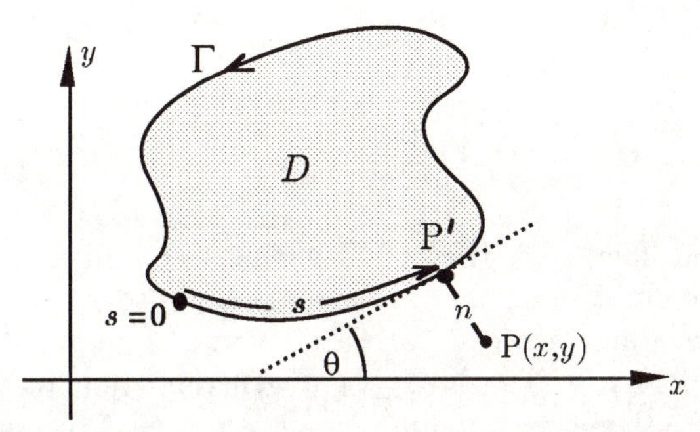

Figure 8.9

Now for many problems it is convenient to express the partial differential equation in one coordinate system, perhaps Cartesian, and the boundary condition in ns coordinates. Clearly, at each point on the boundary Γ of a closed planar region we must consider two first-order partial derivatives:

(i) $\partial/\partial s$: taken along the boundary Γ with the convention that anti-clockwise is taken as positive.

(ii) $\partial/\partial n$: taken along the outward normal.

For simple regions we can relate these derivatives with respect to n and s to derivatives in more familiar coordinates.

- Rectangular region (see Figure 8.10(a)).

$$\Gamma_1: \quad \frac{\partial}{\partial s} \equiv \frac{\partial}{\partial x} \qquad \frac{\partial}{\partial n} \equiv -\frac{\partial}{\partial y}$$

$$\Gamma_2: \quad \frac{\partial}{\partial s} \equiv \frac{\partial}{\partial y} \qquad \frac{\partial}{\partial n} \equiv \frac{\partial}{\partial x}$$

$$\Gamma_3: \quad \frac{\partial}{\partial s} \equiv -\frac{\partial}{\partial x} \qquad \frac{\partial}{\partial n} \equiv \frac{\partial}{\partial y}$$

$$\Gamma_4: \quad \frac{\partial}{\partial s} \equiv -\frac{\partial}{\partial y} \qquad \frac{\partial}{\partial n} \equiv -\frac{\partial}{\partial x}$$

- Wedge shaped region between concentric circles (common centre the origin) (see Figure 8.10(b)).

$$\Gamma_1: \quad \frac{\partial}{\partial s} \equiv \frac{\partial}{\partial x} \qquad \frac{\partial}{\partial n} \equiv -\frac{\partial}{\partial y}$$

$$\Gamma_2: \quad \frac{\partial}{\partial s} \equiv \frac{1}{r_2}\frac{\partial}{\partial \theta} \qquad \frac{\partial}{\partial n} \equiv \frac{\partial}{\partial r}$$

Since $s \equiv r_2\theta + v, n = r + \mu$ where r, θ are polar coordinates and v and μ are constants:

$$\Gamma_3: \quad \frac{\partial}{\partial s} \equiv -\frac{\partial}{\partial r} \qquad \frac{\partial}{\partial n} \equiv -\sin\theta \frac{\partial}{\partial x} + \cos\theta \frac{\partial}{\partial y}$$

$$\Gamma_4: \quad \frac{\partial}{\partial s} \equiv -\frac{1}{r_1}\frac{\partial}{\partial \theta} \qquad \frac{\partial}{\partial n} \equiv -\frac{\partial}{\partial r}$$

We can now, with the aid of n and s coordinates, show how a boundary value problem involving a linear second-order partial differential equation in two independent variables x and y may be recast as a variational problem. The theory is a generalization of Theorem 8.1 (Section 8.4) for the corresponding one-dimensional problem.

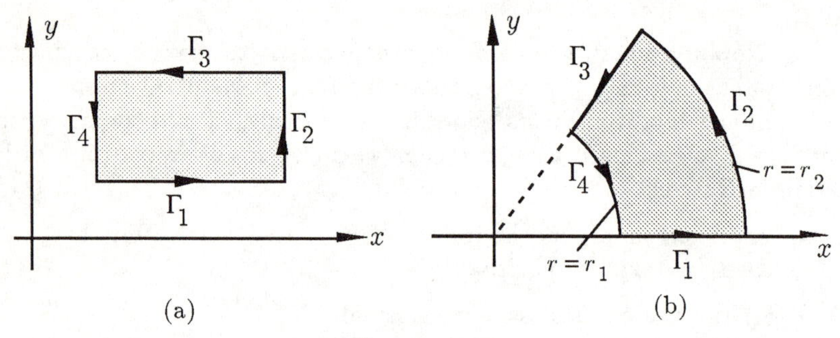

Figure 8.10

Consider the boundary value problem

$$-\frac{\partial}{\partial x}\left(A\frac{\partial u}{\partial x} + B\frac{\partial u}{\partial y}\right) - \frac{\partial}{\partial y}\left(B\frac{\partial u}{\partial x} + C\frac{\partial u}{\partial y}\right) + F\,u = r \qquad \textbf{(8.48a)}$$

in a domain D

$$u = g(s) \quad \text{on} \quad \Gamma_1 \qquad \textbf{(8.48b)}$$

$$a_1(s)\,u + a_2(s)\frac{\partial u}{\partial s} + a_3(s)\frac{\partial u}{\partial n} = a_4(s) \quad \text{on} \quad \Gamma_2 \qquad \textbf{(8.48c)}$$

In other words we have imposed an essential boundary condition on a part, Γ_1, of the boundary curve of D and a suppressible boundary condition on the other part, Γ_2 (see Figure 8.11).

Here A, B, C, F and r are given functions of x and y while g, a_1, a_2, a_3, a_4 are given functions of the arc length parameter s on the boundary. We assume that $a_3(s)$ is non-zero.

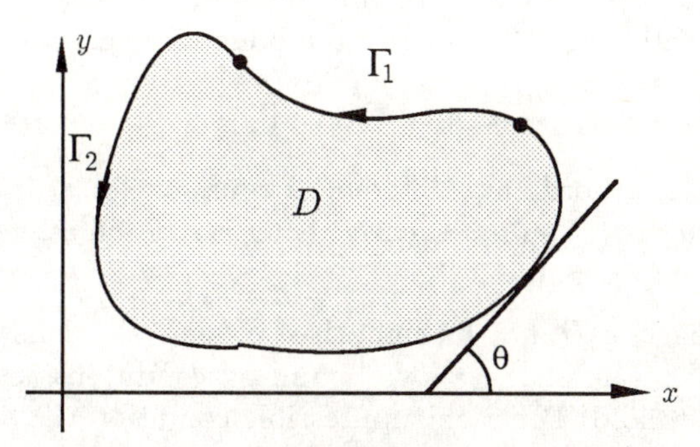

Figure 8.11

Theorem 8.2

The boundary value problem (8.48) may be written as the necessary conditions to be satisfied by the solution of the variational problem

$$J[\phi] = \text{ extremum}$$

where

$$J[\phi] = \iint_D \left\{ A\left(\frac{\partial \phi}{\partial x}\right)^2 + 2\,B\frac{\partial \phi}{\partial x}\frac{\partial \phi}{\partial y} + C\left(\frac{\partial \phi}{\partial y}\right)^2 + F\,\phi^2 - 2\,r\,\phi \right\} dx\,dy + \int_{\Gamma_2} (S\phi^2 + 2\,T\,\phi)ds \qquad \textbf{(8.49)}$$

if and only if the coefficients in (8.48a) satisfy the conditions:

$$\frac{a_2}{a_3} = \frac{(A-C)\alpha\beta + B(\alpha^2 - \beta^2)}{A\alpha^2 - 2B\alpha\beta + C\beta^2} \qquad \textbf{(8.50)}$$

where $\alpha = \cos\theta, \beta = \sin\theta$. (The angle θ is shown in Figure 8.11 and is the angle made by the tangent line at s with the positive x-axis.)

If (8.50) is satisfied then S and T are necessarily related to the other coefficients by

$$S = (A\alpha^2 - 2B\alpha\beta + C\beta^2)\,\frac{a_1}{a_3} \qquad \textbf{(8.51)}$$

$$T = -(A\alpha^2 - 2B\alpha\beta + C\beta^2)\,\frac{a_4}{a_3} \qquad \textbf{(8.52)}$$

The admissible functions $\phi(x)$ must satisfy the essential boundary condition

$$\phi = g(s) \quad \text{on} \quad \Gamma_1$$

Equations (8.49) to (8.52) constitute the basic theorem but there are alternative versions worthy of note:

- If $a_3 \equiv 0$ on Γ_2 then we omit the line integral in (8.49) but must further restrict the functions $\phi(x)$ to satisfy the boundary condition

$$a_1(s)u + a_2(s)\,\frac{du}{ds} = a_4(s) \quad \text{on} \quad \Gamma_2 \tag{8.53}$$

(On Γ_2, (8.53) is an ordinary differential equation for $u(s)$ which can be solved to give $u = u(s)$ on Γ_2. Clearly therefore it is an essential boundary condition.)

- The boundary of D may be comprised of many sections $\Gamma_1, \Gamma_2, \ldots, \Gamma_n$ in each of which one of the above boundary conditions (either essential or suppressible) hold. For each suppressible condition the equality (8.50) must be checked and, if it is satisfied, J will gain a line integral. For each essential boundary condition the ϕ functions in the admissible set must be further restricted to satisfy each such condition. The proof of Theorem 8.2 and the extensions described above are beyond the scope of this text. They can be found in the text by Collatz (1966).

Example 8.7

Determine the variational problem equivalent to the boundary value problem:

$$\frac{\partial^2 u}{\partial x^2} + \frac{\partial^2 u}{\partial y^2} + 2 = 0 \quad \text{within} \quad D$$

$$u = 0 \text{ on } \Gamma \text{ the boundary of } D$$

where D is the rectangle shown in Figure 8.12

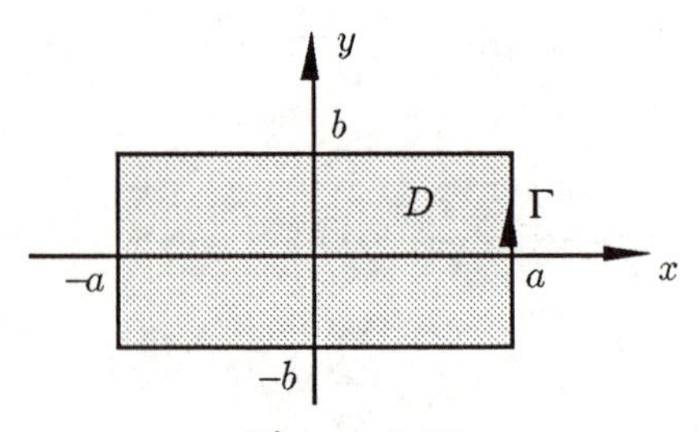

Figure 8.12

Solution

For this problem comparison with (8.48a) gives immediately

$$A = -1, \quad B = 0, \quad C = -1, \quad F = 0, \quad r = -2 \quad \text{and} \quad g(s) = 0$$

There is only one boundary condition which is of 'essential' type. Thus the appropriate variational problem is, by straightforward substitution in (8.49),

$$J[\phi] = \iint_D \left\{ -\left(\frac{\partial \phi}{\partial x}\right)^2 - \left(\frac{\partial \phi}{\partial y}\right)^2 + 4\phi \right\} dx\, dy = \text{extremum}$$

with each admissible function satisfying $\phi = 0$ on Γ.

Apart from an irrelevant sign, this is precisely the functional (8.39a) discussed earlier.

Example 8.8

Determine the variational problem equivalent to the boundary value problem:

$$-\frac{\partial^2 u}{\partial x^2} - \frac{\partial^2 u}{\partial y^2} = 2(x + y) - 4$$

within D (the square shown in Figure 8.13)

$$\text{on} \quad \Gamma_1 : \frac{\partial u}{\partial x} = 2 - 2y - y^2$$

$$\text{on} \quad \Gamma_2 : \frac{\partial u}{\partial y} = 2 - 2x - x^2$$

$$\text{on} \quad \Gamma_3 : u = y^2$$

$$\text{on} \quad \Gamma_4 : u = x^2$$

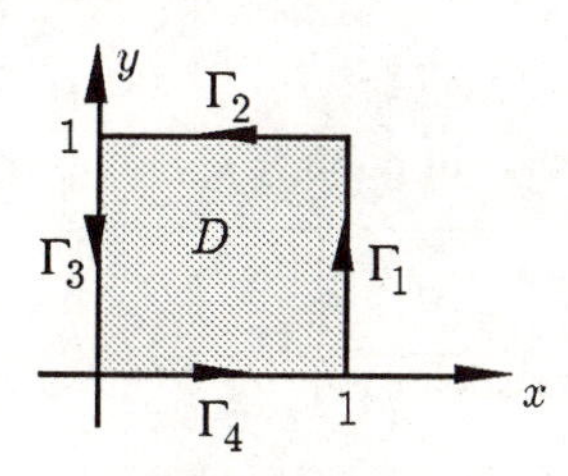

Figure 8.13

Solution

The differential equation part of the boundary value problem is straightforward. We find:

$$A = 1, \quad B = 0, \quad C = 1, \quad F = 0, \quad r = 2\,(x + y) - 4$$

We have to consider each section of the boundary in turn.

(1) Along Γ_1:

In terms of (n, s) coordinates we have here

$$\frac{\partial}{\partial s} \equiv \frac{\partial}{\partial y} \quad \text{and} \quad \frac{\partial}{\partial n} \equiv \frac{\partial}{\partial x}$$

Thus the given boundary condition on Γ_1 may be re-expressed in the form $\partial u/\partial n = 2 - 2y - y^2$ where y is a function of s. This is therefore a suppressible boundary condition. Reading off the coefficients in (8.48c) we have

$$a_1 = 0, \quad a_2 = 0, \quad a_3 = 1, \quad a_4 = 2 - 2y - y^2$$

Condition (8.50) becomes

$$\frac{0}{1} = \frac{(0)\alpha\beta + 0(\alpha^2 - \beta^2)}{\alpha^2 + \beta^2} = 0$$

which is clearly satisfied. Thus for this section of the boundary we have,using (8.51) and (8.52),

$$S = (\alpha^2 + \beta^2)0/a_3 = 0$$

$$T = -(\alpha^2 + \beta^2)(2 - 2y - y^2) = -2 + 2y + y^2$$

Hence, using (8.49) the functional 'includes' a line integral

$$\int_{\Gamma_1} 2(-2 + 2y + y^2)\phi \, ds.$$

(2) Along Γ_2:

Here

$$\frac{\partial}{\partial s} \equiv -\frac{\partial}{\partial x} \quad \text{and} \quad \frac{\partial}{\partial n} \equiv \frac{\partial}{\partial y}$$

Hence the boundary condition on Γ_2 can be written

$$\frac{\partial u}{\partial n} = 2 - 2x - x^2$$

which is also a suppressible condition. Thus for this part of the boundary

$$a_1 = 0, \quad a_2 = 0, \quad a_3 = 1, \quad a_4 = 2 - 2x - x^2$$

Condition (8.50) is again satisfied and so

$$S = 0, \quad T = -(2 - 2x - x^2)/1 = -2 + 2x + x^2$$

Once more the functional $J[\phi]$ includes a line integral of the form

$$\int_{\Gamma_2} 2(-2 + 2x + x^2)\phi \, ds.$$

(3) Along sections Γ_3 and Γ_4 the boundary conditions are essential.

We obtain from (8.49) the final form for the variational problem:

$$J[\phi] = \text{extremum}$$

where

$$J[\phi] = \iint_D \left\{ \left(\frac{\partial \phi}{\partial x}\right)^2 + \left(\frac{\partial \phi}{\partial y}\right)^2 - 2(2(x+y)-4)\,\phi \right\} dx\,dy$$
$$+ \int_{\Gamma_1} 2(-2+2y+y^2)\phi\, ds + \int_{\Gamma_2} 2(-2+2x+x^2)\phi\, ds$$

and the ϕ functions must satisfy

$$\phi = y^2 \quad \text{on} \quad \Gamma_3 \quad \text{and} \quad \phi = x^2 \quad \text{on} \quad \Gamma_4$$

8.11 The Ritz method in two dimensions

The one-dimensional Ritz approximation procedure can be applied to two-dimensional problems with very little change. The one-dimensional approximation (8.23) generalizes to

$$\tilde{u}_k = \theta_0(x,y) + \sum_{i=1}^{k} a_i\,\theta_i\,(x,y) \tag{8.54}$$

where, as we would expect, θ_0 is chosen to satisfy the essential boundary conditions of the given problem and the coordinate functions $\theta_i(x,y), i = 1, 2, \ldots, k$, are chosen to satisfy the homogeneous form of the essential boundary conditions.

The Ritz procedure by itself is of use only if the domain of the boundary value problem is sufficiently regular to permit simple choices for the coordinate functions. However, for irregularly shaped domains the Ritz procedure can be amended by using piecewise polynomial functions – in other words a two-dimensional finite element method – so as to deal with the domain irregularity.

We demonstrate firstly an application of the Ritz method over a regular region which effectively is treated as one element.

Example 8.9

Determine a Ritz approximation to the boundary value problem:

$$-\frac{\partial^2 u}{\partial x^2} - \frac{\partial^2 u}{\partial y^2} = 2(x+y) - 4 \quad \text{in} \quad D \tag{8.55a}$$

$$u = y^2 \text{ on } \Gamma_3 \qquad u = x^2 \text{ on } \Gamma_4 \tag{8.55b}$$

$$\frac{\partial u}{\partial x} = 2 - 2y - y^2 \text{ on } \Gamma_1 \qquad \frac{\partial u}{\partial y} = 2 - 2x - x^2 \text{ on } \Gamma_2 \tag{8.55c}$$

(see Figure 8.14).

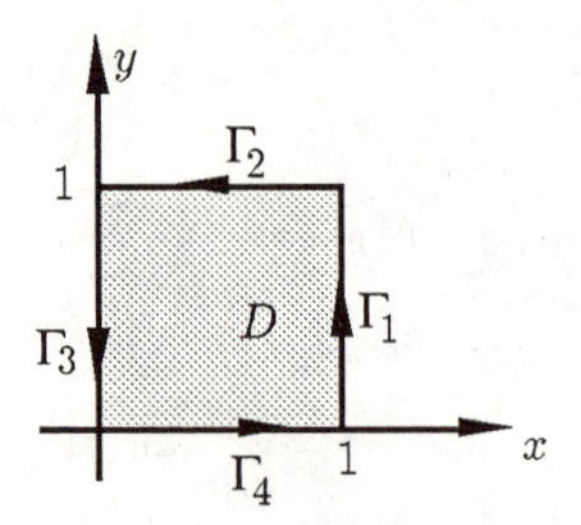

Figure 8.14

Solution

From Example 8.8 the corresponding variational problem is:

$$J[\phi] = \text{extremum}$$

where

$$J[\phi] = \iint_D \left\{ \left(\frac{\partial \phi}{\partial x}\right)^2 + \left(\frac{\partial \phi}{\partial y}\right)^2 - 2(2(x+y)-4)\phi \right\} dx\, dy$$
$$+ \int_{\Gamma_1} 2(-2+2y+y^2)\phi ds + \int_{\Gamma_2} 2(-2+2x+x^2)\phi\, ds \tag{8.56}$$

with $\quad \phi = y^2$ on $\Gamma_3, \qquad \phi = x^2$ on Γ_4

We shall only consider a one-parameter approximation of the form (8.54) viz:

$$\tilde{u}(x,y) = \theta_0(x,y) + a_1\, \theta_1\,(x,y)$$

A function satisfying the two essential boundary conditions (8.55b) is

$$\theta_0(x,y) = x^2 + y^2$$

A simple function satisfying the homogeneous form of these essential boundary conditions is

$$\theta_1(x,y) = x\, y$$

Hence, in this case, (8.54) has the form

$$\tilde{u} = x^2 + y^2 + a_1\, x\, y$$

giving

$$\frac{\partial \tilde{u}}{\partial x} = 2\,x + a_1\, y \qquad \frac{\partial \tilde{u}}{\partial y} = 2\,y + a_1\, x$$

Using this approximation in (8.56)

$$J[\tilde{u}] = \iint_D \{(2\,x + a_1\,y)^2 + (2\,y + a_1\,x)^2 \\ - 2(2(x+y) - 4)(x^2 + y^2 + a_1\,x\,y)\}dx\,dy \\ + \int_{\Gamma_1} 2(-2 + 2\,y + y^2)(x^2 + y^2 + a_1\,x\,y)\,ds \\ + \int_{\Gamma_2} 2(-2 + 2\,x + x^2)(x^2 + y^2 + a_1\,x\,y)\,ds$$

The calculations are easier if we optimise J with respect to a_1 before integrating:

$$\frac{\partial J}{\partial a_1} = \iint_D [4yx + 2a_1y^2 + 4xy + 2a_1x^2 - 2xy(2(x+y) - 4)]dx\,dy \\ + \int_{\Gamma_1} 2xy(-2 + 2y + y^2)ds \\ + \int_{\Gamma_2} 2xy(-2 + 2x + x^2)ds = 0 \qquad (8.57)$$

Now on Γ_1 using a parameter t:

$$x = 1, \quad y = t \qquad 0 < t < 1$$
$$ds = dt$$

$$\therefore \quad \int_{\Gamma_1} 2xy(-2 + 2y + y^2)ds = \int_0^1 2t(-2 + 2t + t^2)dt = -\frac{1}{6}$$

Similarly, on Γ_2, we put

$$y = 1, \quad x = 1 - t \quad 0 < t < 1$$
$$ds = dt$$

and so

$$\int_{\Gamma_2} 2xy(-2 + 2x + x^2)ds = \int_0^1 2(1-t)[-2 + 2(1-t) + (1-t)^2]dt = -\frac{1}{6}$$

The double integral in (8.57) can be re-arranged to

$$\iint_D \{16\,xy - 4\,xy(x+y)\}dx\,dy + a_1 \iint_D 2(y^2 + x^2)dx\,dy$$

Straightforward integration gives

$$\iint_D 16\,x\,y\,dx\,dy = 4 \qquad \iint_D -4\,x\,y(x+y)dx\,dy = -\frac{4}{3}$$

and

$$\iint_D 2(y^2 + x^2)dx\,dy = \frac{4}{3}$$

Hence (8.56) gives

$$4 - \frac{4}{3} + \frac{4}{3}a_1 - \frac{1}{3} = 0$$

or

$$a_1 = -\frac{7}{4}$$

Our one-parameter approximate solution is therefore

$$\tilde{u} = x^2 + y^2 - \frac{7}{4} x\, y$$

This gives a value for the functional of

$$J[\tilde{u}] = \frac{11}{120} = 0.092$$

The exact solution to this problem may be shown to be

$$u = x^2 + y^2 - xy(x + y)$$

with a corresponding functional value

$$J[u] = -\frac{7}{45} = -0.092$$

Finally, in this chapter, we re-work Example 7.3 by the piecewise Ritz method. The boundary value problem was

$$\frac{\partial^2 u}{\partial x^2} + \frac{\partial^2 u}{\partial y^2} + 2 = 0 \quad \text{within} \quad D$$

$$u = 0 \quad \text{on the boundary of} \quad D$$

D being the ellipse shown in Figure 7.13(a). As discussed in the solution to Example 7.3, symmetry considerations give the reduced boundary value problem:

$$\frac{\partial^2 u}{\partial x^2} + \frac{\partial^2 u}{\partial y^2} + 2 = 0 \quad \text{within } A, \text{ the quarter ellipse} \tag{8.58a}$$

$$\frac{\partial u}{\partial x} = 0 \quad \text{on boundary} \quad C_1 : \ x = 0 \tag{8.58b}$$

$$\frac{\partial u}{\partial y} = 0 \quad \text{on boundary} \quad C_2 : \ y = 0 \tag{8.58c}$$

Using Theorem 8.2 the variational problem equivalent to (8.58) is easily found to be

$$J[\phi] = \text{extremum}$$

where

$$J[\phi] = \iint_A \left\{ -\left(\frac{\partial \phi}{\partial x}\right)^2 - \left(\frac{\partial \phi}{\partial y}\right)^2 + 4\,\phi \right\} dx\, dy \tag{8.59a}$$

and all admissible functions must satisfy

$$\phi = 0 \tag{8.59b}$$

on the bounding quarter ellipse. (In this case the suppressible boundary conditions on C_1 and C_2 do not give rise to additional line integral terms in J such as arise in (8.49).)

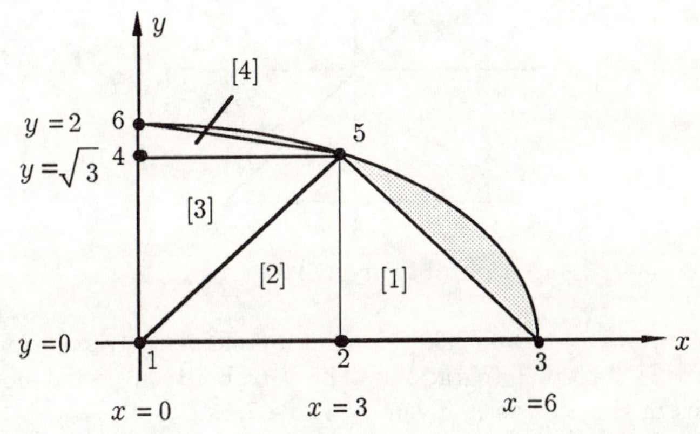

Figure 8.15

To illustrate the piecewise Ritz approach we use the same segmentation as in Example 7.3 viz. four three-noded triangles. See Figure 8.15, a copy of Figure 7.14. Most of the necessary results have already been obtained in Example 7.3 and we will use them freely here. Thus we have for the functional (8.59a)

$$J[\phi] = \sum_{e=1}^{4} \iint_{[e]} \left\{ -\left(\frac{\partial \phi}{\partial x}\right)^2 - \left(\frac{\partial \phi}{\partial y}\right)^2 + 4\,\phi \right\} dx\, dy \tag{8.60}$$

where, within each triangular element, which use the linear approximation (7.55)

$$\tilde{u} = \sum_{i=1}^{3} \tilde{u}_i^{[e]}\, T_i\,(\xi, \eta) \tag{8.61}$$

where ξ and η are related to x and y by the coordinate transformation (7.56)

$$x = \sum_{i=1}^{3} x_i^{[e]}\, T_i\,(\xi, \eta) \tag{8.62a}$$

$$y = \sum y_i^{[e]} T_i(\xi, \eta) \tag{8.62b}$$

This transformation maps each element in turn into the standard triangle ST shown again in Figure 8.16.

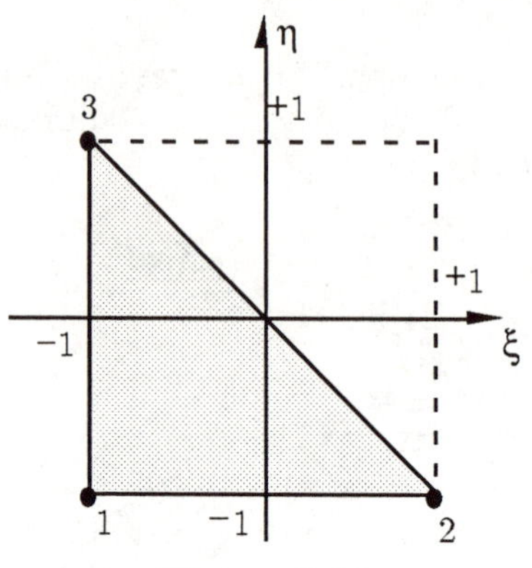

Figure 8.16

We now evaluate (8.60) element by element, transforming each element to ST before integration. The link between global nodes and local nodes in the ST is still given by Table 7.4.

- **Element [1]**

Equations (8.62) become

$$x = \frac{9}{2} + \frac{3}{2}\xi \qquad y = \frac{\sqrt{3}}{2}(1+\eta)$$

giving a Jacobian for the transformation

$$J = \begin{vmatrix} \dfrac{\partial x}{\partial \xi} & \dfrac{\partial x}{\partial \eta} \\ \dfrac{\partial y}{\partial \xi} & \dfrac{\partial y}{\partial \eta} \end{vmatrix} = \begin{vmatrix} \dfrac{3}{2} & 0 \\ 0 & \dfrac{\sqrt{3}}{2} \end{vmatrix} = \frac{3\sqrt{3}}{4}$$

(where J should not be confused with the functional $J[u]$). Also within this element, because of the boundary conditions,

$$\begin{aligned} \tilde{u}(x,y) &= \tilde{u}_2 T_1(\xi, \eta) \\ &= \tilde{u}_2 \left(-\frac{1}{2}(\xi + \eta) \right) \end{aligned}$$

We also need to evaluate derivatives with respect to x and y for use in (8.60). We have

$$\begin{aligned}\frac{\partial \tilde{u}}{\partial x} &= \frac{\partial \tilde{u}}{\partial \xi}\frac{\partial \xi}{\partial x} + \frac{\partial \tilde{u}}{\partial \eta}\frac{\partial \eta}{\partial x} \\ &= \frac{\partial \tilde{u}}{\partial \xi}\left(\frac{2}{3}\right) + 0 \\ &= -\frac{\tilde{u}_2}{3}\end{aligned}$$

$$\begin{aligned}\frac{\partial \tilde{u}}{\partial y} &= \frac{\partial \tilde{u}}{\partial \xi}\frac{\partial \xi}{\partial y} + \frac{\partial \tilde{u}}{\partial \eta}\frac{\partial \eta}{\partial y} \\ &= 0 + \frac{\partial \tilde{u}}{\partial \eta}\left(\frac{2}{\sqrt{3}}\right) \\ &= -\frac{1}{\sqrt{3}}\,\tilde{u}_2\end{aligned}$$

Hence the contribution of element [1] to the functional is

$$\begin{aligned}J_1[\tilde{u}] &= \iint_{ST}\left\{-\frac{\tilde{u}_2^2}{9} - \frac{\tilde{u}_2^2}{3} + 4\,\tilde{u}_2\left(-\frac{1}{2}(\xi+\eta)\right)\right\}\frac{3\sqrt{3}}{4}d\xi d\eta \\ &= -\frac{2\sqrt{3}}{3}\,\tilde{u}_2^2 + 2\sqrt{3}\,\tilde{u}_2\end{aligned} \tag{8.63}$$

where we have used (6.20) for evaluating the integrals.

- **Element [2]**

Here the transformation equations (8.62) are

$$x = 3 + \frac{3}{2}\,(\xi+\eta) \qquad y = \frac{\sqrt{3}}{2}\,(1+\eta)$$

giving a Jacobian $J = 3\sqrt{3}/4$.

The trial solution (8.61) becomes

$$\begin{aligned}\tilde{u} &= \tilde{u}_1\,T_1\,(\xi,\eta) + \tilde{u}_2\,T_2\,(\xi,\eta) \\ &= \tilde{u}_1\left[-\frac{1}{2}(\xi+\eta)\right] + \tilde{u}_2\left[\frac{1}{2}(1+\xi)\right]\end{aligned}$$

Differentiating the transformation equations with respect to x we find, respectively,

$$1 = \frac{3}{2}\left(\frac{\partial \xi}{\partial x} + \frac{\partial \eta}{\partial x}\right) \qquad 0 = \frac{\sqrt{3}}{2}\frac{\partial \eta}{\partial x}$$

from which $\partial\eta/\partial x = 0$ and $\partial\xi/\partial x = 2/3$.

Similarly, differentiating both with respect to y,

$$0 = \frac{3}{2}\left(\frac{\partial \xi}{\partial y} + \frac{\partial \eta}{\partial y}\right) \qquad 1 = \frac{\sqrt{3}}{2}\frac{\partial \eta}{\partial y}$$

from which

$$\frac{\partial \eta}{\partial y} = \frac{2}{\sqrt{3}} \quad \text{and} \quad \frac{\partial \xi}{\partial y} = -\frac{2}{\sqrt{3}}$$

Hence the derivatives needed for (8.60) are

$$\begin{aligned}\frac{\partial \tilde{u}}{\partial x} &= -\frac{1}{2}\tilde{u}_1\left(\frac{\partial \xi}{\partial x} + \frac{\partial \eta}{\partial x}\right) + \frac{1}{2}\tilde{u}_2\frac{\partial \xi}{\partial x} \\ &= -\frac{\tilde{u}_1}{3} + \frac{\tilde{u}_2}{3} \\ \frac{\partial \tilde{u}}{\partial y} &= -\frac{1}{2}\tilde{u}_1\left(\frac{\partial \xi}{\partial y} + \frac{\partial \eta}{\partial y}\right) + \frac{\tilde{u}_2}{2}\frac{\partial \xi}{\partial y} \\ &= 0 - \frac{1}{\sqrt{3}}\tilde{u}_2\end{aligned}$$

Finally the Jacobian for the transformation from element [2] to the standard element is such that its reciprocal is

$$J^{-1} = \begin{vmatrix} \dfrac{\partial \xi}{\partial x} & \dfrac{\partial \xi}{\partial y} \\ \dfrac{\partial \eta}{\partial x} & \dfrac{\partial \eta}{\partial y} \end{vmatrix} = \frac{4}{3\sqrt{3}}$$

Hence the contribution of element [2] to the functional is

$$\begin{aligned}J_2[\tilde{u}] = \iint_{ST}\Bigg\{ &-\left(\frac{-\tilde{u}_1 + \tilde{u}_2}{3}\right)^2 - \frac{\tilde{u}_2{}^2}{3} \\ &+\left(- 4\tilde{u}_1\frac{1}{2}(\xi + \eta) + 4\,\tilde{u}_2\frac{1}{2}(1 + \xi)\right)\Bigg\} 3\frac{\sqrt{3}}{4}\,d\xi\,d\eta\end{aligned}$$

which gives, again using (6.20),

$$\begin{aligned}J_2[\tilde{u}] = \tilde{u}_1^2\left(-\frac{\sqrt{3}}{6}\right) + \tilde{u}_2^2\left(-2\frac{\sqrt{3}}{3}\right) + \tilde{u}_1\,\tilde{u}_2\left(\frac{\sqrt{3}}{3}\right) \\ + \tilde{u}_1(2\sqrt{3}) + \tilde{u}_2(2\sqrt{3})\end{aligned} \tag{8.64}$$

- **Element [3]**

Here

$$x = -\frac{3}{2}(\xi + \eta) \quad y = \frac{\sqrt{3}}{2}(1 + \eta)$$

are the transformation equations and the trial solution is

$$\begin{aligned}\tilde{u} &= \tilde{u}_1 \, T_3 + \tilde{u}_4 \, T_2 \\ &= \frac{\tilde{u}_1}{2}(1+\eta) + \frac{\tilde{u}_4}{2}(1+\xi)\end{aligned}$$

Calculations similar to those performed for elements [1] and [2] give

$$\frac{\partial \tilde{u}}{\partial x} = -\frac{1}{3}\,\tilde{u}_4 \quad \text{and} \quad \frac{\partial \tilde{u}}{\partial y} = \frac{1}{\sqrt{3}}\,(\tilde{u}_1 - \tilde{u}_4)$$

Substitution into (8.60) and evaluation of the integrals gives a contribution to the functional

$$J_3[\tilde{u}] = \tilde{u}_1^2\left(-\frac{\sqrt{3}}{2}\right) + \tilde{u}_4^2\left(-2\frac{\sqrt{3}}{3}\right) + \tilde{u}_1\,\tilde{u}_4\sqrt{3} + \tilde{u}_1\,2\sqrt{3} + \tilde{u}_4\,2\sqrt{3} \quad \textbf{(8.65)}$$

- **Element [4]**

Here

$$x = \frac{3}{2}(1+\xi) \quad y = \frac{(2-\sqrt{3})}{2}\eta + \frac{(2+\sqrt{3})}{2}$$

$$\tilde{u} = \tilde{u}_4\,T_1 = \tilde{u}_4\left(-\frac{1}{2}(\xi+\eta)\right)$$

from which we obtain a contribution to the functional

$$J_4[\tilde{u}] = \tilde{u}_4^2\left\{\frac{2\sqrt{3}-8}{3(2-\sqrt{3})}\right\} + \tilde{u}_4(4 - 2\sqrt{3}) \qquad \textbf{(8.66)}$$

Adding the four equations (8.63) to (8.66) we obtain

$$\begin{aligned}J[\tilde{u}] = \tilde{u}_1^2\left[-2\frac{\sqrt{3}}{3}\right] + \tilde{u}_2^2\left[-4\frac{\sqrt{3}}{3}\right] + \tilde{u}_4^2\left[\frac{-10-6\sqrt{3}}{3}\right] \\ + \tilde{u}_1\,\tilde{u}_2\left[\frac{\sqrt{3}}{3}\right] + \tilde{u}_1\,\tilde{u}_4\sqrt{3} + \tilde{u}_1\,4\sqrt{3} + \tilde{u}_2\,4\sqrt{3} + 4\,\tilde{u}_4\end{aligned}$$

To optimize we set $\partial J/\partial \tilde{u}_1$, $\partial J/\partial \tilde{u}_2$, $\partial J/\partial \tilde{u}_4$ equal to zero giving respectively

$$\begin{aligned}
-\frac{4\sqrt{3}}{3}\tilde{u}_1 + \frac{\sqrt{3}}{3}\tilde{u}_2 + \sqrt{3}\,\tilde{u}_4 + 4\sqrt{3} &= 0 \\
\frac{\sqrt{3}}{3}\tilde{u}_1 - 8\frac{\sqrt{3}}{3}\tilde{u}_2 + 4\sqrt{3} &= 0 \\
\sqrt{3}\,\tilde{u}_1 - \frac{4}{3}(5+3\sqrt{3})\tilde{u}_4 + 4 &= 0
\end{aligned}$$

These are the same equations as those obtained in Example 7.3, illustrating again the equivalence of the Ritz technique to the Galerkin method.

The reader is invited to show that the Ritz and modified Galerkin formulations are equivalent (for the same choice of coordinate functions) for the general linear two-dimensional boundary value problem described in (8.48). In many ways, it is because of this equivalence that the Ritz formulation is less widely used than the modified Galerkin method in connection with finite elements and is the reason why we have concentrated on the latter in this text. However, there are a number of situations in which a variational problem arises naturally and, in those situations, a Ritz formulation is preferred. One such application occurs linear elasticity theory as discussed in Chapter 5 and in Section 7.7.

EXERCISES

8.1 (a) By considering the set of admissible functions

$$\phi(x) = u(x) + \varepsilon\eta(x)$$

(with the usual properties) determine the boundary value problem satisfied by the function $u(x)$ which makes the functional

$$J[\phi] = \int_0^1 \{(\phi')^2 - \phi^2 - 2x^2\phi\}\, dx$$

an extremum. Here $\phi(0) = \phi(1) = 0$.

(b) Show that, had $u(x)$ not been specified at $x = 1$, then the function which optimizes $J[\phi]$ is

$$u(x) = -2\cos x + (2 - 2\sin 1)\frac{\sin x}{\cos 1} - x^2 + 2$$

8.2 Show that

$$\int_0^1 \{(\phi')^2 + 2x\phi\}\, dx + \phi^2(0) + \phi^2(1) \geq -\frac{41}{270}$$

and that the equality is achieved with the function

$$u(x) = \frac{x^3}{6} - \frac{2}{9}(x+1)$$

8.3 Use the Euler–Lagrange equation to show that the function $u(x)$ which optimizes the functional

$$\int_0^{\pi/2} \{-(\phi')^2 + \phi^2\}\, dx$$

subject to $\phi(0) = 1,\ \phi(\pi/2) = 1$ also satisfies the boundary value problem:

$$\frac{d^2u}{dx^2} + u = 0 \qquad 0 \leq x \leq \frac{\pi}{2}$$

$$u(0) = 1 \qquad u(\frac{\pi}{2}) = 1$$

(b) Develop a piecewise linear approximation to $u(x)$ in the above boundary value problem using three equal length elements. Show that the approximation to $u(x)$ is such that (using an obvious numbering system) $\tilde{u}_2 = \tilde{u}_3 = 1.355$. How does your approximation compare with the exact solution?
(c) Repeat (b) using the matrix formulation of Section 8.7.

8.4 Determine corresponding variational problems for the following one-dimensional boundary value problems:

(a) $$\frac{d^2u}{dx^2} + (1+x^2)u + 1 = 0 \qquad -1 \leq x \leq 1$$

$$u(-1) = 0, \quad u(1) = 0$$

(b) $$-\frac{d^2u}{dx^2} + u = 2\cos x \qquad \frac{\pi}{2} \leq x \leq \pi$$

$$u'(\frac{\pi}{2}) + 3u(\frac{\pi}{2}) = -1$$

$$u'(\pi) + 4u(\pi) = -4$$

(c) $$x\frac{d^2u}{dx^2} + \lambda u = 0 \qquad 0 \leq x \leq 1$$

$$u(0) = u(1) = 0$$

8.5 Determine corresponding variational problems for the following two-dimensional boundary value problems:

(a) $$-\frac{\partial^2 u}{\partial x^2} - \frac{\partial^2 u}{\partial y^2} = 1 \qquad \text{within } D$$

where D is the triangular region bounded by the lines $x = 0$, $y = 0$ and $y = 1 - x$

$$u = 0 \quad \text{on } x = 0,\ y = 0$$

$$\frac{\partial u}{\partial n} = 0 \quad \text{on } y = 1 - x$$

(b) $$\frac{\partial^2 u}{\partial x^2} + \frac{\partial^2 u}{\partial y^2} = -1 \qquad \text{within } D$$

$$\frac{\partial u}{\partial n} + u = 0 \qquad \text{on the boundary of } D$$

where D is the square $-1 \leq x,\ y \leq 1$.

8.6 Show, for the following linear boundary value problem that the equations pertinent to the Ritz and modified Galerkin procedures using an n-parameter approximation are equivalent:

$$\frac{\partial^2 u}{\partial x^2} + \frac{\partial^2 u}{\partial y^2} = f(x, y) \qquad \text{within a region } D$$

$$\frac{\partial u}{\partial n} + u = 0 \qquad \text{on the boundary of } D$$

8.7 Reformulate the boundary value problem described in Q2.10 as a variational problem. Hence, using the Ritz method, with two parameters, verify that (for the same choice of coordinate functions) the same equations as those arising from use of the modified Galerkin method are obtained.

8.8 Using an n-parameter trial solution:

$$\tilde{u} = \theta_0(x) + \sum_{i=1}^{n} a_i \theta_i(x)$$

formulate a matrix approach for the Ritz procedure (as in Section 8.7) to determine the equations satisfied by the parameters $a_1, a_2, \ldots, a_n$ for the boundary value problem:

$$-\frac{d}{dx}\left(p(x)\frac{du}{dx}\right) + q(x)u = r(x) \qquad a \leq x \leq b$$

$$u(a) = \gamma_1 \qquad u(b) = \gamma_2$$

8.9 Apply the matrix procedure developed in Q8.8 to obtain an approximate solution to the boundary value problem

$$\frac{d^2u}{dx^2} + u = e^x \qquad 0 \le x \le 1$$

$$u(0) = 0 \qquad u(1) = 0$$

Use a trial solution of the form:

$$\tilde{u}(x) = \alpha_0 + \alpha_1 x + \alpha_2 x^2 + \alpha_3 x^3$$

8.10 Repeat Q4.5 of Chapter 4 using a variational formulation.

8.11 Repeat Q7.3 of Chapter 7 using a variational formulation.

8.12 Repeat Q7.8 of Chapter 7 using a variational formulation. The results (7.48), (7.49) and similar relations obtained in Q7.7 may, of course, be used.

SUPPLEMENT

8S.1 Matrix differentiation

In this supplement we obtain certain results concerning 'matrix differentiation' that have been used explicitly in this chapter in connection with the Ritz method.

Let $[A]$ be an $n \times m$ matrix whose components a_{ij} may be functions of a single parameter u. The derivative of $[A]$ with respect to u is written $d[A]/du$ and is an $n \times m$ matrix whose components are da_{ij}/du.

For example, if

$$[A] = \begin{bmatrix} u & 1 & \sin u & u \\ 0 & u^2 + u & \cos u & u^2 \\ 1 & 2 & u^3 & u^3 \end{bmatrix}$$

then

$$\frac{d[A]}{du} = \begin{bmatrix} 1 & 0 & \cos u & 1 \\ 0 & 2u + 1 & -\sin u & 2u \\ 0 & 0 & 3u^2 & 3u^2 \end{bmatrix}$$

It follows from this definition that the matrix derivative possesses the usual properties of all derivative operators:

$$\text{(a)} \qquad \frac{d}{du}([A]+[B]) = \frac{d[A]}{du} + \frac{d[B]}{du}$$

$$\text{(b)} \qquad \frac{d}{du}([A][B]) = [A]\frac{d[B]}{du} + \frac{d[A]}{du}[B]$$

whenever the product $[A][B]$ is defined. The proof of (a) is obvious. The proof of (b), not surprisingly, utilizes the definition of the matrix product. If $[A]$ is an $n \times k$ matrix with components a_{ij} and $[B]$ is a $k \times m$ matrix with components b_{ij} then $[C] = [A][B]$ is an $n \times m$ matrix with components c_{ij} given by

$$c_{ij} = \sum_{p=1}^{k} a_{ip} b_{pj}$$

Hence

$$\frac{dc_{ij}}{du} = \sum_{p=1}^{k} \left(\frac{da_{ip}}{du} b_{pj} + a_{ip} \frac{db_{pj}}{du} \right)$$

that is

$$\frac{d[C]}{du} = \frac{d[A]}{du}[B] + [A]\frac{d[B]}{du}$$

Let $\{u\}$ be an $n \times 1$ column vector with components u_i, $i = 1, 2, \ldots, n$. We define the symbol $\partial/\partial\{u\}$ to denote the $n \times 1$ column vector whose 'components' are

$$\frac{\partial}{\partial u_i} \qquad i = 1, 2, \ldots, n$$

Employing this notation it follows that if K is a scalar function of $u_1, u_2, \ldots, u_n$ then the matrix product of $\partial/\partial\{u\}$ with K, written $\partial K/\partial\{u\}$, is

$$\frac{\partial K}{\partial\{u\}} = \left\{ \frac{\partial K}{\partial u_1}, \frac{\partial K}{\partial u_2}, \ldots, \frac{\partial K}{\partial u_n} \right\}^T$$

For example, if $K = u_1^2 + u_2^2 + u_1 u_3$ and if $\{u\} = \{u_1, u_2, u_3\}^T$ we have

$$\frac{\partial K}{\partial\{u\}} = \{2u_1 + u_3, 2u_2, u_1\}^T$$

More generally, if $\{u\}$ is an $n \times 1$ column vector with components u_i, $i = 1, 2, \ldots, n$, then **any** linear expression in these u_i components may be written in matrix form:

$$K_L = \{u\}^T \{F\}$$

where $\{F\}$ is a column vector all of whose terms are constant. For example if $K_L = 3u_1 + 2u_2 + 4u_3$ then $K_L = \{u\}^T\{F\}$ in which $\{F\} = \{3, 2, 4\}^T$. In such cases, using the normal rules of matrix multiplication:

$$\frac{\partial K_L}{\partial \{u\}} = \frac{\partial}{\partial \{u\}} \left(\{u\}^T \{F\}\right) = \frac{\partial \{u\}^T}{\partial \{u\}} \{F\}$$

$$= \{F\}$$

since $\dfrac{\partial \{u\}^T}{\partial \{u\}} = \mathrm{I}$, the $n \times n$ identity matrix

The above expression for K, viz $K = u_1^2 + u_2^2 + u_1 u_3$ is called a **quadratic form** in the variables u_1, u_2, u_3. A quadratic form can be expressed in matrix form:

$$K = \{u\}^T [A] \{u\}$$

where $[A]$ is **chosen** to be a symmetric matrix. In this case

$$[A] = \begin{bmatrix} 1 & 0 & 1/2 \\ 0 & 1 & 0 \\ 1/2 & 0 & 0 \end{bmatrix}$$

Now a general result of considerable use in the text (see Section 8.7) is:

$$\frac{\partial K}{\partial \{u\}} = \frac{\partial}{\partial \{u\}} \left(\{u\}^T [A] \{u\}\right) = 2[A]\{u\}$$

The proof employs the 'component' construction of K. If a_{ij} are the components of the symmetric matrix $[A]$ (so that $a_{ij} = a_{ji}$) then any quadratic form may be expressed as:

$$K = \sum_{j=1}^{n} \sum_{i=1}^{n} a_{ij} u_i u_j$$

Now

$$\frac{\partial K}{\partial u_k} = \sum_{j=1}^{n} \sum_{i=1}^{n} (a_{ij} \delta_{ik} u_j + a_{ij} u_i \delta_{jk})$$

where

$$\delta_{ik} = \begin{cases} 1 & \text{if } i = k \\ 0 & \text{if } i \neq k \end{cases}$$

and similarly for δ_{jk}.
Hence

$$\begin{aligned} \frac{\partial K}{\partial u_k} &= \sum_{j=1}^{n} a_{kj} u_j + \sum_{i=1}^{n} a_{ik} u_i \\ &= 2 \sum_{j=1}^{n} a_{kj} u_j \end{aligned}$$

since $[A]$ is symmetric. Thus the components of $\partial K / \partial \{u\}$ are the components of $2[A]\{u\}$.

We may apply these results to a functional similar to the one considered in Section 8.7:

$$J = \frac{1}{2} \{u\}^T [A] \{u\} + \{u\}^T \{F\}$$

In optimizing with respect to the variables $\{u\}$ we obtain

$$\frac{\partial J}{\partial \{u\}} = [A]\{u\} + \{F\} = 0$$

an $n \times n$ system of equations.

For example if

$$J = 2u_1^2 + 4u_1 u_2 + u_3^2 + 2u_1 u_3 + 2u_1 + u_2 + u_3 \tag{8S.1}$$

then, in matrix form:

$$J = \{u\}^T [A] \{u\} + \{u\}^T \{F\}$$

where

$$\{u\} = \begin{Bmatrix} u_1 \\ u_2 \\ u_3 \end{Bmatrix} \qquad [A] = \begin{bmatrix} 2 & 2 & 1 \\ 2 & 0 & 0 \\ 1 & 0 & 1 \end{bmatrix} \qquad \{F\} = \begin{Bmatrix} 2 \\ 1 \\ 1 \end{Bmatrix}$$

Hence by the results proved in this supplement

$$\frac{\partial J}{\partial \{u\}} = 2[A]\{u\} + \{F\} = \begin{Bmatrix} 4u_1 + 4u_2 + 2u_3 + 2 \\ 4u_1 + 1 \\ 2u_1 + 2u_3 + 1 \end{Bmatrix}$$

which is clearly correct since, by differentiation of (8S.1),

$$\frac{\partial J}{\partial u_1} = 4u_1 + 4u_2 + 2u_3 + 2$$

$$\frac{\partial J}{\partial u_2} = 4u_1 + 1$$

$$\frac{\partial J}{\partial u_3} = 2u_1 + 2u_3 + 1$$

9

Pre- and Post-processing, and Solution Assembly

PREVIEW

This chapter introduces techniques and procedures used in the generation of finite element meshes and in the graphical display of finite element output. We find that properties of elements and shape functions are applied in these important aspects of finite element analysis. Consequently material introduced in earlier chapters is used in this chapter. We also introduce two practical approaches, the banded matrix technique and the frontal method, to the assembly and consequent solution of finite element equations.

9.1 Mesh generation

Any realistic use of the finite element method for solving boundary value problems involves the collating together of a large amount of information. This information can be divided, quite naturally, into two blocks which comprise the basic input data to a finite element computer program:

- Information concerned with the mathematical description of the problem: the differential equation and the boundary conditions.

- Element 'topology': number of nodes, number of elements, nodal coordinate positions and global node numbers associated with each element.

The topological data can be very extensive indeed. Consider, for example, a two-dimensional region meshed with 16 eight-noded elements, a very modest number compared with the hundreds or thousands of elements often used in practice. For the 16 elements we would need to log

(a) the x and y coordinates of each node (of which there are 65) i.e. 130 pieces of information

(b) the node numbers attached to each element which involves recording eight numbers in the correct order for each of the sixteen elements

(c) the node numbers attached to the boundary elements.

All this information would have to be put into a data file for input into a computer program. Clearly, if the information is determined manually and then transferred manually to a data file, errors of calculation and/or transcription can easily occur.

What is required is an 'automatic' procedure for generating all the information in the element topology block. This is the information which is supplied by a '**mesh generator**' and is part of the 'preprocessing' required in a practical implementation of finite elements. We now show how simple mesh generators can be written for both one and two-dimensional problems.

9.1.1 One-dimensional mesh

In this case the domain over which the boundary value problem is defined is a section, say $a \leq x \leq b$, of the x-axis (see Figure 9.1).

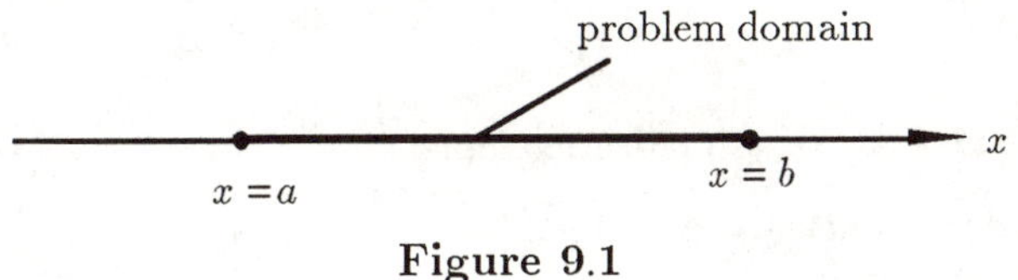

Figure 9.1

For the sake of example, we will mesh the region using three quadratic elements of equal length. For generality we let *nel* denote the total number of elements and *nde* denote the total number of nodes ($nde =$ 7 and $nel = 3$ here). The information required for a finite element analysis is detailed in Figure 9.2.

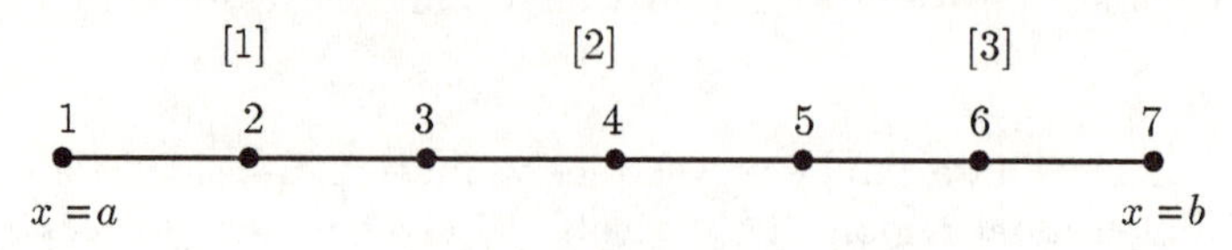

Figure 9.2

We can conveniently store the global node number information for each element in an array $nenn\,(i,j)$ where $i = 1, 2, \ldots, nel$ and $j = 1, 2, 3$ for three-noded elements.

For the simple mesh of Figure 9.2 the terms of this array are

$$nenn\,(1,1) = 1 \qquad nenn\,(1,2) = 2 \qquad nenn\,(1,3) = 3$$
$$nenn\,(2,1) = 3 \qquad nenn\,(2,2) = 4 \qquad nenn\,(2,3) = 5$$
$$nenn\,(3,1) = 5 \qquad nenn\,(3,2) = 6 \qquad nenn(3,3) = 7$$

The nodal coordinates can be stored in an array $xnode\ (i)$, $i = 1, 2, \ldots,$ nde. For the above mesh we have

$$xnode\,(1) = a$$
$$xnode\,(2) = a + (b-a)/6$$
$$xnode\,(3) = a + 2(b-a)/6$$
$$xnode\,(4) = a + 3(b-a)/6$$
$$xnode\,(5) = a + 4(b-a)/6$$
$$xnode\,(6) = a + 5(b-a)/6$$
$$xnode\,(7) = a + 6(b-a)/6 = b$$

Our immediate purpose is to outline an algorithm which will specify nde, $nenn\,(i,j)$ and $xnode\,(i)$ for any given value of nel and any given domain $a \le x \le\ b$.

The coordinate information is easy to obtain in general form. If $\ell = (b-a)/nel$ is the element length then

$$xnode\,(i) = a + (i-1)\ell/2 \tag{9.1}$$

It is also clear that for a quadratic element

$$nde = 2(nel) + 1 \tag{9.2}$$

This last statement, (9.2), gives us an easy way to determine $nenn\,(e,1)$. Because there are $(e-1)$ elements to the left of element $[e]$ (see Figure 9.3) we find that the node number of the first node of element $[e]$ is

$$nenn\,(e,1) = 2(e-1)+1 = 2e-1$$

Hence

$$nenn\,(e,2) = 2e$$

$$nenn\,(e,3) = 2e+1 \tag{9.3}$$

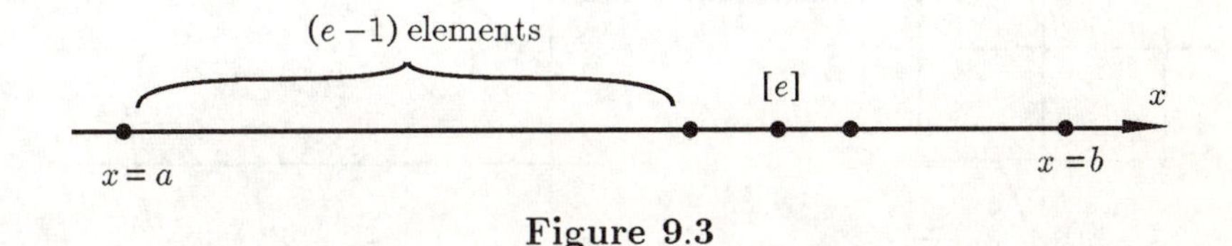

Figure 9.3

This completes the element topology required for any one-dimensional finite element analysis (using three-noded elements).

The reader should not find it too taxing to consider the effect on these formulae of accommodating a variable element length ℓ (which affects only (9.1)), to consider other element types (linear, cubic etc) or indeed to produce a mesh incorporating different element types.

9.1.2 Two-dimensional mesh

In two dimensions the problem of generating topological data is slightly more complex than in one dimension. As well as a meshing task we have also to consider geometric factors since the domain over which the boundary value problem is defined must also be approximated.

The geometric problem is relatively easily solved, particularly if we consider only simple regions. We have seen in Section 6.5 how, for example, a standard eight-noded element can be transformed into a quadrilateral with four parabolic edges. In this section we will assume that the domain under consideration can be modelled as such a quadrilateral (see Figure 9.4). Each 'point' in the standard domain is transformed into a point in the problem domain using

$$x = \sum_{i=1}^{8} X_i\, Q_i(\xi,\eta) \qquad y = \sum_{i=1}^{8} Y_i\, Q_i(\xi,\eta) \tag{9.4}$$

where $Q_i(\xi,\eta)$ are the shape functions (6.25) appropriate to an eight-noded element. Here (X_i, Y_i), $i = 1, 2, \ldots, 8$, are 'control points' on the boundary of the problem domain chosen so that the four parabolic edges model this boundary accurately. (We note that if the boundary cannot be modelled in this way with sufficient accuracy then the problem domain can be split into a number of sub-regions, each of which can be accurately modelled as described in this section. We leave this situation as an extension of the basic approach which the reader can carry through.)

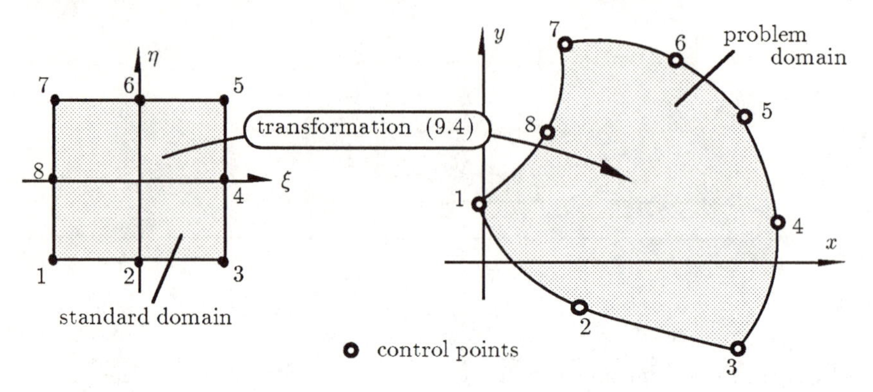

Figure 9.4

It is clear that in order to mesh the actual problem domain it will be sufficient to mesh the standard domain. This is because the coordinate transformation (9.4) will then transform the standard mesh so as to cover the problem domain.

To be specific we shall use eight-noded quadrilateral elements, and number both nodes and elements from left to right as we proceed from lower 'levels' to higher 'levels' (see Figure 9.5).

In this mesh we have chosen n elements in the ξ–direction and m elements in the η–direction. Our purpose is to determine the element topology for a general element $[e]$ which is highlighted in the figure. If the elements are graded evenly in both directions then the width w and height h are $w = 2/n$ and $h = 2/m$ respectively, remembering that the standard domain is a square of side 2.

From what we have learnt in analysing one-dimensional regions it is obvious that the odd-numbered levels contain $(2n + 1)$ nodes whilst clearly the even-numbered levels have $(n+1)$ nodes. Also, if element $[e]$ is located at the intersection of column N and row M (both N and M are odd-numbered levels) then there are $(M - 1)$ full even-rows and $(M - 1)$ full odd-rows below $[e]$. This gives $(M-1)(2n+1)+(M-1)(n+1)$ nodes arising from the 'full' levels. Also there are $2(N - 1) + 1$ nodes arising from the nodes on row M itself. By simple addition, the node number of the node marked A (Figure 9.5) is

$$3Mn + 2(M + N) - 3n - 3$$

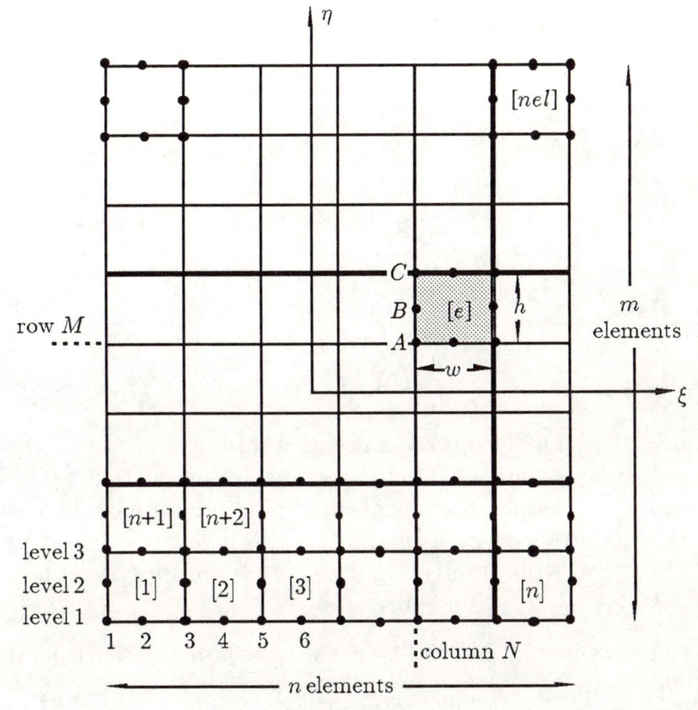

Figure 9.5

The reader should use similar arguments to deduce that the nodes marked B and C are numbered

$$3Mn + 2M + N - n - 1 \qquad \text{and} \qquad 3Mn + 2(M + N) - 1$$

respectively.

The full nodal information for element $[e]$ can then be recorded in an array $nenn\,(e, j)$, $j = 1, 2, \ldots, 8$, as follows:

$$\begin{aligned}
nenn\,(e,1) &= 3Mn + 2(M+N) - 3n - 3 \\
nenn\,(e,2) &= nenn\,(e,1) + 1 \\
nenn\,(e,3) &= nenn\,(e,1) + 2 \\
nenn\,(e,4) &= 3Mn + 2M + N - n \\
nenn\,(e,5) &= 3Mn + 2(M+N) + 1 \\
nenn\,(e,6) &= nenn\,(e,5) - 1 \\
nenn\,(e,7) &= nenn\,(e,5) - 2 \\
nenn\,(e,8) &= nenn\,(e,4) - 1
\end{aligned}$$

where we have used the local node numbering convention shown in Figure 9.4. All that remains is to outline an algorithm for finding the values of

M and N for a given $[e]$. This is accomplished in the following pseudo-program segment:

given e, n, m
initialise $N = 1$ and $M = 1$
if $(e - n) > 0$ **then**
$M = M + 1$
$e = e - n$
else
$N = e$
end if

This routine continually subtracts multiples of n from e until e is reduced to a number less than n. Each time a subtraction takes place M is increased by 1 corresponding to moving up an odd-numbered level. N is clearly the final value of $(e-n)$ just before this quantity becomes negative.

In all, over the standard domain, there are $(m+1)$ even rows and m odd rows giving a total number of nodes:

$$nde = (m+1)(2n+1) + m(n+1) = 3mn + 2(n+m) + 1 \quad \textbf{(9.5)}$$

and, obviously, a total number of elements:

$$nel = mn \quad \textbf{(9.6)}$$

Having deduced element/node number relations the nodal coordinates can be obtained relatively easily. It is clear that the nodal coordinates of the node marked A in Figure 9.5 are

$$\xi = -1 + Nw, \qquad \eta = -1 + Mh \quad \textbf{(9.7)}$$

From this, the nodal coordinates of all other nodes attached to element $[e]$ may be deduced without difficulty, so that arrays $xnode\,(i)$, $ynode\,(i)$, $i = 1, 2, \ldots, nde$, which store the problem domain nodal coordinates can be constructed using (9.4).

In some boundary value problems, as we saw in Chapter 7, line integrals need to be evaluated and, in these cases, boundary elements along the edge of the problem domain need to be considered. Fortunately, as the mesh on the boundary is directly dependent upon the mesh chosen for the interior, little additional work is needed. We record the boundary element node numbers in the array $nennb(b, j)$ where $j = 1, 2, 3$ (since eight-noded 'interior' elements induce three-noded elements on the boundary) and b is an index which ranges over the number of elements on the boundary. Considering each side of the standard domain separately we number the boundary elements in an anti-clockwise fashion as shown in Figure 9.6.

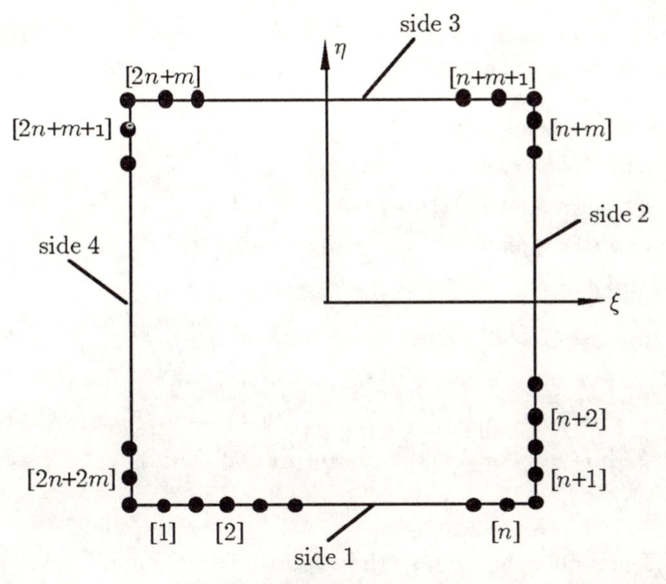

Figure 9.6

Clearly on side 1

$$\begin{aligned}
nennb(1,1) &= nenn\,(1,1) = 1\\
nennb(1,2) &= nenn\,(1,2) = 2\\
nennb(1,3) &= nenn\,(1,3) = 3\\
&\vdots\\
nennb(2,3) &= nenn\,(2,3) = 5 \qquad \text{etc.}
\end{aligned}$$

and in general on this side

$$\begin{aligned}
nennb(b,1) &= nenn\,(b,1)\\
nennb(b,2) &= nenn\,(b,2) \qquad 1 \le b \le n\\
nennb(b,3) &= nenn\,(b,3)
\end{aligned}$$

On side 2 the calculation is more complex but is easily organized in the following pseudo-program segment:

for $i = 1$ **to** m **do**
$nennb(n+i,1) = nenn(ni,3)$
$nennb(n+i,2) = nenn(ni,4)$
$nennb(n+i,3) = nenn(ni,5)$
end for

Similarly the following two pseudo-program segments determine the boundary node information on sides 3 and 4 respectively:

for $i = 1$ **to** n **do**
$nennb(n+m+i,1) = nenn\,(nm+1-i,5)$
$nennb(n+m+i,2) = nenn\,(nm+1-i,6)$
$nennb(n+m+i,3) = nenn(nm+1-i,7)$
end for

for $i = 1$ **to** m **do**
$nennb(2n+m+i,1) = nenn\,(nm+1-ni,7)$
$nennb(2n+m+i,2) = nenn\,(nm+1-ni,8)$
$nennb(2n+m+i,3) = nenn(nm+1-ni,1)$
end for

The reader should take the opportunity of validating the formulae developed in this section either manually or (preferably) by writing simple mesh generators for one- and two-dimensional problems.

9.2 Assembly and solution

As we have seen in Chapters 4 and 7 the overall system of equations arising from a finite element analysis is concisely expressed (for linear problems) in the matrix form

$$[K]\{\tilde{u}\} = \{F\} + \{B\}$$

where $[K]$ is a symmetric matrix and $\{\tilde{u}\}$, $\{F\}$ and $\{B\}$ are column vectors. These equations are normally solved using a direct numerical procedure such as Gaussian elimination.

The final forms of $[K]$, $\{F\}$ and $\{B\}$ are obtained by assembling from element contributions as outlined for two-dimensional problems in Sections 7.4 and 7.5. That is,

$$[K] = \sum_{e=1}^{nel} [K]^{[e]} \qquad \{F\} = \sum_{e=1}^{nel} \{F\}^{[e]} \qquad \{B\} = \sum_{b=1}^{nbe} \{B\}^{[b]}$$

where nel and nbe are the total number of interior and boundary elements respectively. Here, generally, each member of $[K]^{[e]}$, $\{F\}^{[e]}$ and $\{B\}^{[b]}$ is obtained by numerical integration. A detailed examination of this procedure was carried through in Examples 7.2 and 7.3 (although in those examples all integrations were performed analytically).

In a typical finite element problem involving large numbers of elements the assembly will, of necessity, be carried out by a computer and some care (or, in computer jargon, 'housekeeping') is needed to ensure that each member of $[K]^{[e]}$, $\{F\}^{[e]}$ and of $\{B\}^{[b]}$ is assigned to the correct position in the global matrices $[K]$, $\{F\}$ and $\{B\}$ respectively. This 'positioning-information' is stored in the arrays $nenn(e,i)$ for the 'interior' elements and in $nennb(b,i)$ for the 'boundary-elements', as discussed in the previous section.

However, there are still problems in storing all the terms of $[K]$ in a computer memory. The core storage of any computer is finite and, for large problems, it is desirable to consider ways of minimizing the storage requirements.

We outline two approaches that have been used – the banded matrix approach (BM) and the frontal method (FM). The BM approach is the easier to understand and to program. The frontal method requires considerable housekeeping programming skills and could only be considered by advanced programmers.

9.2.1 Banded matrices

For many problems (including all those considered in this text) the coefficient matrix $[K]$ is symmetric and hence an obvious way to save storage space is to store only those components in the 'upper-triangle'. Thus, instead of storing the $(nde)^2$ terms of $[K]$ we need only store $\frac{1}{2}nde(nde+1)$ terms. However, for many finite element meshes, we can do considerably better than this since many of the terms of $[K]$ will be zero and so need not be stored. To clarify this point consider the simple mesh shown in Figure 9.7(a).

If we consider element $[c]$, for example, the element matrix $[K]^{[c]}$ will have the following form:

$$\begin{array}{c} \quad\;\; \begin{array}{cccc} 5 & 6 & 8 & 7 \end{array} \\ [K]^{[c]} = \left[\begin{array}{cccc} * & * & 0 & * \\ * & * & * & * \\ 0 & * & * & * \\ * & * & * & * \end{array}\right] \begin{array}{c} 5 \\ 6 \\ 8 \\ 7 \end{array} \end{array}$$

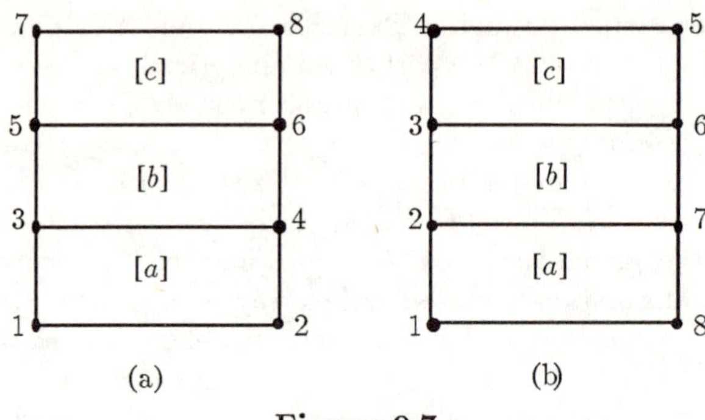

Figure 9.7

in which an asterisk denotes a non-zero term. Some terms of an element matrix may, fortuitously, be zero such as $K^{[c]}_{13} = K^{[c]}_{31} = 0$ here.

As usual we have attached the global node numbers of the element to indicate precisely where each term will be located in the global matrix $[K]$. Of course, these numbers are $nenn(c, i),\ i = 1, 2, 3, 4$.

When all three element matrices are superposed into $[K]$ we obtain:

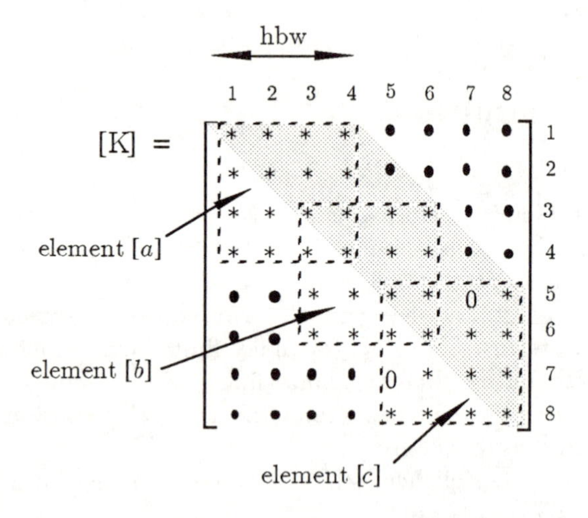

The matrix $[K]$ has a marked **banded structure**. The reader should note that if a different global node numbering scheme is used, for example that shown in Figure 9.7(b), the strongly banded structure is lost.

All the essential non-zero terms are contained within the shaded band. These terms can be stored in an 8×4 rectangular matrix:

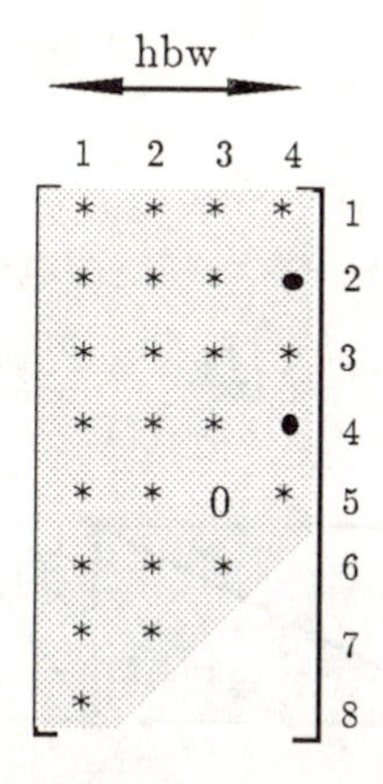

Here we have stored some zeroes and a small 'lower triangle' unnecessarily. However, to eliminate all such non-essential terms would considerably increase the housekeeping programming and is not generally carried out. Nonetheless, we have made a significant saving in memory requirement since all information is now stored in an $nde \times hbw$ rectangular matrix instead of an $\frac{1}{2}nde(nde+1)$ square matrix $[K]$. The quantity hbw is called the **half-band width** and is defined as:

$hbw = 1+$ maximum difference in global node numbers per element

In our example $hbw = 4$, whereas for the global numbering scheme of Figure 9.7(b) we would have $hbw = 8$ implying no saving at all.

There exist general-purpose equation solvers (based on the Gaussian elimination procedure) which can be applied directly to systems of linear equations written in banded form, although, as would be expected, the computer coding is more complex than for the standard procedure. As well as economizing in storage space the BM approach also produces a significant saving in the time taken to execute the solution procedure. The number of operations (multiplications or divisions) in the banded routine can be shown to be approximately $\frac{1}{2}nde(hbw)^2$ compared to approximately $\frac{1}{6}(nde)^3$ for a standard routine applied to a square symmetric matrix. This is particularly useful in practical situations where nde may be many times larger than hbw.

9.2.2 The frontal method

This method makes use of the fact that when in the Gaussian elimination process a variable is eliminated from the system $[K]\{\tilde{u}\} = \{F\}$ many of the terms in $[K]$ and $\{F\}$ remain unaffected. This implies that the solution procedure can begin **before** the process of assembly of $[K]$ and $\{F\}$ is complete. As we shall see, this can lead to a significant reduction in computer storage requirements.

A simple example will illustrate the approach. Consider the triangular mesh shown in Figure 9.8.

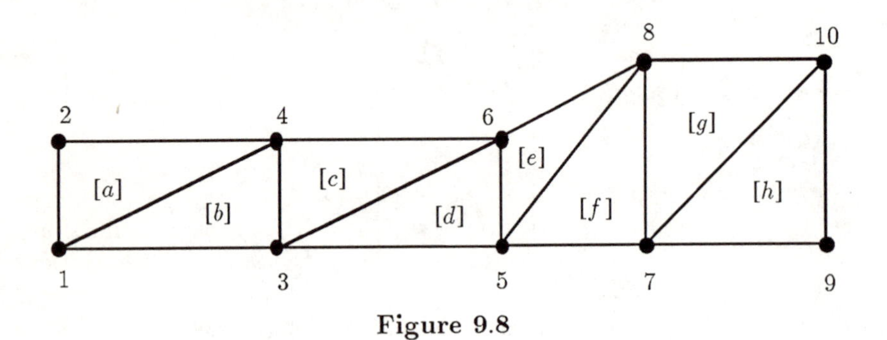

Figure 9.8

Assume, for the sake of argument, that

$$[K]^{[a]} = \begin{array}{c} \begin{array}{ccc} \scriptstyle 1 & \scriptstyle 4 & \scriptstyle 2 \end{array} \\ \left[\begin{array}{ccc} 1 & 1 & 1 \\ 1 & 4 & 1 \\ 1 & 1 & 2 \end{array} \right] \end{array} \begin{array}{c} \scriptstyle 1 \\ \scriptstyle 4 \\ \scriptstyle 2 \end{array} \qquad \{F\}^{[a]} = \left\{ \begin{array}{c} 1 \\ 4 \\ 2 \end{array} \right\} \begin{array}{c} \scriptstyle 1 \\ \scriptstyle 4 \\ \scriptstyle 2 \end{array}$$

The global matrices $[K]$ (a 10×10 symmetric matrix) and $\{F\}$ (a 10×1 column vector) are both initialized with all components at zero. On superposing $[K]^{[a]}$ into $[K]$ and $\{F\}^{[a]}$ into $\{F\}$ we obtain

$$\begin{array}{c} \begin{array}{ccccccc} \scriptstyle 1 & \scriptstyle 2 & \scriptstyle 3 & \scriptstyle 4 & \scriptstyle . & \scriptstyle . & \scriptstyle 10 \end{array} \\ \left[\begin{array}{ccccccc} 1 & 1 & 0 & 1 & . & . & . \\ 1 & 2 & 0 & 1 & . & . & . \\ 0 & 0 & 0 & 0 & . & . & . \\ 1 & 1 & 0 & 4 & . & . & . \\ . & . & . & . & . & . & . \\ . & . & . & . & . & . & . \\ . & . & . & . & . & . & . \end{array} \right] \end{array} \left\{ \begin{array}{c} \tilde{u}_1 \\ \tilde{u}_2 \\ \tilde{u}_3 \\ \tilde{u}_4 \\ . \\ . \\ \tilde{u}_{10} \end{array} \right\} = \left\{ \begin{array}{c} 1 \\ 2 \\ 0 \\ 4 \\ . \\ . \\ . \end{array} \right\}$$

No other element is attached to node 2 so that the second row (and second column) of both $[K]$ and $\{F\}$ will be unaffected by assembly of further elements. Hence **at this stage** we can eliminate the variable $\tilde{u}_2$:

$$\tilde{u}_2 = \frac{1}{2}(2 - \tilde{u}_1 - \tilde{u}_4)$$

Hence, substituting for $\tilde{u}_2$ in the other equations we obtain

$$\rightarrow \begin{bmatrix} 1/2 & \overset{\downarrow}{0} & 0 & 1/2 & . & . & . \\ 1 & 2 & 0 & 1 & . & . & . \\ 0 & 0 & 0 & 0 & . & . & . \\ 1/2 & 0 & 0 & 7/2 & . & . & . \\ . & . & . & . & . & . & . \\ . & . & . & . & . & . & . \\ . & . & . & . & . & . & . \end{bmatrix} \begin{Bmatrix} \tilde{u}_1 \\ \tilde{u}_2 \\ \tilde{u}_3 \\ \tilde{u}_4 \\ . \\ . \\ \tilde{u}_{10} \end{Bmatrix} = \begin{Bmatrix} 0 \\ 2 \\ 0 \\ 3 \\ . \\ . \\ . \end{Bmatrix}$$

For the remainder of the assembly and elimination process the second row (and second column) of this equation could be removed and stored outside of the computer's working memory. These positions (suitably relabelled) could then be used to incorporate new element information. In the interests of clarity we will not do this here but will use an arrow (as shown) to indicate that the corresponding row and column will remain fixed as the assembly and elimination process proceeds.

Adding element [b] in which (say)

$$[K]^{[b]} = \begin{bmatrix} \overset{1}{1} & \overset{3}{1} & \overset{4}{1} \\ 1 & 3 & 1 \\ 1 & 1 & 4 \end{bmatrix} \begin{matrix} 1 \\ 3 \\ 4 \end{matrix} \qquad \{F\}^{[b]} = \begin{Bmatrix} 1 \\ 3 \\ 4 \end{Bmatrix} \begin{matrix} 1 \\ 3 \\ 4 \end{matrix}$$

we obtain

$$\rightarrow \begin{bmatrix} 3/2 & \overset{\downarrow}{0} & 1 & 3/2 & . & . & . \\ 1 & 2 & 0 & 1 & . & . & . \\ 1 & 0 & 3 & 1 & . & . & . \\ 3/2 & 0 & 1 & 15/2 & . & . & . \\ . & . & . & . & . & . & . \\ . & . & . & . & . & . & . \\ . & . & . & . & . & . & . \end{bmatrix} \begin{Bmatrix} \tilde{u}_1 \\ \tilde{u}_2 \\ \tilde{u}_3 \\ \tilde{u}_4 \\ . \\ . \\ \tilde{u}_{10} \end{Bmatrix} = \begin{Bmatrix} 1 \\ 2 \\ 3 \\ 7 \\ . \\ . \\ . \end{Bmatrix}$$

No other element (other than [a] and [b]) is attached to node 1 so that the first row (and column) of [K] and {F} will remain unaffected by further element contributions. Hence the variable $\tilde{u}_1$ can be eliminated at this stage using row 1 viz.

$$\tilde{u}_1 = \frac{2}{3} - \frac{2}{3}\tilde{u}_3 - \tilde{u}_4$$

leading to

$$
\begin{array}{c} \\ \rightarrow \\ \rightarrow \\ \\ \\ \\ \\ \\ \end{array}
\begin{array}{c}
\begin{array}{cccccccc} \downarrow & \downarrow & & & & & & \end{array} \\
\begin{bmatrix}
3/2 & 0 & 1 & 3/2 & . & . & . \\
1 & 2 & 0 & 1 & . & . & . \\
0 & 0 & 7/3 & 0 & . & . & . \\
0 & 0 & 0 & 6 & . & . & . \\
. & . & . & . & . & . & . \\
. & . & . & . & . & . & . \\
. & . & . & . & . & . & .
\end{bmatrix}
\end{array}
\begin{Bmatrix} \tilde{u}_1 \\ \tilde{u}_2 \\ \tilde{u}_3 \\ \tilde{u}_4 \\ . \\ . \\ \tilde{u}_{10} \end{Bmatrix}
=
\begin{Bmatrix} 3/2 \\ 1 \\ 7/3 \\ 6 \\ . \\ . \\ . \end{Bmatrix}
$$

Again the first row can be removed and stored outside of the computer's working memory and brought back later when the full assembly and elimination procedure is complete. Note that if the rows and columns corresponding to eliminated variables are ignored the remaining array is still symmetric.

This procedure of adding each element in turn and eliminating variables whenever possible is continued so that at each stage the working space (non-zero terms in non-arrowed rows and columns) is only a fraction of the full array for $[K]$. Indeed, for this specific example, a 6×1 array is all that is needed to store the information after the assembly of all elements. This should be compared with the requirement of a 10×4 rectangular array using a banded matrix approach. The price to be paid for this saving of storage space is a significant increase in 'housekeeping' which presents a considerable programming task.

It is instructive to record, at each element addition, the variables eliminated and the variables 'in core' – these latter being called **frontal nodes**. Lines connecting the frontal nodes define a 'wave-front' in the mesh as the assembly proceeds.

For our specific example we have

element introduced	nodes eliminated	frontal nodes (wave-front)	storage requirements
[*a*]	2	1,4	6×1
[*b*]	1	3,4	6×1
[*c*]	4	3,6	6×1
[*d*]	3	5,6	6×1
[*e*]	6	5,8	6×1
[*f*]	5	7,8	6×1
[*g*]	8	7,10	6×1
[*h*]	7,9,10		

For the frontal approach the global node numbering system does not affect the efficiency of the procedure. However the order in which the elements are assembled (element numbering) is significant. For example, had our triangular mesh been labelled according to Figure 9.9 we would obtain the following table:

element introduced	nodes eliminated	frontal nodes (wave-front)	storage requirements
[*a*]	2	1,4	6×1
[*b*]	9	1,4,7,10	15×1
[*c*]	1	3,4,7,10	15×1
[*d*]	none	3,4,5,6,8,7,10	28×1
[*e*]	4	3,5,6,8,7,10	28×1
[*f*]	none	3,5,6,8,7,10	21×1
[*g*]	7,8,10	3,5,6	21×1
[*h*]	3,5,6		

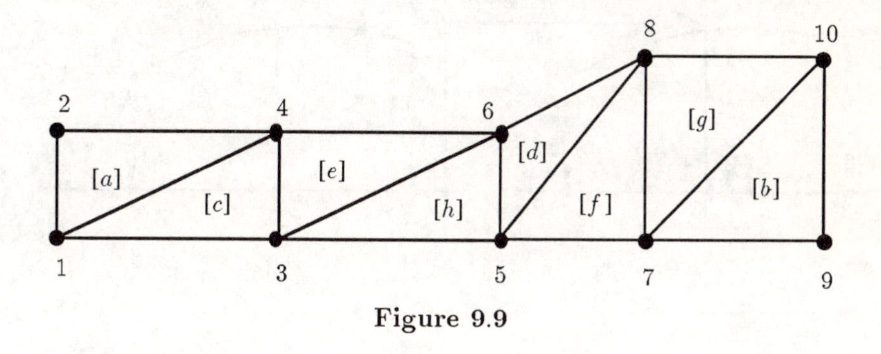

Figure 9.9

We see that the storage requirements are considerably greater than with the previous element numbering scheme. As with banded procedures there exist computer routines to number elements so as to minimize 'front-width'.

9.3 Solution curves

The operation known as post-processing is concerned with extracting as much information as possible from a finite element analysis and (generally) displaying that information graphically.

We consider firstly the post-processing of results from one-dimensional problems. The extension to two-dimensional problems will be considered in the next section.

We assume that, as a result of a finite element analysis, we have obtained the nodal values $\tilde{u}_i$, $i = 1, 2, \ldots, nde$, of a function $\tilde{u}(x)$ which is an approximation to the solution of a boundary value problem. To be precise we consider the situation in which three-noded quadratic elements are used. It follows that, within each element,

$$\tilde{u}(x) = \sum_{i=1}^{3} \tilde{u}_i^{[e]} \, Q_i(\xi) \tag{9.8}$$

in which $\tilde{u}_i^{[e]}$ are the nodal function values in element $[e]$. The variables x and ξ are linearly related by

$$x = x_1^{[e]} \frac{1}{2}(1 - \xi) + x_3^{[e]} \frac{1}{2}(1 + \xi) \tag{9.9}$$

See Figure 9.10.

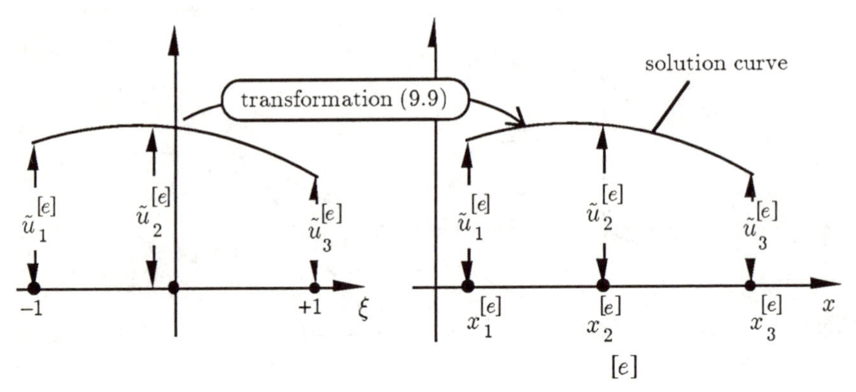

Figure 9.10

To draw the solution curve we need to calculate, for each value of x within element $[e]$, a corresponding value of $\tilde{u}(x)$ using (9.8). Of course, we only determine a **finite** number of closely spaced points $(x, \tilde{u}(x))$ and join them by straight lines. If these points are sufficiently close then the overall appearance will be that of a smooth curve.

With much pre- and post-processing, calculations are carried out with respect to the standard element and the appropriate quantity is then transferred into the problem domain by a coordinate transformation such as (9.9). In this case we begin at the left hand end of the standard domain $(\xi = -1)$ and calculate $x, \tilde{u}(x)$, as ξ is successively incremented in steps, of length say 0.1, until the right hand end $(\xi = +1)$ is reached. This routine is repeated for each element. All the calculated points $(x, \tilde{u}(x))$ are then joined by straight lines producing the solution curve over the whole problem domain.

To draw the curve automatically a computer terminal, capable of displaying graphics, is required. In particular, for this application, all that is required is a machine that can move a pen to a given point (on the screen or drawing pad) and then draw straight lines connecting given points. These are the two basic requirements of any graphics package.

The following algorithm in pseudo-code outlines the major steps to be followed when drawing a solution curve. We assume that nodal coordinates are held in the array $xnode\,(i), i = 1, 2, \ldots, nde$, and that nodal function values are held in the array $\tilde{u}(i),\ i = 1, 2, \ldots, nde$.

move to $(xnode\,(1), \tilde{u}(1))$
for $e = 1$ **to** nel **do**
$x1 = xnode\ (nenn\,(e, 1))$
$x3 = xnode\ (nenn\,(e, 3))$
$\tilde{u}1 = \tilde{u}\,(nenn\ (e, 1))$
$\tilde{u}2 = \tilde{u}\,(nenn\ (e, 2))$
$\tilde{u}3 = \tilde{u}\,(nenn\ (e, 3))$
$\xi = -1$
for $j = 1$ **to** 20 **do**
$\xi = \xi + 0.1$
$x = x1\,\frac{1}{2}\,(1 - \xi) + x3\,\frac{1}{2}\,(1 + \xi)$
$\tilde{u} = \tilde{u}1\,\frac{1}{2}\,\xi\,(\xi - 1) + \tilde{u}2\,(1 - \xi^2) + \tilde{u}3\,\frac{1}{2}\xi(\xi + 1)$
draw line to $(x, \tilde{u})$
end for j
end for e

9.4 Contour curves

'Contouring' is a particular application which is invaluable in post-processing the results of a two-dimensional finite element analysis. The technique is also extremely extremely useful in the display of raw data and in the geometrical analysis of surfaces.

Consider a function of two variables $u(x, y)$ defined over some region A of the xy plane. Geometrically this function defines a surface:

$$z = u(x, y) \qquad (x, y) \text{ within } A$$

(See Figure 9.11.)

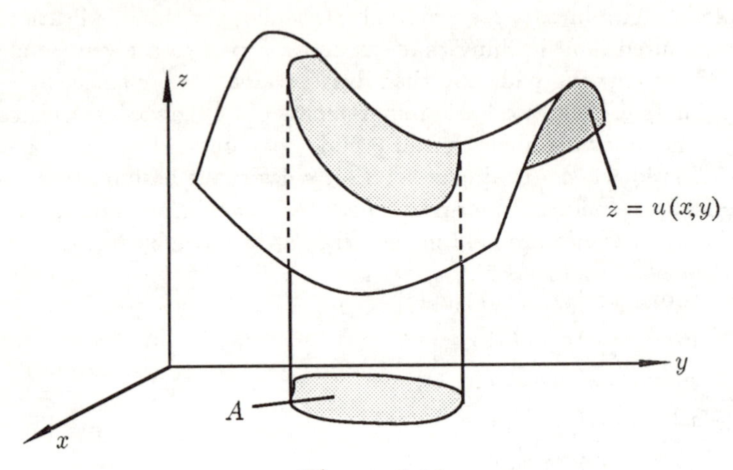

Figure 9.11

There is often a requirement to draw the **level curves (or contour curves)** of $u(x, y)$ so as to visualize more easily the shape of the surface from a two-dimensional representation. In other words, we need to draw the curves, say $y = g(x)$, which are the solutions of the equation

$$u(x, y) = c \tag{9.10}$$

where c is a given constant defining a particular contour curve.

In Figure 9.12 we show a surface and a selection of contour curves.

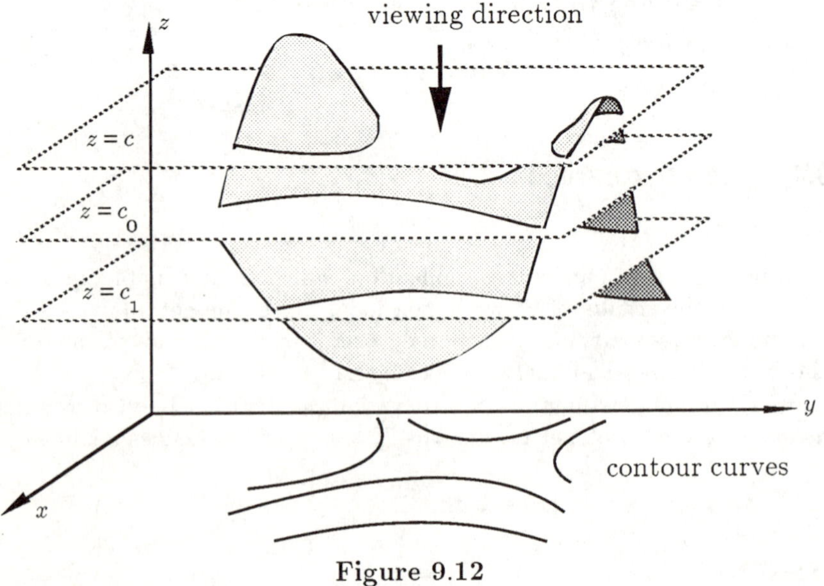

Figure 9.12

The level curve $u(x,y)=c$ is that curve obtained by the intersection of the surface with the plane $z=c$. Other contour curves are obtained by choosing different values of the constant c. The 'contour map' of the surface is the image obtained by viewing the contour curves along the direction of the z-axis.

In practical situations it may be very difficult to solve (9.10) explicitly for y in terms of x (and hence to obtain $g(x)$). We often need to construct the curves numerically. To do this the region A under consideration is searched for a given contour curve $u(x,y)=c$. One way of doing this is to increment the variable x in small steps from a to b (these being the values of x at the extreme left and extreme right of the domain A). For each value (say x_0) of x a value of y (say y_0) is determined by numerically solving the equation $u(x_0,y)=c$. The value of y_0 is tested to see if the point (x_0,y_0) lies in the domain A. If it does then we may calculate other closely related points on the contour as outlined below and hence (by joining these points with straight lines) draw the contour (see Figure 9.13). Closely related points on the contour curve are obtained by incrementing x from x_0 to x_1 and utilizing the Taylor series expansion of $u(x,y)$ in the neighbourhood of (x_0,y_0) namely:

$$u(x_1,y_1)\approx u(x_0,y_0)+(x_1-x_0)\left.\frac{\partial u}{\partial x}\right|_{(x_0,y_0)}+(y_1-y_0)\left.\frac{\partial u}{\partial y}\right|_{(x_0,y_0)} \qquad \mathbf{(9.11)}$$

where we ignore higher order terms.

Now if (x_0,y_0) and (x_1,y_1) lie on the same contour curve then

$$u(x_0,y_0)=c \qquad \text{and} \qquad u(x_1,y_1)=c$$

so that, by (9.11),

$$0=(x_1-x_0)\left.\frac{\partial u}{\partial x}\right|_{(x_0,y_0)}+(y_1-y_0)\left.\frac{\partial u}{\partial y}\right|_{(x_0,y_0)}$$

which gives

$$y_1=y_0-(x_1-x_0)\frac{\left.\dfrac{\partial u}{\partial x}\right|_{(x_0,y_0)}}{\left.\dfrac{\partial u}{\partial y}\right|_{(x_0,y_0)}} \qquad \mathbf{(9.12)}$$

from which, y_1 can be determined. We can then calculate the next point on the contour curve by repeating the analysis starting from (x_1,y_1). We continue in this way until either the domain boundary is reached or the initial point (x_0,y_0) is again obtained (indicating a **closed** contour).

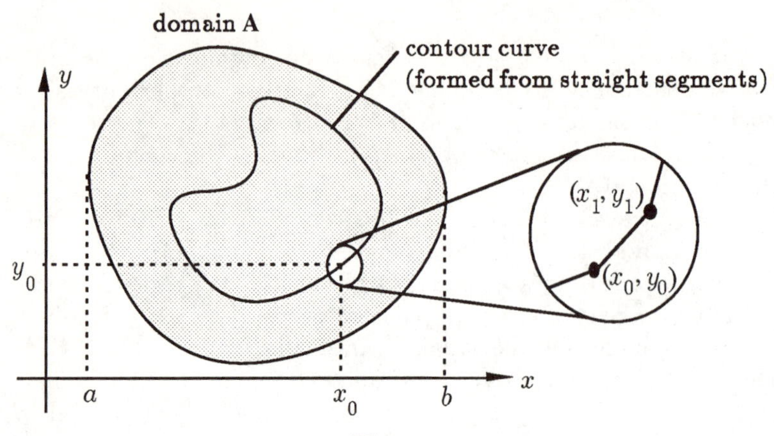

Figure 9.13

There are a number of difficulties associated with this approach.

- The functional form of $u(x, y)$ may be such as to make the numerical solution of $u(x, y) = c$ (for given x) intractible and certainly it would be very difficult to automate the process.

- For arbitrarily shaped domains it is difficult to formulate an algorithm for deciding whether or not a particular point (x, y) lies within that domain. Again the complication lies in attempting to automate the process.

- If the contour is closed then it is highly unlikely (unless the increments are very small) that the above plotting procedure will produce a contour passing through its initial point. In practice the contour will loop indefinitely.

These difficulties can be almost completely eliminated if an element/shape function approach to 'contouring' is used. This involves approximating the domain by a number of elements (say eight-noded quadrilateral elements). At each node of the mesh the function $u(x, y)$ is sampled (or has been obtained by a finite element calculation). Hence, over the complete domain $u(x, y)$ may be approximated by

$$\tilde{u}(x, y) = \sum_{i=1}^{nde} \tilde{u}_i \, R_i \, (x, y)$$

in which, as usual, $\tilde{u}_i$, $i = 1, 2, \ldots, nde$, are the nodal function values and R_i, $i = 1, 2, \ldots, nde$ are the roof functions. This implies that within each element

$$\tilde{u}(x, y) = \sum_{i=1}^{8} \tilde{u}_i^{[e]} \, Q_i \, (\xi, \eta) \qquad \textbf{(9.13)}$$

where

$$x = \sum_{i=1}^{8} x_i^{[e]}\, Q_i\,(\xi, \eta) \qquad \text{and} \qquad y = \sum_{i=1}^{8} y_i^{[e]}\, Q_i\,(\xi, \eta) \tag{9.14}$$

and the $Q_i(\xi, \eta)$ are the shape functions for an eight-noded quadrilateral.

Of course if we are solving a boundary value problem then, as part of the process of finding a solution, we have made all of the assumptions inherent in (9.13) and (9.14).

With this approach the problem of drawing contours in the whole domain A, can be transferred, element by element, to the problem of drawing contour curves within the standard element which (obviously) has well defined boundaries. Moreover, throughout the standard region the approximation to $\tilde{u}$ varies in a well-defined manner throughout the square and quadratically along each edge (see Figure 9.14).

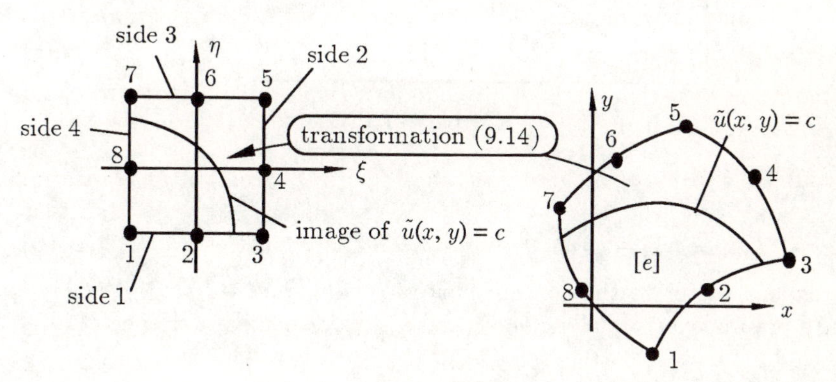

Figure 9.14

A particular algorithm for drawing the contour curve $\tilde{u}(x, y) = c$ in element $[e]$ then proceeds as follows.

The point of intersection (if any) of the contour curve with side 1 of the standard element is first determined. This is relatively easy since, along this side,

$$\begin{aligned}\tilde{u}(x, y) &= \tilde{u}_1^{[e]}\, Q_1\,(\xi, -1) + \tilde{u}_2^{[e]}\, Q_2\,(\xi, -1) + \tilde{u}_3^{[e]}\, Q_3\,(\xi, -1) \\ &= \tilde{u}_1^{[e]}\, \frac{1}{2}\, \xi(\xi - 1) + \tilde{u}_2^{[e]}\, (1 - \xi^2) + \tilde{u}_3^{[e]}\, \frac{1}{2}\, \xi(\xi + 1)\end{aligned}$$

Now, at the point, say $(\xi_0, -1)$, of intersection we know $\tilde{u}(x, y) = c$. Hence ξ_0 satisfies the quadratic equation

$$\xi^2 \left[\frac{1}{2}\, \tilde{u}_1^{[e]} - \tilde{u}_2^{[e]} + \frac{1}{2}\, \tilde{u}_3^{[e]}\right] + \xi \left[\frac{1}{2}\, \tilde{u}_3^{[e]} - \frac{1}{2}\, \tilde{u}_1^{[e]}\right] + \tilde{u}_2^{[e]} - c = 0 \tag{9.15}$$

If this equation has a **real** root lying between -1 and $+1$ then the contour curve does indeed intersect side 1.

The curve can now be traced through the standard element by again using a Taylor expansion. Having found a point $(\xi_{\text{old}}, \eta_{\text{old}})$ on the curve a new point, with coordinates $(\xi_{\text{new}}, \eta_{\text{new}})$, a distance δs away is calculated as follows (see Figure 9.15).

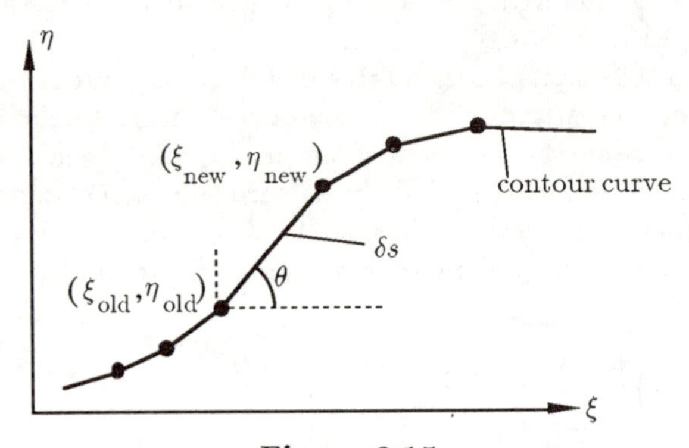

Figure 9.15

Clearly

$$\xi_{\text{new}} = \xi_{\text{old}} + \delta s \cos\theta \qquad \eta_{\text{new}} = \eta_{\text{old}} + \delta s \sin\theta \tag{9.16}$$

A tangent vector to the contour curve is $(-\partial\tilde{u}/\partial\eta\ \hat{\mathbf{i}} + \partial\tilde{u}/\partial\xi\ \hat{\mathbf{j}})$ where $\hat{\mathbf{i}}$ and $\hat{\mathbf{j}}$ are unit vectors along the ξ and η axes respectively. Hence

$$\theta = \tan^{-1}\left(-\frac{\partial\tilde{u}/\partial\xi}{\partial\tilde{u}/\partial\eta}\right)$$

But, within the element, $\tilde{u}$ is a known function, (9.13), of (ξ, η) and hence its derivatives with respect to ξ and η may be evaluated at $(\xi_{\text{old}}, \eta_{\text{old}})$ and hence $(\xi_{\text{new}}, \eta_{\text{new}})$ determined from (9.16). A smooth curve is obtained if δs is chosen to be sufficiently small.

Certain precautions (concerning the signs of $\sin\theta$ and $\cos\theta$) need to be taken to ensure that the contour is traced **into** the element. The tracing is continued as long as ξ_{new} and η_{new} satisfy $-1 \le \xi_{\text{new}} \le 1$, $-1 \le \eta_{\text{new}} \le 1$. As each new point is calculated within the standard element the corresponding (x, y) coordinates are determined using (9.14).

Proceeding in this way for the other three sides of the standard element, each intersection of the contour $\tilde{u}(x, y) = c$ will be determined and the contour curve drawn.

With this rather simple algorithm each of the contour curves will be drawn twice. For example a contour curve impinging on side 1 and side 4 (as in Figure 9.14) will be drawn from side 1 until it meets side 4 but then, on proceeding with the algorithm, it will be re-drawn from side 4 until it meets side 1. This double drawing can be eliminated but is, in fact, quite a good check on the accuracy of the drawing routine (in particular on the size of δs).

The algorithm outlined above for contouring surfaces is relatively easily implemented and examples of its use occur throughout this text, particularly in Chapter 7 (see Figures 7.4, 7.16(b),(c)).

10

Fourth-order Boundary Value Problems

PREVIEW

This chapter examines the application of weighted residual procedures and the Ritz variational technique to fourth-order boundary value problems in both one and two dimensions. We utilise Hermite-type shape functions for one-dimensional problems. We show how complicated shape functions may be avoided by reformulating a fourth-order differential equation as a system of two second-order differential equations.

10.1 One-dimensional fourth-order problems

One important example of a fourth-order boundary value problem arises in the theory of the transverse deformation of beams.

Consider a beam of length ℓ as shown in Figure 10.1. As indicated in the figure the cross section of the beam may be variable and the x-axis is chosen to lie along the line of centroids of each cross section. The beam is supported at its extremities $x = 0$ and $x = \ell$ so as to prevent it moving as a rigid body. These supports can take many forms, simple and clamped supports being the simplest. The original configuration of the beam is altered by the action of transverse deforming forces which may be point forces placed at various positions along the length and/or distributed forces acting along a portion of the surface.

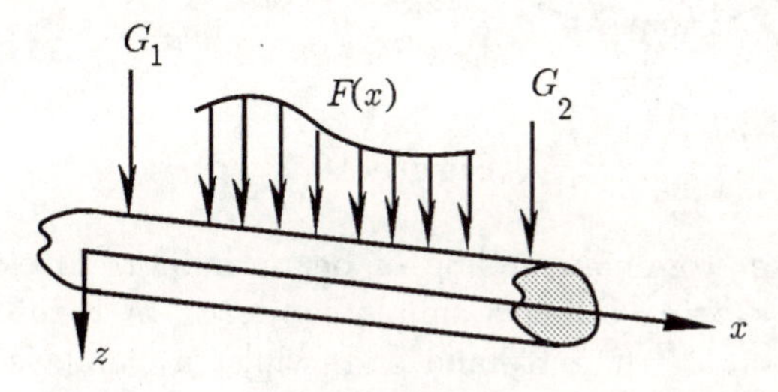

Figure 10.1

For long slender beams the Euler–Bernouilli theory of beam bending in which attention is focused on particles along the centre-line of the beam (the x-axis) is generally regarded as an adequate model. The theory predicts that the curve $z = u(x)$ assumed by the centre-line as a result of the deforming forces is a solution of the **fourth-order differential equation**:

$$\frac{d^2}{dx^2}\left(EI(x)\frac{d^2u}{dx^2}\right) = F(x) \qquad 0 \le x \le \ell \tag{10.1}$$

in which E is the Young's Modulus characterising the material from which the beam is formed, and $I(x)$ is the (possibly variable) second moment of area of the cross section. We have assumed in (10.1) that the deforming forces $F(x)$ are distributed. (Point forces can be regarded as a special case of distributed forces.)

Equation (10.1) is a fourth-order linear equation and consequently four boundary conditions, two at each end of the beam, must be specified before a unique solution can be obtained. For example, if the beam is clamped rigidly at both ends as in Figure 10.2(a) the boundary conditions are

$$u(0) = 0, \quad \frac{du}{dx}(0) = 0, \quad u(\ell) = 0, \quad \frac{du}{dx}(\ell) = 0 \tag{10.2}$$

On the other hand if the beam is clamped at $x = 0$ and unconstrained (free) at $x = \ell$ as in Figure 10.2(b) then the boundary conditions are

$$u(0) = 0, \ \frac{du}{dx}(0) = 0, \quad \frac{d^2u}{dx^2}(\ell) = 0, \quad \frac{d}{dx}\left(EI\frac{d^2u}{dx^2}\right)(\ell) = 0 \qquad \textbf{(10.3)}$$

The boundary conditions at $x = \ell$ imply the vanishing at this point of the bending monent $M(\equiv EI\, d^2u/dx^2)$ and of the shear force S $(\equiv -d/dx(EI\, d^2u/dx^2))$ respectively.

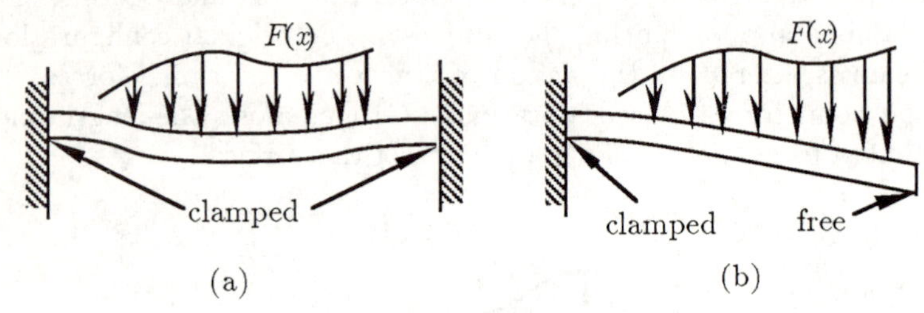

Figure 10.2

Another common method of beam support involves the use of hinges, a hinge being called a simple support. In Figure 10.3 there is a simple support at the left-hand end, whilst a clamp is placed at the right-hand end.

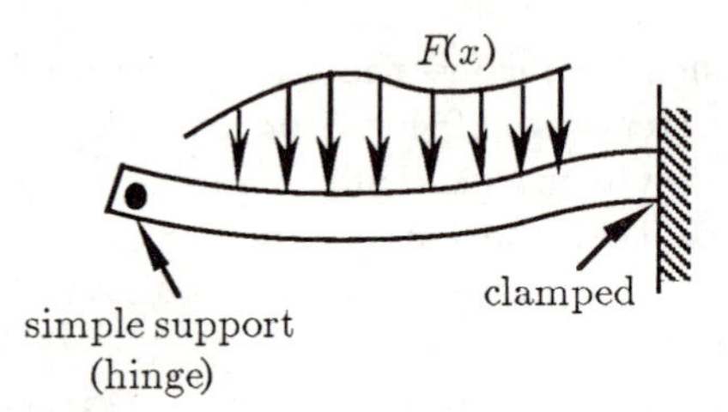

Figure 10.3

The boundary conditions in this case are

$$u(0) = 0, \quad \frac{d^2u}{dx^2}(0) = 0, \quad u(\ell) = 0, \quad \frac{du}{dx}(\ell) = 0 \qquad \textbf{(10.4)}$$

A hinge cannot support or exert any bending moment. Hence the bending moment must vanish, which leads to the second boundary condition in (10.4).

The differential equation (10.1) together with four boundary conditions (such as (10.2), (10.3) or (10.4)) form a **one-dimensional fourth-order boundary value problem.**

As with second-order boundary value problems the boundary conditions may be classified as either 'essential' or 'suppressible'. However in the fourth order case a boundary condition is called 'suppressible' if it contains derivatives of order **two or more**. This is because, as we shall see, the information contained in such conditions can be utilized in full in the transition from a standard to a modified Galerkin approach and the trial solution need not satisfy such conditions. On the other hand, boundary conditions involving derivatives of **order one or zero** cannot be fully utilized at this stage and trial solutions must satisfy any such conditions. Hence the classification 'essential'.

It is interesting to note that, in the case of beam deformation, essential-type boundary conditions could also be described as 'geometric' because they are obvious from the geometrical configuration of the beam. On the other hand suppressible-type boundary conditions are directly related to 'forces' acting in the system and are less obvious. A detailed physical knowledge of the Euler–Bernoulli model is needed before one can be confident about specifying suppressible boundary conditions. They certainly cannot be deduced from the geometrical configuration alone.

The relation (10.1) is, in fact, a very simple example of a fourth-order differential equation since a unique solution $u(x)$ can be found by integrating four times and imposing the particular boundary conditions of the problem.

Example 10.1

Determine the deflection of a uniform cantilever beam (clamped at its left hand end) deformed by a tip force P as shown in Figure 10.4.

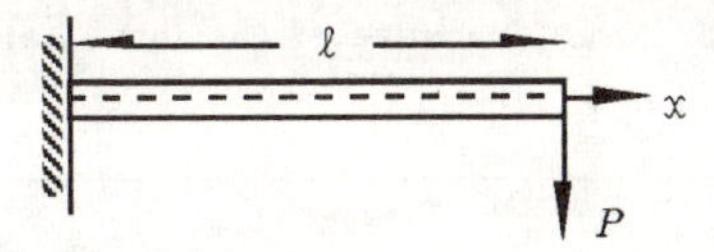

Figure 10.4

Solution

A point force P can be modelled by a Dirac delta-function. However, since in this case P is applied at the end-point $x = \ell$, it is easier to absorb it into the boundary conditions. Thus the boundary value problem describing this physical situation is

$$\frac{d^2}{dx^2}\left(EI\,\frac{d^2u}{dx^2}\right) = 0 \quad 0 \leq x \leq \ell \tag{10.5a}$$

$$u(0) = 0, \quad \frac{du}{dx}(0) = 0 \tag{10.5b}$$

$$\frac{d^2u}{dx^2}(\ell) = 0, \quad -\frac{d}{dx}\left(EI\frac{d^2u}{dx^2}\right)(\ell) = P \tag{10.5c}$$

The solution is easily obtained by successively integrating (10.5a):

$$\frac{d}{dx}\left(EI\frac{d^2u}{dx^2}\right) = C_1 \qquad 0 \le x \le \ell$$

where $C_1 = -P$ due to the second equation of (10.5c).

Integrating again

$$EI\frac{d^2u}{dx^2} = -Px + C_2$$

where $C_2 = P\ell$ due to the first equation of (10.5c).

Continuing, we find (since I is constant for a uniform beam)

$$EI\frac{du}{dx} = -P\frac{x^2}{2} + P\ell x + C_3$$

where $C_3 = 0$ due to the second equation of (10.5b).

Finally

$$EI\,u = -P\frac{x^3}{6} + P\ell\frac{x^2}{2} + C_4 \tag{10.6}$$

But $C_4 = 0$ due to the first equation of (10.6) so that

$$u(x) = \frac{P}{EI}\left(-\frac{x^3}{6} + \frac{x^2}{2}\ell\right) \tag{10.7}$$

In realistic problems the deflection (10.7) would be hardly noticeable (unless P is very large) because of the large value of Young's modulus E for most structural materials.

More complicated fourth-order boundary value problems than (10.5) can easily be constructed. For example a beam elastically supported along its entire length can be modelled by the differential equation

$$\frac{d^2}{dx^2}\left(E_b\,I(x)\frac{d^2u}{dx^2}\right) + E_s\,u = F(x) \tag{10.8}$$

where E_b, E_s are the Young's modulus of the beam and support materials respectively.

Both (10.1) and (10.8) are special cases of the general linear fourth-order equation in self-adjoint form:

$$\frac{d^2}{dx^2}\left(p(x)\frac{d^2u}{dx^2}\right) - \frac{d}{dx}\left(q(x)\frac{du}{dx}\right) + r(x)u = s(x) \quad a \le x \le b \tag{10.9}$$

where $p(x)$, $q(x)$, $r(x)$ and $s(x)$ are given functions. Many (but not all) linear fourth-order differential equations can be written in this form. This is because many fourth order differential equations in problems of physical interest can be derived using a variational principle from which the self-adjoint form naturally arises. This is particularly true in linear elasticity theory which is the source of many fourth-order boundary value problems both in one and two dimensions.

A brief inspection of (10.9), and indeed of the special cases (10.1) and (10.8), suggests that finding exact solutions to fourth-order boundary value problems is not possible unless, as in Example 10.1, the coefficient functions $p(x), q(x), r(x)$ and $s(x)$ have very simple forms. Thus, as in the treatment of second-order boundary value problems, we are led to consider numerical approximations and, in particular, the applicability of the finite element technique.

We begin, as in the second-order case, by reworking the boundary value problems using an integral formulation provided either by the Galerkin method or an appropriate variational principle.

10.2 Galerkin procedures

We shall not consider fourth-order boundary value problems in full generality because of the complexity of the attendant algebra. We restrict attention to the special case of (10.9) with $q(x) \equiv 0$ and also restrict the boundary conditions to commonly occuring forms.

As usual, we begin with the Galerkin approximation

$$\tilde{u}(x) = \theta_0(x) + \sum_{i=1}^{n} a_i\,\theta_i(x)$$

in which, at the outset, $\theta_0(x)$ satisfies **all** the boundary conditions of the problem and $\theta_i(x)$, $i = 1, 2, \ldots, n$, satisfies the corresponding homogeneous form of all the boundary conditions.

The Galerkin equations are then obtained by the familiar weighted residual method using the coordinate functions $\theta_i(x)$ as weighting functions:

$$\int_a^b \left\{ \frac{d^2}{dx^2}\left(p(x)\frac{d^2\tilde{u}}{dx^2} \right) + r(x)\tilde{u} - s(x) \right\} \theta_i\,dx = 0 \quad i = 1, 2, \ldots, n \quad \textbf{(10.10)}$$

For reasons similar to those discussed for second-order problems we prefer to reduce as far as possible the order of the derivatives appearing in the integrand of (10.10). In the fourth-order case we must integrate the fourth derivative terms by parts **twice**:

$$\int_a^b \frac{d^2}{dx^2}\left(p(x)\frac{d^2\tilde{u}}{dx^2}\right)\theta_i\,dx = -\int_a^b \frac{d}{dx}\left(p(x)\frac{d^2\tilde{u}}{dx^2}\right)\frac{d\theta_i}{dx}\,dx + \left[\frac{d}{dx}(p(x)\frac{d^2\tilde{u}}{dx^2})\,\theta_i\right]_a^b$$

$$= \int_a^b p(x)\frac{d^2\tilde{u}}{dx^2}\,\frac{d^2\theta_i}{dx^2}\,dx + \left[\frac{d}{dx}\left(p(x)\frac{d^2\tilde{u}}{dx^2}\right)\theta_i - p(x)\frac{d^2\tilde{u}}{dx^2}\,\frac{d\theta_i}{dx}\right]_a^b$$

$$i = 1, 2, \ldots, n \qquad \textbf{(10.11)}$$

Continued use of the integration by parts technique would increase the order of the highest derivative of the coordinate functions occurring in the integrand and this is undesirable. In the original form (10.10) the coordinate functions must be (at least) three times differentiable because the integrand must be piecewise continuous. However, in (10.11) only differentiable coordinate functions need be considered. Obviously functions which are only once differentiable are easier to construct than those for which higher order differentiability is required.

Any further simplification of (10.11) is dependent on the boundary conditions imposed. For example, if all four boundary conditions are essential then the coordinate functions $\theta_i(x)$ must satisfy the corresponding homogeneous forms:

$$\theta_i(a) = 0, \quad \frac{d\theta_i}{dx}(a) = 0, \quad \theta_i(b) = 0, \quad \frac{d\theta_i}{dx}(b) = 0 \quad i = 1, 2, .., n$$

and so, in this case, we can write (10.11) as

$$\int_a^b \frac{d^2}{dx^2}\left(p(x)\frac{d^2\tilde{u}}{dx^2}\right)\theta_i\,dx = \int_a^b p(x)\frac{d^2\tilde{u}}{dx^2}\,\frac{d^2\theta_i}{dx^2}\,dx$$

The standard Galerkin approach thus gives rise to a modified Galerkin procedure in which the parameters a_i, $i = 1, 2, \ldots, n$, are, by (10.10), solutions of the n (linear) equations:

$$\int_a^b \left\{p(x)\frac{d^2\tilde{u}}{dx^2}\,\frac{d^2\theta_i}{dx^2} + (r(x)\tilde{u} - s(x))\theta_i\right\}dx = 0 \quad i = 1, 2, \ldots, n \qquad \textbf{(10.12)}$$

in which $\tilde{u}$ satisfies all (essential) boundary conditions of the problem.

On the other hand, if the boundary conditions at $x = a$ are suppressible whilst those at $x = b$ are essential, then at $x = a$ we have

$$\frac{d^2\tilde{u}}{dx^2}(a) = k_1, \qquad \frac{d^3\tilde{u}}{dx^3}(a) = k_2$$

whilst

$$\theta_i\,(b) = 0, \qquad \frac{d\theta_i}{dx}\,(b) = 0$$

In this case (10.11) simplifies to

$$\int_a^b \frac{d^2}{dx^2}\left(p(x)\frac{d^2\tilde{u}}{dx^2}\right)\theta_i\,dx = \int_a^b p(x)\frac{d^2\tilde{u}}{dx^2}\,\frac{d^2\theta_i}{dx^2}\,dx$$
$$- p'(a)\,k_1\,\theta_i(a) - p(a)\,k_2\theta_i\,(a) + p(a)k_1\frac{d\theta_i}{dx}(a)$$

The modified Galerkin equations (10.10) become

$$\int_a^b \left\{p(x)\frac{d^2\tilde{u}}{dx^2}\,\frac{d^2\theta_i}{dx^2} + (r(x)\tilde{u} - s(x))\theta_i\right\}dx$$
$$-p'(a)\,k_1\theta_i\,(a) - p(a)\,k_2\,\theta_i(a) + p(a)k_1\,\frac{d\theta_i}{dx}(a) = 0$$
$$i = 1, 2, \ldots, n \qquad \textbf{(10.13)}$$

in which $\tilde{u}$ is chosen to satisfy only the essential boundary conditions at $x = b$. The condition that $\tilde{u}$ satisfies the suppressible conditions may be relaxed as we have already used these conditions in obtaining (10.13).

Other combinations of boundary conditions can be used in a similar way to simplify (10.11) and obtain appropriate modified Galerkin equations.

Example 10.2

Use the modified Galerkin method to solve the boundary value problem:

$$\frac{d^4u}{dx^4} + u = 1 \qquad 0 \le x \le 1$$

$u(0) = u(1) = 0$ (essential)
$u''(0) = u''(1) = 0$ (suppressible)

Choose a two-parameter trial solution of the form

$$\tilde{u} = a_1 \sin \pi x + a_2 \sin 3\pi x$$

Compare this approximation with the exact solution.

Solution

The Galerkin equations

$$\int_0^1 \left\{\frac{d^4\tilde{u}}{dx^4} + \tilde{u} - 1\right\}\theta_i\,dx = 0 \qquad i = 1, 2$$

are modified by integrating the fourth-order derivative term by parts (twice). After inserting the suppressible boundary conditions we obtain

$$\int_0^1 \left\{ \frac{d^2\tilde{u}}{dx^2}\frac{d^2\theta_i}{dx^2} + (\tilde{u}-1)\theta_i \right\} dx = 0 \qquad i = 1, 2 \tag{10.14}$$

Note that even though (strictly) the approximation $\tilde{u}$ in (10.14) needs to satisfy only the essential boundary conditions of the problem the actual approximation chosen does in fact satisfy **all** the boundary conditions.

In this case we have

$$\theta_1 = \sin \pi x \qquad \theta_2 = \sin 3\pi x$$

and

$$\frac{d^2\theta_1}{dx^2} = -\pi^2 \sin \pi x \qquad \frac{d^2\theta_2}{dx^2} = -9\pi^2 \sin 3\pi x$$

so

$$\frac{d^2\tilde{u}}{dx^2} = -a_1\,\pi^2\,\sin \pi x - 9\,\pi^2\,a_2\,\sin 3\pi x$$

Hence for $i = 1$ in (10.14) we obtain

$$\int_0^1 \left\{ a_1\pi^4 \sin^2 \pi x + 9\pi^4 a_2 \sin 3\pi x \sin \pi x \right.$$
$$\left. + (a_1 \sin \pi x + a_2 \sin 3\pi x - 1) \sin \pi x \right\} dx = 0$$

giving, after straightforward integration,

$$\frac{\pi^4 a_1}{2} + \frac{a_1}{2} - 2 = 0$$

Hence

$$a_1 = 4/(\pi^4 + 1)$$

Similarly for $i = 2$ we have

$$\int_0^1 \left\{ \pi^4\,a_1\,9\sin 3\pi x\,\sin\,\pi x + 81\,\pi^4\,a_2\,\sin^2 3\pi x \right.$$
$$\left. + (a_1 \sin \pi x + a_2 \sin 3\pi x - 1) \sin 3\pi x \right\} dx = 0$$

which gives

$$81\frac{\pi^4 a_2}{2} + \frac{a_2}{2} - 2 = 0$$

that is

$$a_2 = 4/(81\pi^4 + 1)$$

We thus obtain a two-parameter approximation:

$$\tilde{u} = \frac{4}{(\pi^4 + 1)} \sin \pi x + \frac{4}{(81\pi^4 + 1)} \sin 3\pi x$$

The exact solution to this problem is

$$u = 1.3798 \cos \frac{x}{\sqrt{2}} \sinh \frac{x}{\sqrt{2}} - 0.3215 \sin \frac{x}{\sqrt{2}} \cosh \frac{x}{\sqrt{2}} - \exp \frac{x}{\sqrt{2}} \cos \frac{x}{\sqrt{2}} + 1$$

The two solutions are compared in Figure 10.5. The apparently poor correspondence between approximate and exact solutions is illusory, being due only to the scaling on the vertical axis.

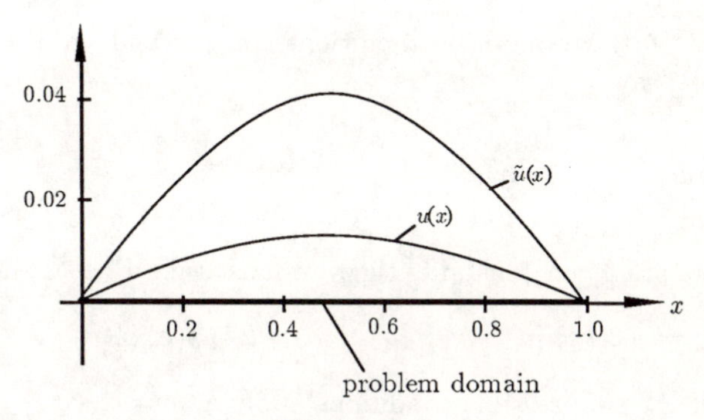

Figure 10.5

10.3 Variational formulation

In this section we again consider the special case $q(x) \equiv 0$ of (10.9).

In a similar manner to Chapter 8 (Theorems 8.1 and 8.2) it is possible to recast linear fourth-order boundary value problems in variational form. The prescription for the procedure is as follows.

Theorem 10.1

The fourth-order linear differential equation:

$$\frac{d^2}{dx^2}\left(p(x)\frac{d^2u}{dx^2}\right) + r(x)u = s(x) \quad a \le x \le b \qquad \textbf{(10.15)}$$

can be written as the necessary conditions to be satisfied by the solution of the variational problem

$$J[\phi] = \text{extremum}$$

where

$$J[\phi] = \int_a^b \left\{ p(x) \left(\frac{d^2\phi}{dx^2} \right)^2 + r(x)\phi^2 - 2\, s(x)\phi \right\} dx + B_a + B_b \qquad \textbf{(10.16)}$$

The following cases should be noted:

• Case 1. If all the boundary conditions are essential, that is

$$u(a) = \alpha, \quad \frac{du}{dx}(a) = \beta, \quad u(b) = \gamma, \quad \frac{du}{dx}(b) = \delta$$

where $\alpha, \beta, \gamma, \delta$ are given constants then we can take $B_a \equiv B_b \equiv 0$ but the admissible functions $\phi(x)$ **must** satisfy all the boundary conditions.

• Case 2. If the boundary conditions at one end, say $x = a$, are suppressible, say

$$\frac{d^2u}{dx^2}(a) = k_1, \quad \frac{d^3u}{dx^3}(a) = k_2$$

where k_1, k_2 are given constants then we must take $B_b \equiv 0$ and

$$B_a = -2(k_2\, p(a) + k_1\, p'(a))\, \phi(a) + 2\, k_1\, p(a)\, \phi'(a)$$

Similarly if the suppressible conditions apply at $x = b$, say

$$\frac{d^2u}{dx^2}(b) = h_1 \qquad \frac{d^3u}{dx^3}(b) = h_2$$

then we would take $B_a \equiv 0$ and

$$B_b = 2\,(h_2\, p(b) + h_1\, p'(b))\, \phi(b) - 2\, h_1\, p(b)\, \phi'(b)$$

In both these situations $\phi(x)$ need only satisfy the essential conditions.

• Case 3. If the boundary conditions at both ends are suppressible, of the form given in case 2 above, then B_a and B_b must both be taken as in case 2 but in this case the admissible functions $\phi(x)$ need not satisfy any boundary conditions.

The Ritz technique for determining approximations to second-order boundary value problems can be applied without change to fourth-order problems. An approximation of the form

$$\tilde{u}(x) = \theta_0(x) + \sum_{i=1}^{n} a_i\, \theta_i(x)$$

is sought with the usual constraint that $\theta_0(x)$ satisfies the essential boundary conditions of the problem and $\theta_i(x)$, $i = 1, 2, \ldots, n$, satisfies the corresponding homogeneous form of the essential conditions.

We illustrate with a simple example.

Example 10.3

Determine a two-parameter polynomial approximation to the solution of the boundary value problem:

$$\frac{d^4u}{dx^4} = P\delta(x-\ell) + P\delta(x-2\ell) \qquad 0 \leq x \leq 2\ell$$

$$u(0) = \frac{du}{dx}(0) = 0, \qquad \frac{d^2u}{dx^2}(2\ell) = \frac{d^3u}{dx^3}(2\ell) = 0$$

Physically, this is a model for a cantilevered beam (clamped at $x = 0$) deformed by two point forces at $x = \ell$ and $x = 2\ell$. These forces are being modelled using Dirac delta-functions. The constants E, I are chosen such that the product $EI = 1$.

Solution

Here the appropriate variational problem is (from (10.16) case 2) with $s(x) = P\delta(x-\ell) + P\delta(x-2\ell)$:

$$J[\phi] = \text{extremum}$$

in which

$$J[\phi] = \int_0^{2\ell} \left(\frac{d^2\phi}{dx^2}\right)^2 dx - 2\,P\,\phi(\ell) - 2\,P\,\phi(2\ell) \qquad \textbf{(10.17)}$$

In obtaining (10.17) we have used the well-known 'sampling' property of the delta-function viz.

$$\int \phi(x)\delta(x-k)dx = \phi(k).$$

The term B_b is zero in (10.17) due to the simple form of the suppressible boundary conditions. The admissible functions ϕ must satisfy the essential boundary conditions of the problem: $\phi(0) = \phi'(0) = 0$.

The simplest two-parameter approximation consistent with the essential boundary conditions is (choosing $\tilde{u}$ as a **particular** ϕ):

$$\tilde{u}(x) = a_1\,x^2 + a_2\,x^3$$

giving

$$\tilde{u}'' = 2\,a_1 + 6\,a_2 x$$

Hence from (10.17)

$$J[\tilde{u}] = \int_0^{2\ell} \{4\,a_1^2 + 36\,a_2^2\,x^2 + 24\,a_1\,a_2\,x\}\,dx$$

$$-\,2P(a_1\,\ell^2 + a_2\,\ell^3) - 2P(4\,a_1\ell^2 + 8\,a_2\,\ell^3)$$

$$= 4\,a_1^2(2\ell) + 36\,a_2^2\left(\frac{8\ell^3}{3}\right) + 24\,a_1\,a_2\,(2\ell^2)$$

$$-\,2P(a_1\,\ell^2 + a_2\,\ell^3) - 2P(4\,a_1\ell^2 + 8\,a_2\,\ell^3)$$

Optimizing with respect to a_1, a_2:

$$\frac{\partial J}{\partial a_1} = 16\,\ell\,a_1 + 48\,\ell^2\,a_2 - 10P\ell^2 = 0$$

$$\frac{\partial J}{\partial a_2} = 48\,\ell^2\,a_1 + 192\,\ell^3\,a_2 - 18P\ell^3 = 0$$

Solving:

$$a_1 = \frac{11}{8}P\ell \qquad a_2 = -\frac{P}{4}$$

giving

$$\tilde{u}(x) = P\ell\left(\frac{11}{8}\,x^2 - \frac{1}{4\ell}\,x^3\right)$$

as the two-parameter approximation.

The exact solution to this problem is

$$u(x) = P\ell\left[-\frac{x^3}{3\ell} + \frac{3x^2}{2} + \frac{(x-\ell)^3}{6\ell}H(x-\ell) + \frac{(x-2\ell)^3}{6\ell}H(x-2\ell)\right]$$

where $H(x)$ is the Heaviside unit step function:

$$H(x) = \begin{cases} 0 & \text{if } x < 0 \\ 1 & \text{if } x \geq 0 \end{cases}$$

As an indication of the accuracy of the approximate solution we find, after some algebra,

$$J[u] = -2.333P^2\,\ell^3 \quad \text{whereas} \quad J[\tilde{u}] = -2.3125P^2\,\ell^3$$

10.4 Finite element procedures

The finite element method can be applied to fourth-order boundary value problems in much the same way as to second-order boundary value problems. However there are some important differences of detail which we now examine.

In second-order problems the simplest element used was the two-noded linear element. However this choice is inappropriate for fourth or higher order boundary value problems.

To illustrate the difficulties that arise we consider again the simple beam problem of Example 10.1 (with $EI \equiv 1$ for convenience):

$$\frac{d^4u}{dx^4} = 0, \quad 0 \leq x \leq \ell \tag{10.18a}$$

$$u(0) = 0, \quad u'(0) = 0, \qquad u''(\ell) = 0, \quad -u'''(\ell) = P \tag{10.18b}$$

Replacing ϕ by a **particular** $\tilde{u}$ in (10.16) the corresponding variational problem is

$$J[\tilde{u}] = \int_0^\ell \left(\frac{d^2\tilde{u}}{dx^2}\right)^2 dx - 2P\tilde{u}(\ell) = \text{extremum}$$
$$\tilde{u}(0) = 0, \qquad \tilde{u}'(0) = 0 \tag{10.19}$$

Attempting to apply a piecewise linear approximation to this functional will fail, for if we use just two elements (see Figure 10.6a) then the derivative $d\tilde{u}/dx$ takes the form shown in Figure 10.6b.

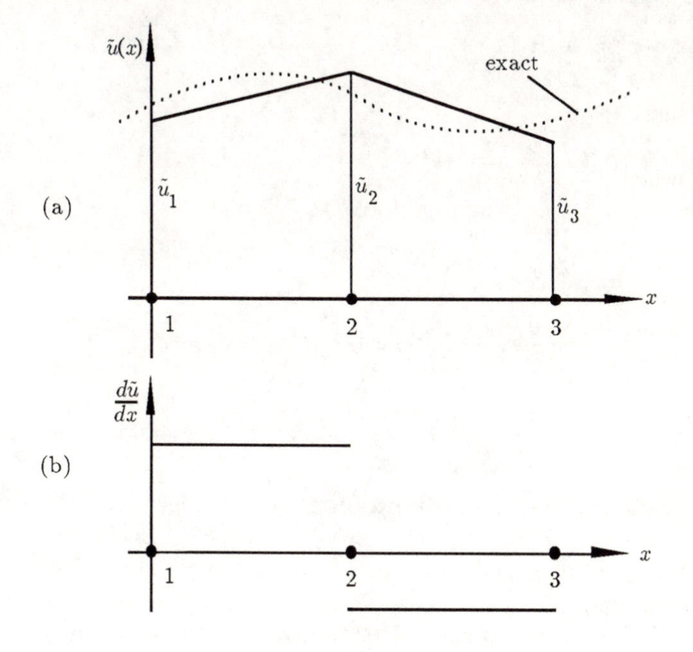

Figure 10.6

That is (as drawn in Figure 10.6a) $\tilde{u}$ has a positive (constant) gradient in element [1] and a negative (constant) gradient in element [2]. Taking a second derivative produces $\tilde{u}'' = 0,\ 0 \le x \le \ell$, except at $x = \ell/2$ where $\tilde{u}''$ is not defined due to the lack of continuity in $\tilde{u}'$. This causes substantial mathematical problems in attempting to assign a meaning to the integral term in (10.19).

There is another difficulty with the use of piecewise linear functions in a problem of this kind. Even if we could obtain an expression for $J[\tilde{u}]$ in terms of the nodal values $\tilde{u}_1, \tilde{u}_2, \tilde{u}_3$ there is no obvious way of imposing the essential boundary conditions in (10.18b). Clearly $\tilde{u}(0) = 0$ implies $\tilde{u}_1 = 0$ but $\tilde{u}'(0) = 0$ does not have a direct correspondence with any of the nodal values.

It is clear then that a piecewise linear function is not a good model to describe the solutions to fourth-order boundary value problems. A better model is one in which both the displacement $u(x)$ and the first derivative du/dx are treated on an equal footing.

We need to develop an expression for $\tilde{u}(x)$ similar to the piecewise linear approximation but which is specified by **four** parameters (corresponding to two degrees of freedom per node). These parameters are $\tilde{u}_1, \tilde{u}_2$ (the approximations to $u(x)$ at each node) and $\tilde{u}_1', \tilde{u}_2'$ (the approximations to du/dx at each node).

The simplest polynomial specified by four parameters is a cubic. For an element extending from $x = a$ to $x = b$ we find:

$$\tilde{u}(x) = \tilde{u}_1\, H_1(x) + \tilde{u}_1'\, H_2(x) + \tilde{u}_2\, H_3(x) + \tilde{u}_2'\, H_4(x) \tag{10.20}$$

in which

$$H_1(x) = \frac{(b-x)}{(b-a)^3}\,(b^2 + xb - 3ab + 3ax - 2x^2)$$

$$H_2(x) = \frac{(x-a)}{(b-a)^2}\,(b-x)^2$$

$$H_3(x) = \frac{(a-x)}{(a-b)^3}\,(a^2 + xa - 3ab + 3bx - 2x^2)$$

$$H_4(x) = \frac{(x-b)}{(a-b)^2}\,(a-x)^2 \tag{10.21}$$

These four cubic **Hermite shape functions** are derived in the Appendix for the standard element with $a = -1, b = +1$. The shape functions (10.21) may be obtained from the standard functions by a simple linear transformation.

The cubic expansion, (10.20), has the desired properties:

$$\tilde{u}(a) = \tilde{u}_1, \quad \tilde{u}'(a) = \tilde{u}_1', \quad \tilde{u}(b) = \tilde{u}_2, \quad \tilde{u}'(b) = \tilde{u}_2'$$

This approximation ensures that $d\tilde{u}/dx$ is continuous at the point where two elements join and so does not suffer from the problem encountered with piecewise linear functions. To examine it further we shall apply it to obtain an approximate solution of the boundary value problem (10.18) whose corresponding variational problem is (10.19). For simplicity we split the domain of the problem into just two equal length elements. We then have

$$J[\tilde{u}] = \int_0^{\ell/2} \left(\frac{d^2\tilde{u}}{dx^2}\right)^2 dx + \int_{\ell/2}^{\ell} \left(\frac{d^2\tilde{u}}{dx^2}\right)^2 dx - 2\,P\,\tilde{u}(\ell)$$

In element [1], using (10.20) with $a = 0, b = \ell/2$,

$$\frac{d^2\tilde{u}}{dx^2} = \frac{8}{\ell^3}\left[-3(\ell - 4x)(\tilde{u}_1 - \tilde{u}_2) + \tilde{u}_1'(3x - \ell)\ell + \frac{1}{2}\,\tilde{u}_2'(6x - \ell)\ell\right]$$

and in element [2], with $a = \ell/2, b = \ell$,

$$\frac{d^2\tilde{u}}{dx^2} = \frac{8}{\ell^3}\left[-3(3\ell - 4x)(\tilde{u}_2 - \tilde{u}_3) + \frac{1}{2}\,\tilde{u}_2'\,(6x - 5\ell)\ell + \tilde{u}_3'(3x - 2\ell)\ell\right]$$

Hence

$$\begin{aligned}\int_0^{\ell/2} \left(\frac{d^2\tilde{u}}{dx^2}\right)^2 dx = {} & \frac{96}{\ell^3}\,\tilde{u}_1^2 + \frac{96}{\ell^3}\,\tilde{u}_2^2 + \frac{8}{\ell}\,(\tilde{u}_1')^2 + \frac{8}{\ell}(\tilde{u}_2')^2 - \frac{192}{\ell^3}\,\tilde{u}_1\,\tilde{u}_2 \\ & + \frac{48}{\ell^2}(\tilde{u}_1 - \tilde{u}_2)\tilde{u}_1' + \frac{48}{\ell^2}(\tilde{u}_1 - \tilde{u}_2)\,\tilde{u}_2' + \frac{8}{\ell}\tilde{u}_1'\,\tilde{u}_2'\end{aligned}$$

and

$$\begin{aligned}\int_{\ell/2}^{\ell} \left(\frac{d^2\tilde{u}}{dx^2}\right)^2 dx = {} & \frac{96}{\ell^3}\,\tilde{u}_2^2 + \frac{96}{\ell^3}\,\tilde{u}_3^2 + \frac{8}{\ell}\,(\tilde{u}_2')^2 + \frac{8}{\ell}(\tilde{u}_3')^2 - \frac{192}{\ell^3}\,\tilde{u}_2\,\tilde{u}_3 \\ & + \frac{48}{\ell^2}(\tilde{u}_2 - \tilde{u}_3)\tilde{u}_2' + \frac{48}{\ell^2}(\tilde{u}_2 - \tilde{u}_3)\,\tilde{u}_3' + \frac{8}{\ell}\tilde{u}_2'\,\tilde{u}_3'\end{aligned}$$

The expression for $J[\tilde{u}]$ can be written directly in matrix form:

$$J = \{\tilde{u}\}^T[K]\{\tilde{u}\} + 2\{\tilde{u}\}^T\{F\} \tag{10.22}$$

where

$$[K] = [K^{[1]}] + [K^{[2]}], \qquad \{F\} = \{0, 0, 0, 0, -P, 0\}^T$$

with $\{\tilde{u}\} = \{\tilde{u}_1, \tilde{u}_1', \tilde{u}_2, \tilde{u}_2', \tilde{u}_3, \tilde{u}_3'\}^T$

$$[K^{[1]}] = \frac{1}{\ell^3}\begin{bmatrix} 96 & 24\ell & -96 & 24\ell & 0 & 0 \\ 24\ell & 8\ell^2 & -24\ell & 4\ell^2 & 0 & 0 \\ -96 & -24\ell & 96 & -24\ell & 0 & 0 \\ 24\ell & 4\ell^2 & -24\ell & 8\ell^2 & 0 & 0 \\ 0 & 0 & 0 & 0 & 0 & 0 \\ 0 & 0 & 0 & 0 & 0 & 0 \end{bmatrix}$$

$$[K^{[2]}] = \frac{1}{\ell^3}\begin{bmatrix} 0 & 0 & 0 & 0 & 0 & 0 \\ 0 & 0 & 0 & 0 & 0 & 0 \\ 0 & 0 & 96 & 24\ell & -96 & 24\ell \\ 0 & 0 & 24\ell & 8\ell^2 & -24\ell & 4\ell^2 \\ 0 & 0 & -96 & -24\ell & 96 & -24\ell \\ 0 & 0 & 24\ell & 4\ell^2 & -24\ell & 8\ell^2 \end{bmatrix}$$

Optimizing J with respect to $\{\tilde{u}\}$ (see supplement to Chapter 8)

$$\frac{\partial J}{\partial\{\tilde{u}\}} = 0$$

giving

$$[K]\{\tilde{u}\} + \{F\} = 0 \tag{10.23}$$

As we have noted elsewhere not all of the equations from the matrix system (10.23) are valid. Only those obtained from $\partial J/\partial\tilde{u}_2$, $\partial J/\partial\tilde{u}'_2$, $\partial J/\partial\tilde{u}_3$, $\partial J/\partial\tilde{u}'_3$, are meaningful as $\tilde{u}_1$ and $\tilde{u}'_1$ are fixed by the two essential boundary conditions in (10.18b).

Thus, as usual, we ignore the invalid equations of (10.23) (the first two) and impose the essential boundary conditions; we find that the system to solve is

$$\frac{1}{\ell^3}\begin{bmatrix} 192 & 0 & -96 & 24\ell \\ 0 & 16\ell^2 & -24\ell & 4\ell^2 \\ -96 & -24\ell & 96 & -24\ell \\ 24\ell & 4\ell^2 & -24\ell & 8\ell^2 \end{bmatrix}\begin{Bmatrix} \tilde{u}_2 \\ \tilde{u}'_2 \\ \tilde{u}_3 \\ \tilde{u}'_3 \end{Bmatrix} = \begin{Bmatrix} 0 \\ 0 \\ P \\ 0 \end{Bmatrix} \tag{10.24}$$

with solution

$$\tilde{u}_2 = \frac{5}{48}\ell^3 P, \quad \tilde{u}'_2 = \frac{3}{8}P\ell^2, \quad \tilde{u}_3 = \frac{1}{3}P\ell^2, \quad \tilde{u}'_3 = \frac{1}{2}P\ell^2 \tag{10.25}$$

These results are in fact exact. This should be no surprise as the exact solution in this particular case is a cubic in x (see Example 10.1). Our aim here has been to show that once an **appropriate** choice has been made for the shape functions then the finite element procedures used earlier for second order boundary value problems can be applied, almost without change, to fourth-order problems.

10.5 Alternative treatment of fourth-order problems

It is obvious, from the previous section, that the treatment of fourth-order boundary value problems employing Hermite shape functions can be a somewhat tortuous process. In this section we shall show how it is feasible to treat such problems using the simpler one-dimensional shape functions from Chapters 3 and 4. This simpler approach depends on the fact that a fourth-order boundary value problem can be reformulated as a system of two second-order differential equations together with appropriate boundary conditions. We explain this approach by again considering the boundary value problem (10.18):

$$\frac{d^4u}{dx^4} = 0 \qquad 0 \leq x \leq \ell \tag{10.26a}$$

$$u(0) = u'(0) = u''(\ell) = 0, \qquad u'''(\ell) = -P \tag{10.26b}$$

If we introduce a new function say $v(x)$ such that

$$v = \frac{d^2u}{dx^2} \tag{10.27}$$

then (10.26a) is clearly equivalent to two second-order differential equations

$$\begin{aligned} &\frac{d^2v}{dx^2} = 0 \\ &\frac{d^2u}{dx^2} - v = 0 \end{aligned} \tag{10.28}$$

with boundary conditions

$$u(0) = u'(0) = 0, \quad v(\ell) = 0, \quad v'(\ell) = -P \tag{10.29}$$

We can now attempt approximations to both $u(x)$ and $v(x)$ of the usual Galerkin form:

$$\begin{aligned} \tilde{u}(x) &= \theta_0(x) + \sum_{i=1}^{m} a_i\,\theta_i(x) \\ \tilde{v}(x) &= \phi_0(x) + \sum_{i=1}^{n} b_i\,\phi_i(x) \end{aligned} \tag{10.30}$$

Note that the number of parameters m and n in these approximations may be different.

Naturally, we impose the usual constraints that $\theta_0(x), \phi_0(x)$ and the coordinate functions $\theta_i(x), \phi_i(x)$ are chosen such that $\tilde{u}, \tilde{v}$ satisfy all the boundary conditions regardless of the values of a_i, b_i.

Using these approximations we can define two residual functions:

$$\varepsilon_u(x) \equiv \frac{d^2\tilde{u}}{dx^2} - \tilde{v}, \qquad \varepsilon_v(x) \equiv \frac{d^2\tilde{v}}{dx^2}$$

and, correspondingly, two sets of Galerkin equations:

$$\int_0^\ell \frac{d^2\tilde{v}}{dx^2}\phi_i(x)dx = 0 \qquad i = 1, 2, \ldots, n$$

$$\int_0^\ell \left\{ \frac{d^2\tilde{u}}{dx^2} - \tilde{v} \right\} \theta_i(x)dx = 0 \qquad i = 1, 2, \ldots, m \tag{10.31}$$

These equations can be converted to modified Galerkin form in the usual way to obtain:

$$-\int_0^\ell \frac{d\tilde{v}}{dx}\frac{d\phi_i}{dx}\, dx - \frac{d\tilde{v}}{dx}(0)\, \phi_i\, (0) = 0 \qquad i = 1, 2, \ldots, n \tag{10.32a}$$

$$-\int_0^\ell \left\{ \frac{d\tilde{u}}{dx}\frac{d\theta_i}{dx} + \tilde{v}\, \theta_i \right\} dx + \frac{d\tilde{u}}{dx}\, (\ell)\theta_i\, (\ell) = 0 \qquad i = 1, 2, \ldots, m \tag{10.32b}$$

As an example of this approach we use the following simple approximations for $\tilde{u}(x), \tilde{v}(x)$:

$$\begin{aligned} \tilde{u}(x) &= a_1\, x^2 \\ \tilde{v}(x) &= -P(x - \ell) + b_1\, (x - \ell)^2 \end{aligned}$$

giving from (10.32a)

$$-\int_0^\ell \{-P + 2\, b_1(x - \ell)\}\, 2(x - \ell)dx - [-P - 2b_1\, \ell]\, \ell^2 = 0$$

which leads to $b_1 = 0$. Similarly from (10.32b) we find

$$-\int_0^\ell \left\{4\, a_1\, x^2 + (-P(x - \ell) + b_1(x - \ell)^2)x^2\right\} dx + 2\, a_1\, \ell^3 = 0$$

leading to $a_1 = P\ell/8$.

Thus a first approximation to the solution of the boundary value problem (10.26) is $\tilde{u}(x) = P\ell x^2/8$. Although this is not particularly accurate we do have, in this alternative approach, an independent approximation for the bending moment (the second derivative) viz.

$$\tilde{v}(x) = -P(x - \ell)$$

which is, in fact, the exact result. The reader is given the opportunity in the end of chapter exercises to show that, as might be expected, the exact solution for the displacement would be obtained with the three-parameter approximation:

$$\begin{aligned} \tilde{u}(x) &= a_1\, x^2 + a_2\, x^3 \\ \tilde{v}(x) &= -P(x - \ell) + b_1(x - \ell)^2 \end{aligned}$$

More complicated examples of fourth-order boundary value problems in one dimension can be treated similarly. Indeed, by reformulating fourth order equations as a system of two second-order equations finite element procedures may be used to obtain approximations without the need to consider Hermite-type shape functions.

10.6 Fourth-order boundary value problems in two dimensions

Perhaps the simplest example of a fourth-order boundary value problem in two dimensions arises in the mathematical modelling of the transverse deformation of a thin plate.

We consider a thin plate of uniform thickness with the x- and y-axes chosen to lie along the mid-plane (see Figure 10.7).

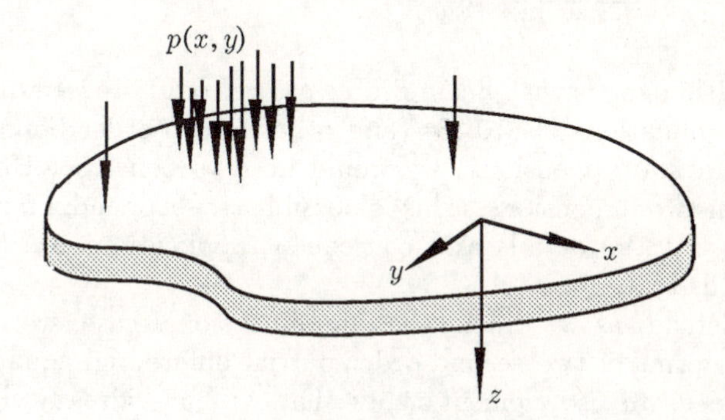

Figure 10.7

The plate is assumed to be supported by forces applied along its edge which can be a general curve C. The deformation of the plate is characterised by the displacement $u(x, y)$ of points on the mid-plane. The plate is assumed to be deformed by transverse forces $p(x, y)$ which may be a combination of distributed and/or point forces.

Using assumptions consistent with those of linear elasticity theory (see Chapter 5) it can be shown that after deformation the equation of the surface of the mid-plane, $z = u(x, y)$, satisfies the fourth-order partial differential equation:

$$\frac{\partial^4 u}{\partial x^4} + 2\frac{\partial^4 u}{\partial x^2 \partial y^2} + \frac{\partial^4 u}{\partial y^4} = p(x, y) \qquad \textbf{(10.33)}$$

When $p(x, y) = 0$ this important differential equation is known as the **biharmonic equation**. The boundary conditions associated with (10.33) depend on the supports which are applied at the edge of the plate. For a general curve C the boundary conditions may be quite complex. The simplest conditions are those pertaining to so-called clamped conditions. These take the form

$$u = 0, \quad \frac{\partial u}{\partial n} = 0 \quad \text{on} \quad C \qquad \textbf{(10.34)}$$

in which n is a coordinate measured along the outwardly drawn normal to C (see Section 8.10).

If, instead of clamped conditions, the edge is simply supported (i.e. the displacement and the bending moment are zero) then we find

$$u = 0, \quad \kappa\nu \frac{\partial u}{\partial n} + \frac{\partial^2 u}{\partial n^2} = 0 \quad \text{on} \quad C \tag{10.35}$$

Here κ is the curvature along the edge and ν is an elastic constant (Poisson's ratio) characteristic of the plate material. Along a straight edge κ is zero and in this case it may be shown that

$$\frac{\partial^2 u}{\partial n^2} = \frac{\partial^2 u}{\partial x^2} + \frac{\partial^2 u}{\partial y^2}$$

Although, as with beam problems, we could proceed directly with integral formulations of (10.33) (and other fourth-order boundary value problems in two dimensions) we would need to construct Hermite-type elements in two dimensions. This is possible (see Appendix for an outline approach) but the details and consequent applications are beyond the scope of this text.

Instead, as we did in the one-dimensional case, we reformulate (10.33) in terms of two second-order partial differential equations. This will enable us to use element types that we have already discussed in earlier chapters.

We introduce a new variable:

$$v = \frac{\partial^2 u}{\partial x^2} + \frac{\partial^2 u}{\partial y^2} \tag{10.36}$$

from which it follows readily that (10.33) is equivalent to the coupled system:

$$\begin{aligned} \frac{\partial^2 u}{\partial x^2} + \frac{\partial^2 u}{\partial y^2} - v &= 0 \\ \frac{\partial^2 v}{\partial x^2} + \frac{\partial^2 v}{\partial y^2} &= p(x, y) \end{aligned} \tag{10.37}$$

Clamped conditions are as before ($u = 0$, $\partial u/\partial n = 0$ along the boundary), whereas along a simply supported straight edge (along which the curvature κ is zero) the appropriate boundary conditions are

$$u = 0, \qquad v = 0$$

The Galerkin approach may now be applied to the system (10.37) by selecting separate approximations for $u(x, y)$ and $v(x, y)$:

$$\tilde{u}(x,y) = \theta_0(x,y) + \sum_{i=1}^{m} a_i\,\theta_i\,(x,y)$$
$$\tilde{v}(x,y) = \phi_0(x,y) + \sum_{i=1}^{n} b_i\,\phi_i\,(x,y) \tag{10.38}$$

(Compare (10.30) for the one-dimensional case.)

We can now proceed as in Section 10.5 to derive a system of modified Galerkin equations to determine the parameters a_i and b_i. This is left as an exercise for the reader to carry through, using the techniques outlined in Chapter 7.

EXERCISES

10.1 Determine the boundary value problem which results from minimizing the functional

$$J[\phi] = \int_0^{\ell} \{EI(\phi'')^2 - 2q\phi\}\,dx$$

over the class of functions satisfying $\phi(0) = \phi(\ell) = 0$. Here E is a constant and I, q are variable. Explain why the boundary conditions split into either 'essential' or 'suppressible' conditions and classify the boundary conditions in this case.

10.2 A cantilevered beam of length ℓ is fixed at $x = 0$ and free at $x = \ell$. When the beam is deformed to a curve $\phi(x)$ by a uniform load p per unit length the potential energy is

$$J[\phi] = \int_0^{\ell} \frac{1}{2}EI\left(\frac{d^2\phi}{dx^2}\right)^2\,dx - \int_0^{\ell} p\phi\,dx$$

J is minimized by the actual deformation $u(x)$. Using the Ritz method with a single parameter and a coordinate function

$$\theta_1 = \frac{x^2}{2} - \frac{2\ell}{\pi}\left[x - \frac{2\ell}{\pi}\sin\frac{\pi x}{2\ell}\right]$$

obtain an approximation to the deflection u at $x = \ell$ and compare with the exact solution:

$$u(x) = \frac{1}{24}p(x^4 - 4\ell x^3 + 6\ell^2x^2)$$

Note that the boundary conditions for this problem are:

$$u(0) = u'(0) = 0, \qquad u''(\ell) = u'''(\ell) = 0$$

and that the coordinate function satisfies **all** these conditions.

10.3 A circular plate of radius β with a concentric hole of radius α is clamped along its edges and is subjected to a uniform pressure p. The displacement satisfies the boundary value problem

$$\frac{d^2}{dr^2}\left(r\frac{d^2u}{dr^2}\right) - \frac{d}{dr}\left(\frac{1}{r}\frac{du}{dr}\right) = p \qquad \alpha \le r \le \beta$$

$$u(\alpha) = u(\beta) = 0 \qquad \frac{du}{dr}(\alpha) = \frac{du}{dr}(\beta) = 0$$

(a) Show, from first principles, that the equivalent variational problem is

$$J[\phi] = \int_\alpha^\beta \left\{ r(\phi'')^2 + \frac{1}{r}(\phi')^2 - 2p\phi \right\} dr = \text{extremum}$$

over the admissible class of functions $\phi(r)$ for which

$$\phi(\alpha) = \phi(\beta) = 0 \qquad \frac{d\phi}{dr}(\alpha) = \frac{d\phi}{dr}(\beta) = 0$$

(b) If $\alpha = 1$, $\beta = 2$ and $p = 1$ determine a two-parameter Ritz approximation using the trial solution

$$\phi = (r^2 - 1)^2(r^2 - 4)^2(a_1 + a_2 r^2)$$

and compare with the exact solution:

$$u = (-0.2661 - 0.2792r^2)\ln r + 0.2414r^2 + 0.01563r^4 - 0.2570$$

10.4 Develop modified Galerkin equations for the boundary value problem of Q10.3, and show that, for the same approximating function, these are equivalent to the equations obtained by the Ritz method.

10.5 Using two equal elements determine an approximation to the deformation $u(x)$ of a uniform beam clamped at each end and deformed by a point load at its centre, as shown in Figure Q10.5.

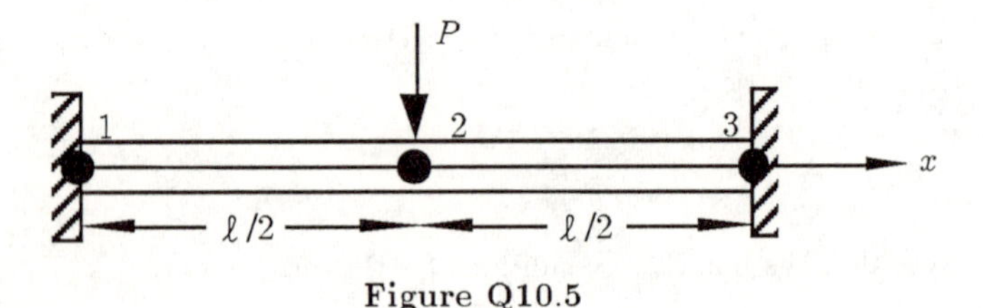

Figure Q10.5

10.6 Using two equal length elements of the type developed in (10.20) determine an approximation to the displacement of the 'stepped' cantilevered beam, subjected to the tip force P, shown in Figure Q10.6.

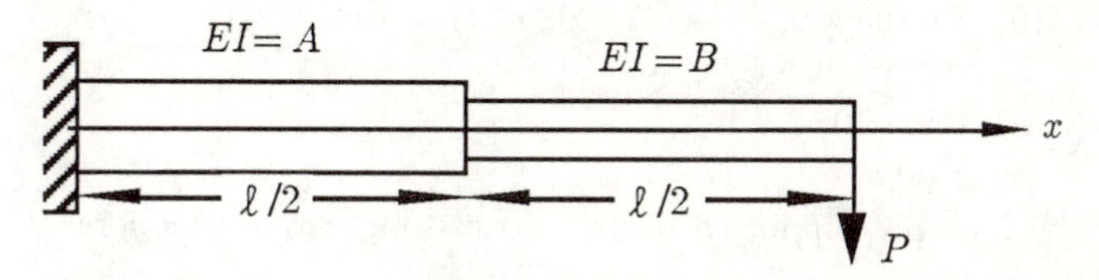

Figure Q10.6

10.7 For the Hermite shape functions introduced in (10.20) verify the following identities:

$$H_1(x) + H_3(x) = 1$$

$$aH_1(x) + H_2(x) + bH_3(x) + H_4(x) = x$$

$$a^2H_1(x) + 2aH_2(x) + b^2H_3(x) + 2bH_4(x) = x^2$$

$$a^3H_1(x) + 3a^2H_2(x) + b^3H_3(x) + 3b^2H_4(x) = x^3$$

10.8 Find a two-parameter Galerkin approximation (one for each variable) to the system of differential equations:

$$\frac{d^2u}{dx^2} - v = 1 \qquad 0 \le x \le 1$$

$$\frac{d^2v}{dx^2} = -1 \qquad 0 \le x \le 1$$

with boundary conditions

$$u(0) = \frac{du}{dx}(0) = 0 \qquad v(1) = \frac{dv}{dx}(1) = 1$$

Compare your results with the exact solution.

10.9 Show that the three-parameter expressions:

$$\tilde{u}(x) = a_1x^2 + a_2x^3 \qquad \tilde{v}(x) = -P(x-\ell) + b_1(x-\ell)^2$$

provide an exact solution to the system of differential equations:

$$\frac{d^2u}{dx^2} - v = 0$$

$$\frac{d^2v}{dx^2} = 0 \qquad 0 \le x \le \ell$$

with the boundary conditions

$$u(0) = u'(0) = 0, \qquad v(\ell) = 0,\ v'(\ell) = -P$$

(Cf. Section 10.5.)

10.10 For a simply supported beam subject to a compressive force T it can be shown that:

$$\frac{d^4u}{dx^4} + k^2\frac{d^2u}{dx^2} = 0$$

in which $u(x)$ is the transverse deflection and $k^2 = -T/EI$. The boundary conditions are:

$$u(0) = u(\ell) = 0 \qquad \frac{d^2u}{dx^2}(0) = \frac{d^2u}{dx^2}(\ell) = 0$$

(This eigenvalue problem governs the onset of buckling in the beam.)
(a) Using trial solutions

(i) $\tilde{u}(x) = a_1x(x-\ell) + a_2x^2(x-\ell)$

(ii) $\tilde{u}(x) = a_1 \sin\dfrac{\pi x}{\ell} + a_2 \sin\dfrac{2\pi x}{\ell}$

find approximations to the lowest value of T at which buckling occurs. Use an appropriate weighted residual or variational procedure.

(b) Repeat for a beam clamped at each end.

Appendix

Shape Functions

The use of shape functions is central to the finite element method. In the text we have concentrated mainly on **using** shape functions. In this appendix we discuss the **derivation** of shape functions for commonly used elements in both one and two dimensions.

We have briefly covered the derivation of one-dimensional shape functions in Chapter 3. Some of that material will be repeated in the interests of clarity and to make this appendix largely self-contained.

A1.1 Definite integrals and elements

We have seen that in using finite elements to determine approximations to the solution of boundary value problems a reformulation, utilizing integrals, is carried out. This is true irrespective of whether the Galerkin technique is used or whether the problem is reformulated according to some variational principle. It is the definite integral in its various guises and its treatment using elements and shape functions that is at the heart of the finite element technique.

The 'integral' under consideration may be an ordinary integral, in the case of one-dimensional boundary value problems or, for two-dimensional problems, a double integral, possibly supplemented by a line integral. (In three dimensions, triple integrals, surface integrals and line integrals arise.)

It is relatively easy to determine an approximation to the value of a definite integral. We simply divide the range of the integral into a large number of segments and then, depending on the degree of approximation required, choose to replace the integrand within each segment by a simpler function such as, in one dimension, a polynomial of low degree.

Unfortunately, the integrals which arise in the reformulation of a boundary value problem have an integrand which contains an unknown function – the solution of that problem. It is this function that is of prime interest – not the integral. However, we do know what 'form' the integrand takes and it is this 'form' that is replaced by a more convenient expression.

In finite elements there are three major steps to obtaining an approximation to the unknown function of interest.

- Obtaining an 'element' representation for the domain of the problem.
- Developing a suitable representation for the unknown function in terms of nodal function values.
- Prescribing a suitable algorithm by which the parameters defining the unknown function may be determined.

In this appendix we examine the first two aspects.

A1.2 Domain representation

For one-dimensional problems the simplest approach is to divide the range $a \le x \le b$ (Figure A1.1a) into a number of segments each connected together at nodes (Figure A1.1b).

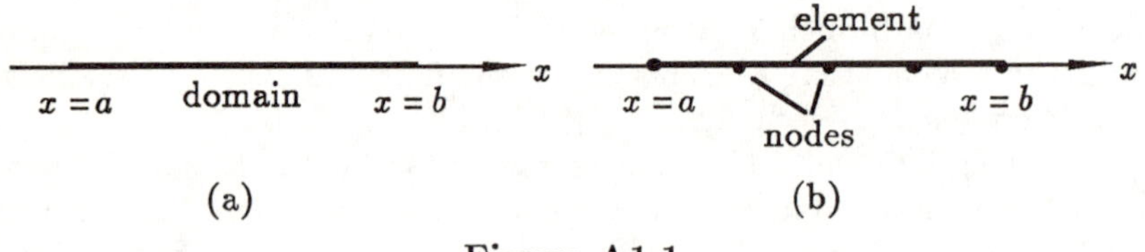

Figure A1.1

As we have seen in Chapter 3, more sophisticated one-dimensional elements, with interior nodes may also be employed.

For two-dimensional problems (Figure A1.2), although there are other possibilities, in practice only two 'shapes' are used to approximate to the domain, triangular and quadrilateral. See Figure A1.3 for three-noded triangular and four-noded quadrilateral elements, each edge being straight with two nodes.

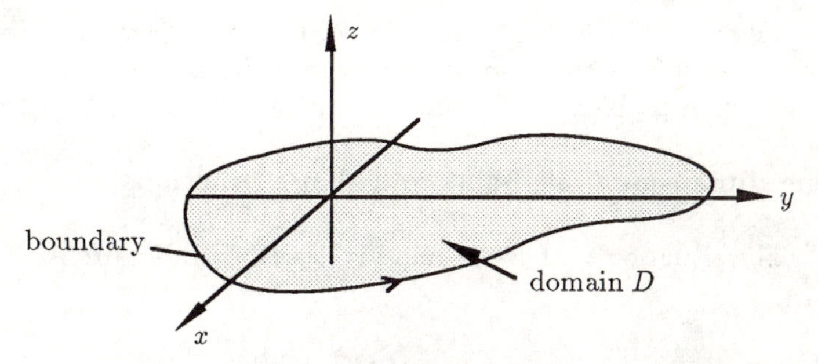

Figure A1.2

We see that for irregularly shaped regions the boundary cannot be represented exactly although clearly we can get as close as we desire to the actual boundary by using small straight-edged elements or alternatively elements with curved edges.

Approximation of a region using a number of elements inevitably gives rise to errors, known as **discretization errors**

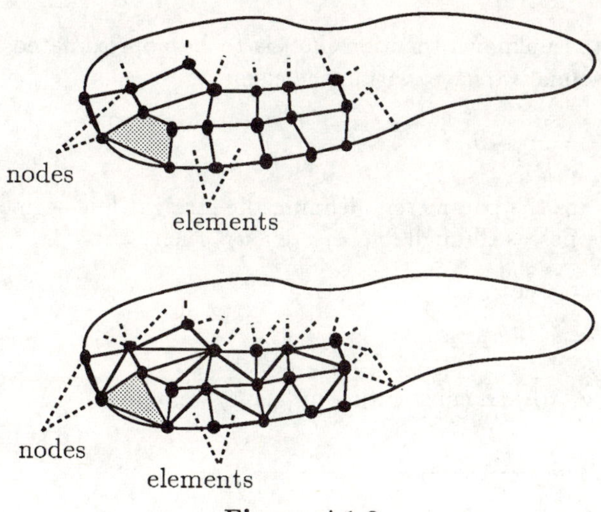

Figure A1.3

A1.3 Representation of the unknown function

Having chosen to divide the domain of the boundary value problem into a number of elements we must now decide how to approximate the unknown function within each element. We generally choose a simple approximation to the unknown function within each element. For one-dimensional problems this is usually a polynomial in one variable x, for two-dimensional problems a multinomial in two variables x, y and so on.

- **One dimensional elements and shape functions**

The situation we have in mind is described in Figure A1.4.

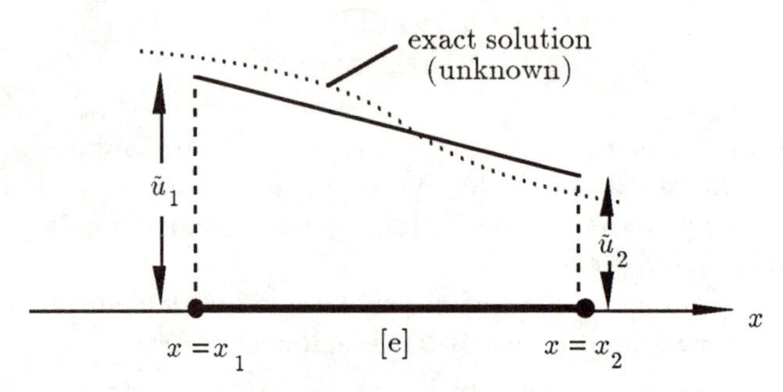

Figure A1.4

Assuming the unknown function $u(x)$ is to be approximated by a first-order polynomial we have, within the element,

$$\tilde{u}(x) = a_1 + a_2\, x \qquad \textbf{(A1.1)}$$

Here a_1, a_2 are the parameters defining the straight line segment. If this straight line passes through $(x_1, \tilde{u}_1); (x_2, \tilde{u}_2)$ then

$$\begin{aligned} \tilde{u}_1 &= a_1 + a_2\, x_1 \\ \tilde{u}_2 &= a_1 + a_2\, x_2 \end{aligned} \qquad \textbf{(A1.2)}$$

Solving for a_1, a_2 and substituting in (A1.1) we obtain

$$\tilde{u}(x) = \frac{\tilde{u}_1\, x_2 - \tilde{u}_2 x_1}{x_2 - x_1} + \left(\frac{\tilde{u}_2 - \tilde{u}_1}{x_2 - x_1} \right) x \qquad \textbf{(A1.3)}$$

This is the simplest approach. In finite elements we complicate matters slightly by emphasizing the part played by the nodal values $\tilde{u}_1, \tilde{u}_2$ of the unknown function $\tilde{u}(x)$. The main reason for doing this is that the nodal values $\tilde{u}_1, \tilde{u}_2$ have an obvious interpretation, whereas the significance of the parameters a_1, a_2 is less obvious. Thus, we rearrange (A1.3) in the equivalent form:

$$\tilde{u}(x) = \tilde{u}_1 \left\{ \frac{x_2 - x}{x_2 - x_1} \right\} + \tilde{u}_2 \left\{ \frac{x - x_1}{x_2 - x_1} \right\} \qquad \textbf{(A1.4)}$$

As far as the mathematics is concerned it is irrelevant whether we choose to use a_1, a_2 or $\tilde{u}_1, \tilde{u}_2$ as the two parameters in this linear approximation. We need only note that instead of multiplying a_1, a_2 by the polynomial terms '1' and 'x' respectively to obtain (A1.1) we must multiply $\tilde{u}_1$ and $\tilde{u}_2$ by slightly more complicated polynomials, $(x_2 - x)/(x_2 - x_1)$ and $(x - x_1)/(x_2 - x_1)$ respectively, to obtain (A1.4).

The polynomials

$$L_1(x) = \frac{x_2 - x}{x_2 - x_1} \qquad L_2(x) = \frac{x - x_1}{x_2 - x_1} \tag{A1.5}$$

are called linear shape functions.

These functions have the obvious property:

$$L_i = \begin{cases} 1 & \text{at node } i \\ 0 & \text{at the other element node} \end{cases} \tag{A1.6a}$$

together with (less obviously)

$$\begin{aligned} L_1 + L_2 &= 1 \qquad \text{for all } x \\ \text{and} \quad x_1 L_1 + x_2 L_2 &= x \qquad \text{for all } x \end{aligned} \tag{A1.6b}$$

It is worthwhile at this stage to examine the meaning of the identities listed in (A1.6b). In fact their existence confirms that both a constant term '1' and a linear term 'x' are present in the polynomial expansion (A1.4) of $\tilde{u}(x)$. This fact can be shown more generally. For example, if we assume as an approximation within the element a polynomial of degree $(n-1)$ then, corresponding to (A1.1), we would have

$$\tilde{u}(x) = a_1 + a_2\, x + \ldots + a_n\, x^{n-1} \tag{A1.7}$$

We now rewrite this expression in nodal value/shape function form. To do this we need n nodes, at $x_1, x_2, \ldots, x_n$, at which to sample $\tilde{u}(x)$, giving nodal values $\tilde{u}_1, \tilde{u}_2, \ldots, \tilde{u}_n$ respectively. We can then reformulate (A1.7) as

$$\tilde{u}(x) = \tilde{u}_1\, S_1(x) + \tilde{u}_2\, S_2(x) + \ldots + \tilde{u}_n\, S_n(x) \tag{A1.8}$$

where $S_1, S_2, \ldots, S_n$ are the shape functions appropriate to this n-noded element.

Now the right-hand side of (A1.8) is simply a re-arrangement of the polynomial in (A1.7). As such, the expression (A1.8) **must** reduce to the polynomial in (A1.7) no matter what values are assigned to the parameters $a_1, a_2, \ldots a_n$. Because of this the shape functions cannot be independent but must be inter-related and satisfy a certain number of easily deduced identities. For example if, in (A1.7), we choose

$$a_1 = 1, \; a_2 = a_3 = \ldots = a_n = 0$$

(so that the nth order polynomial has degenerated into a constant $\tilde{u} = 1$) then, from the definition of $\tilde{u}_i \equiv \tilde{u}(x = x_i)$ we must have

$$\tilde{u}_1 = 1, \qquad \tilde{u}_2 = 1, \ldots, \tilde{u}_n = 1$$

and so (A1.8) gives

$$1 = S_1(x) + S_2(x) + \ldots + S_n(x) = \sum_{i=1}^{n} S_i(x) \qquad \textbf{(A1.9)}$$

In other words if the reformulation of (A1.7) in the form (A1.8) is valid then the identity (A1.9) must result.

In fact, n similar identities can be deduced. For example, if we choose

$$a_1 = 0, \qquad a_2 = 1, \qquad a_3 = a_4 = \ldots = a_n = 0$$

(implying that $\tilde{u}(x) \equiv x$) then we must have from the definition of nodal function values

$$\tilde{u}_1 = \tilde{u}(x_1) = x_1, \qquad \tilde{u}_2 = x_2, \ldots, \tilde{u}_n = x_n$$

It follows from (A1.8) that

$$x = x_1 \, S_1(x) + x_2 \, S_2(x) + \ldots + x_n \, S_n(x) = \sum_{i=1}^{n} x_i \, S_i(x) \qquad \textbf{(A1.10)}$$

By similar reasoning and choosing, in turn, $\tilde{u} = 1, x, x^2, \ldots, x^{n-1}$ we find that the n identities satisfied by the shape functions of an n-noded element are:

$$\sum_{i=1}^{n} x_i^k \, S_i(x) = x^k \qquad k = 0, 1, \ldots, n-1 \qquad \textbf{(A1.11)}$$

For reference purposes these will be called the '**order-properties**' of shape functions. Equations (A1.6b) are of course the two order-properties for linear shape functions.

Each of the identities (A1.11) can be regarded as confirming the existence of the corresponding polynomial x^k in the expansion (A1.7). If one of the polynomial terms in that expansion is absent then the corresponding property in (A1.11) will fail. For example, if

$$\tilde{u}(x) \equiv a_1 + a_2 \, x^3$$

(implying a two-noded element as there are two parameters) then

$$\tilde{u}_1 = a_1 + a_2\, x_1^3$$
$$\tilde{u}_2 = a_1 + a_2\, x_2^3$$

from which

$$a_1 = \frac{\tilde{u}_1\, x_2^3 - \tilde{u}_2\, x_1^3}{x_2^3 - x_1^3} \qquad a_2 = \frac{\tilde{u}_2 - \tilde{u}_1}{x_2^3 - x_1^3}$$

and so

$$\tilde{u}(x) = \tilde{u}_1 \left[\frac{x_2^3 - x^3}{x_2^3 - x_1^3}\right] + \tilde{u}_2 \left[\frac{x^3 - x_1^3}{x_2^3 - x_1^3}\right]$$

In this case the shape functions are clearly

$$S_1(x) = \frac{x_2^3 - x^3}{x_2^3 - x_1^3} \qquad S_2(x) = \frac{x^3 - x_1^3}{x_2^3 - x_1^3}$$

The following relations are easily verified:

$$S_1 + S_2 = 1 \qquad \text{('1' present in polynomial)}$$
$$x_1\, S_1 + x_2\, S_2 \neq x \qquad \text{('}x\text{' term absent in polynomial)}$$
$$x_1^2\, S_1 + x_2^2\, S_2 \neq x^2 \qquad \text{('}x^2\text{' term absent in polynomial)}$$
$$x_1^3\, S_1 + x_2^3\, S_2 = x^3 \qquad \text{('}x^3\text{' term present in polynomial)}$$

There are other properties satisfied by shape functions, akin to those of (A1.6a) in the linear case, that warrant examination. These are obtained directly from (A1.8).

By using the definition of the nodal function values:

$$\tilde{u}_i = \tilde{u}(x = x_i) \qquad i = 1, 2, \ldots, n$$

we have at a particular node $x = x_k$

$$\tilde{u}_k = \tilde{u}_1\, S_1(x_k) + \ldots + \tilde{u}_k S_k\,(x_k) + \ldots + \tilde{u}_n\, S_n(x_k) \qquad \textbf{(A1.12)}$$

However, since the parameters $\tilde{u}_i, i = 1, 2, \ldots n$, are independent there can be no relation between them. Hence, we can deduce from (A1.12)

$$S_k(x_k) = 1$$
$$S_i(x_k) = 0 \qquad i \neq k$$

But these relations are true for **each** node x_k and so we can conclude:

$$S_i(x) = \begin{cases} 1 & \text{at node } x = x_i \\ 0 & \text{at every other node within the element} \end{cases} \qquad \textbf{(A1.13)}$$

These relations will be called the '**interpolating-properties**' of shape functions.

A1.4 Shape function derivation

We have seen in Chapter 3 and also in the previous section how shape functions may be derived algebraically by reformulating the standard polynomial expansion (cf. (A1.1) and (A1.7)) into a form which incorporates the nodal function values as parameters (cf. (A1.4) and (A1.8)). Unfortunately the algebra involved in this process, particularly for high order one-dimensional elements and for two- and three-dimensional elements, is somewhat complicated and not directly amenable to automatic calculation using a computer.

However, there does exist a technique, employing the properties of determinants, which can be used to derive all the element shape functions used in this text and which can readily be developed into an algorithm for computer implementation. The technique uses the order-properties of shape functions, (A1.11), and the interpolating properties (A1.13).

The technique is loosely based on the alternative derivation of shape functions outlined in Section 3.6, in that certain 'forms' (involving determinants) are introduced which manifestly satisfy the interpolation properties of shape functions. As we shall see, it is less obvious that the shape functions satisfy the order-properties. We outline the new approach by introducing an appropriate notation.

Let G and H_i be $(1 \times m)$ row vectors with individual terms $g_i,\ i = 1, 2, \ldots, m$ and $h_{ij},\ j = 1, 2, \ldots, m$ respectively.
That is,

$$G = \{g_1, g_2, \ldots, g_m\}$$
$$H_i = \{h_{i1}, h_{i2}, \ldots, h_{im}\} \qquad i = 1, 2, \ldots, m$$

Let D be the determinant of a matrix whose rows are $H_1, H_2, \ldots, H_m$ taken in order:

$$D = det\ [H_1, H_2, \ldots, H_m]^T$$
$$= det \begin{bmatrix} h_{11} & h_{12} & \ldots & h_{1m} \\ h_{21} & h_{22} & \ldots & h_{2m} \\ \vdots & \vdots & \vdots & \vdots \\ h_{m1} & h_{m2} & \ldots & h_{mm} \end{bmatrix}$$

Finally, let D_{iG} be the determinant obtained from D by replacing the ith row H_i by the row vector G.

Example A1.1

If

$$G = \{1 \quad x \quad x^2\}, \qquad H_1 = \{1 \quad 2 \quad 3\}$$
$$H_2 = \{3 \quad 4 \quad 5\}, \qquad H_3 = \{0 \quad -1 \quad 2\}$$

(i) determine D, D_{1G}, D_{2G}, D_{3G}.

(ii) verify the result

$$\sum_{j=1}^{3} h_{ji}\, D_{jG} = g_i D \qquad i = 1, 2, 3$$

Solution

(i) Clearly

$$D = \begin{vmatrix} 1 & 2 & 3 \\ 3 & 4 & 5 \\ 0 & -1 & 2 \end{vmatrix} = -8$$

and

$$D_{1G} = \begin{vmatrix} 1 & x & x^2 \\ 3 & 4 & 5 \\ 0 & -1 & 2 \end{vmatrix} = 13 - 6x - 3x^2$$

$$D_{2G} = \begin{vmatrix} 1 & 2 & 3 \\ 1 & x & x^2 \\ 0 & -1 & 2 \end{vmatrix} = -7 + 2x + x^2$$

$$D_{3G} = \begin{vmatrix} 1 & 2 & 3 \\ 3 & 4 & 5 \\ 1 & x & x^2 \end{vmatrix} = -2 + 4x - 2x^2$$

(ii) For $i = 1$ consider

$$h_{11}\, D_{1G} + h_{21}\, D_{2G} + h_{31}\, D_{3G}$$

We obtain

$$1(13 - 6x - 3x^2) + 3(-7 + 2x + x^2) + 0(-2 + 4x - 2x^2)$$
$$= -8 = g_1 D$$

By similar calculations we can readily show

$$\sum_{j=1}^{3} h_{j2}\, D_{jG} = g_2\, D$$

and $$\sum_{j=1}^{3} h_{j3}\, D_{jG} = g_3\, D$$

In the above example we have verified in one simple case what is a general result viz.

$$\sum_{j=1}^{m} h_{ji} D_{jG} = g_i D \qquad i = 1, 2, \ldots, m \tag{A1.14}$$

(for the case where G has m elements). This is a little known property of these determinants and one that is of fundamental importance in the construction of shape functions. We now apply the above formalism to such construction. Consider firstly a one-dimensional nth order finite element, that is, we assume within the element a trial solution of the form

$$\tilde{u}(x) = a_1 + a_2\, x + \ldots + a_n\, x^{n-1} = \sum_{i=1}^{n} \tilde{u}_i\, S_i(x) \tag{A1.15}$$

where $S_i(x)$, $i = 1, 2, \ldots, n$, are the shape functions we seek. The nodes are located at positions $x = x_i$, $i = 1, 2, \ldots, n$, within the element and $\tilde{u}_i = \tilde{u}(x = x_i)$, $i = 1, 2, \ldots, n$.

The polynomial expansion on the left-hand side of (A1.15) suggests we consider the row vector

$$G = \{1, x, x^2, \ldots, x^{n-1}\} \tag{A1.16}$$

We also define n further row vectors:

$$H_i \equiv G(x = x_i) \qquad i = 1, 2, \ldots, n \tag{A1.17}$$

that is, the terms in H_i are formed by evaluating the terms of G at each node in turn. We then have

$$D = \begin{vmatrix} 1 & x_1 & x_1^2 & \ldots & x_1^{n-1} \\ 1 & x_2 & x_2^2 & \ldots & x_2^{n-1} \\ \vdots & \vdots & \vdots & \vdots & \vdots \\ 1 & x_n & x_n^2 & \ldots & x_n^{n-1} \end{vmatrix}$$

and

$$D_{iG} = \begin{vmatrix} 1 & x_1 & x_1^2 & \dots & x_1^{n-1} \\ 1 & x_2 & x_2^2 & \dots & x_2^{n-1} \\ \vdots & \vdots & \vdots & \vdots & \\ 1 & x & x^2 & \dots & x^{n-1} \\ \vdots & \vdots & \vdots & \vdots & \\ 1 & x_n & x_n^2 & \dots & x_n^{n-1} \end{vmatrix} \leftarrow i\text{th row}$$

It is easily verified that the shape functions are given by the formula

$$S_i(x) = \frac{D_{iG}}{D} \qquad i = 1, 2, \dots, n \tag{A1.18}$$

For example, for a two-noded linear element

$$G = \{1, x\}, \quad H_1 = \{1, x_1\}, \quad H_2 = \{1, x_2\}$$

$$D = \begin{vmatrix} 1 & x_1 \\ 1 & x_2 \end{vmatrix} = x_2 - x_1$$

$$D_{1G} = \begin{vmatrix} 1 & x \\ 1 & x_2 \end{vmatrix} = x_2 - x \text{ and } D_{2G} = \begin{vmatrix} 1 & x_1 \\ 1 & x \end{vmatrix} = x - x_1.$$

Hence from (A1.18)

$$S_1(x) = \frac{D_{1G}}{D} = \frac{x_2 - x}{x_2 - x_1}$$

$$S_2(x) = \frac{D_{2G}}{D} = \frac{x - x_1}{x_2 - x_1}$$

which we recognize as the linear shape functions $L_1(x)$, $L_2(x)$ for a two-noded element.

We can also readily show that the interpolation-properties (A1.13) and the order-properties (A1.11) of shape functions follow from (A1.18).

• Interpolating-properties

From (A1.18), if j is a node other than i then

$$S_i(x = x_j) = \frac{D_{iG}(x = x_j)}{D}$$

which is zero since two rows of the determinant D_{iG} will now be identical.

Also, at node i itself:

$$S_i(x = x_i) = \frac{D_{iG}(x = x_i)}{D} = 1$$

since $D_{iG}(x = x_i)$ is identical to the determinant D.

- **Order-properties**

Using the fact that h_{ji} is the ith member of the jth row of D whilst g_i is the ith component of G it follows that if $i = k$ then

$$h_{jk} = x_j^{k-1} \qquad g_k = x^{k-1}$$

Hence the determinant property (A1.14) implies

$$\sum_{j=1}^{n} x_j^{k-1}\left(\frac{D_{jG}}{D}\right) = x^{k-1}$$

or

$$\sum_{j=1}^{n} x_j^{k-1} S_j(x) = x^{k-1} \qquad k = 1, 2, \ldots, n$$

which of course are the order-properties expected of these shape functions.

Example A1.2

Derive shape functions for a standard three-noded quadratic element with nodes at $x = -1, 0, +1$.

Solution

Here $G = \{1, x, x^2\}$

$$H_1 = G(x = -1) = \{1, -1, 1\}$$
$$H_2 = G(x = 0) = \{1, 0, 0\}$$
$$H_3 = G(x = 1) = \{1, 1, 1\}$$

Therefore

$$D = \begin{vmatrix} 1 & -1 & 1 \\ 1 & 0 & 0 \\ 1 & 1 & 1 \end{vmatrix} = 2$$

$$D_{1G} = \begin{vmatrix} 1 & x & x^2 \\ 1 & 0 & 0 \\ 1 & 1 & 1 \end{vmatrix} = -x + x^2$$

$$D_{2G} = \begin{vmatrix} 1 & -1 & 1 \\ 1 & x & x^2 \\ 1 & 1 & 1 \end{vmatrix} = x - x^2 - (-2) + (-x^2 - x)$$

$$= -2x^2 + 2 = -2(x-1)(x+1)$$

$$D_{3G} = \begin{vmatrix} 1 & -1 & 1 \\ 1 & 0 & 0 \\ 1 & x & x^2 \end{vmatrix} = x^2 + x$$

Hence

$$\frac{D_{1G}}{D} = \frac{1}{2}x(x-1) \qquad \frac{D_{2G}}{D} = (1-x)(1+x) \qquad \frac{D_{3G}}{D} = \frac{1}{2}x(x+1)$$

which are the quadratic shape functions $Q_1(x), Q_2(x)$ and $Q_3(x)$ first derived (using a variable ξ) in Chapter 3.

A1.5 Two-dimensional elements

• Triangular elements

The simplest two-dimensional element is a three-noded triangle. In such an element we seek an approximation of the form:

$$\tilde{u}(x,y) = a_1 + a_2\,x + a_3\,y$$

which, as in the one-dimensional situation, is conveniently re-expressed in terms of nodal function values and shape functions:

$$\tilde{u}(x,y) = \tilde{u}_1 T_1(x,y) + \tilde{u}_2 T_2(x,y) + \tilde{u}_3 T_3(x,y)$$

where $T_i(x,y),\ i = 1,2,3$ are called linear shape functions for a three-noded triangular element. The interpolation properties (A1.13) of one-dimensional shape functions have obvious two-dimensional equivalents:

$$T_i(x,y) = \begin{cases} 1 & \text{at node } i : x = x_i, y = y_i \\ 0 & \text{at the other two element nodes} \end{cases}$$

The order-properties (A1.9), (A1.10) can also be extended in an obvious way:

$$\sum_{i=1}^{3} T_i(x,y) = 1 \qquad \sum_{i=1}^{3} x_i\, T_i(x,y) = x \qquad \sum_{i=1}^{3} y_i\, T_i(x,y) = y$$

The determinant approach can be applied to find the explicit form of two-dimensional shape functions. For the linear triangular element we define row vectors

$$G = \{1, x, y\}$$

$$H_i = G(x = x_i, y = y_i) = \{1, x_i, y_i\} \qquad i = 1, 2, 3$$

(cf. (A1.16) and (A1.17) in the one-dimensional case) and so

$$D = \begin{vmatrix} 1 & x_1 & y_1 \\ 1 & x_2 & y_2 \\ 1 & x_3 & y_3 \end{vmatrix}$$

To be precise let us derive shape functions for a standard three-noded triangular element with nodes at $(-1,-1), (1,-1), (-1,1)$.

Then

$$H_1 = \{1, -1, -1\}, \quad H_2 = \{1, 1, -1\}, \quad H_3 = \{1, -1, 1\}$$

$$D = \begin{vmatrix} 1 & -1 & -1 \\ 1 & 1 & -1 \\ 1 & -1 & 1 \end{vmatrix} = 4$$

$$D_{1G} = \begin{vmatrix} 1 & x & y \\ 1 & 1 & -1 \\ 1 & -1 & 1 \end{vmatrix} = -2(x+y)$$

$$D_{2G} = \begin{vmatrix} 1 & -1 & -1 \\ 1 & x & y \\ 1 & -1 & 1 \end{vmatrix} = 2(1+x)$$

$$D_{3G} = \begin{vmatrix} 1 & -1 & -1 \\ 1 & 1 & -1 \\ 1 & x & y \end{vmatrix} = 2(1+y)$$

Hence, using (A1.18)

$$T_1(x,y) = \frac{D_{1G}}{D} = -\frac{1}{2}(x+y)$$

$$T_2(x,y) = \frac{D_{2G}}{D} = \frac{1}{2}(1+x)$$

$$T_3(x,y) = \frac{D_{3G}}{D} = \frac{1}{2}(1+y)$$

which were first used in Chapter 6 (where they were written in terms of variables ξ, η).

These procedures may be extended to situations in which $\tilde{u}(x,y)$ is assumed to be a higher order multinomial. Consider a quadratic variation within the triangular element:

$$\tilde{u}(x,y) = a_1 + a_2\,x + a_3\,y + a_4\,x^2 + a_5\,xy + a_6\,y^2$$

We require three additional nodes in order to be able to determine uniquely the parameters a_i, $i = 1,\ldots,6$. These extra nodes are normally positioned on the edges of the element (see Figure A1.5).

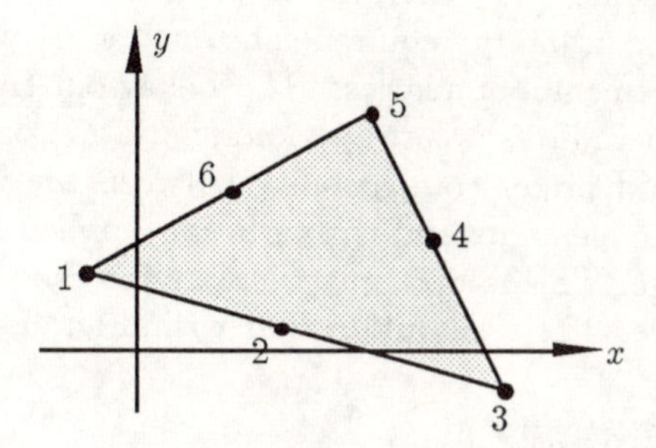

Figure A1.5

The shape functions are determined as above but in this case using the basic row vectors

$$G = \{1, x, y, x^2, xy, y^2\}$$

$$H_i = G(x = x_i, y = y_i) \quad i = 1,2,\ldots,6 \qquad \textbf{(A1.19)}$$

The extension to higher-order triangular elements is now clear. For triangular elements the choice of the multinomial terms occurring in the approximation of the unknown function is obtained most easily by reference to the **Pascal Triangle** (see Figure A1.6).

		total number of terms above level
1		
$x \quad y$	level 1	3
$x^2 \quad xy \quad y^2$	level 2	6
$x^3 \quad x^2y \quad xy^2 \quad y^3$	level 3	10
$x^4 \quad x^3y \quad x^2y^2 \quad xy^3 \quad y^4$	level 4	15

Figure A1.6

Thus for a linear element all terms on and above level 1 are chosen, as we have seen, whilst for a cubic element all terms on and above level 3 would be required. We note that for elements being constructed the distribution of nodes within the triangle is as specified by the 'geometrical' shape of the Pascal Triangle. A quadratic element has all its nodes on the edges, as in Figure A1.5; a cubic element has one interior node; a quartic element three interior nodes and so on. The reason for this is related to the fact that elements will generally be connected together along their sides and certain continuity requirements on $\tilde{u}(x,y)$ will be demanded across common element boundaries. The behaviour in $\tilde{u}(x,y)$ assumed within the element induces a corresponding variation on the edges of the element. We would prefer compatibility between the induced variation on the edges of the elements and the variation implied by the number of nodes on that edge. This is certainly the case for the three noded linear element seen above. The assumption that within an element $[e]$:

$$\tilde{u}(x,y) = a_1 + a_2\,x + a_3\,y$$

implies linear variation on the edges of the element. This is compatible with the occurrence of two nodes per edge (a unique straight line can be drawn through two points). Any other three-noded element attached to element $[e]$ (see Figure A1.7) will have exactly the same variation along the common edge and so the function $\tilde{u}(x,y)$ will be continuous across the element boundary. For, along the common boundary, approaching through element $[e]$:

$$\begin{aligned}\tilde{u}^{[e]} &= \tilde{u}_1 T_1^{[e]} + \tilde{u}_2 T_2^{[e]} + \tilde{u}_3 T_3^{[e]} \\ &= \tilde{u}_2 T_2^{[e]} + \tilde{u}_3 T_3^{[e]}\end{aligned}$$

since $T_1^{[e]} \equiv 0$ along the common edge. Because the shape functions are linear, it follows that $\tilde{u}^{[e]}$ varies linearly along the common edge taking a value $\tilde{u}_2$ at node 2 and a value $\tilde{u}_3$ at node 3.

However, if the edge is approached through element $[f]$:

$$\begin{aligned}\tilde{u}^{[f]} &= \tilde{u}_2\, T_1^{[f]} + \tilde{u}_4\, T_2^{[f]} + \tilde{u}_3\, T_3^{[f]} \\ &= \tilde{u}_2\, T_1^{[f]} + \tilde{u}_3\, T_3^{[f]}\end{aligned}$$

since $T_2^{[f]} \equiv 0$ along the common edge. Again the variation along the common edge is linear, varying between the values $\tilde{u}_2, \tilde{u}_3$ at nodes 2 and 3 respectively. We conclude that $\tilde{u}^{[e]} \equiv \tilde{u}^{[f]}$ along the common edge confirming that $\tilde{u}$ is continuous across element boundaries.

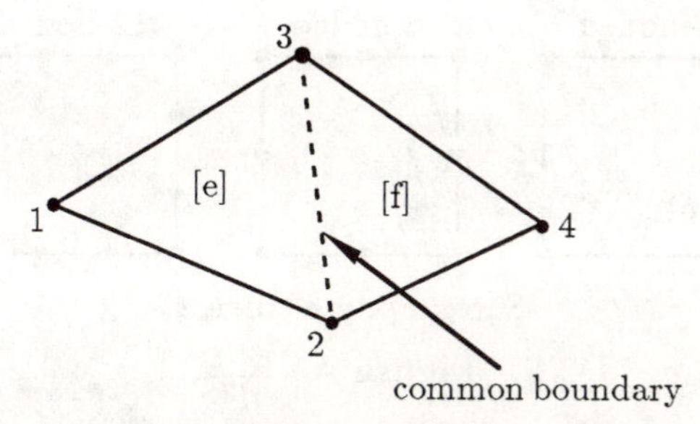

Figure A1.7

On the other hand, had we considered a cubic element and chosen to distribute the nodes so that we had two internal nodes and a side with just three nodes (instead of the normal four) then the cubic variation on the three noded side would depend on values within the element and would not be fully defined by the edge information alone. Thus any other (similar) element joined along the three-noded side would not necessarily have the same variation in $\tilde{u}(x, y)$ on the common boundary and $\tilde{u}(x, y)$ would not necessarily be continuous across element boundaries.

• **Rectangular elements**

The Pascal Triangle is also of considerable use in the construction of shape functions for rectangular elements of which there are two main types: the **Lagrangian** elements shown in Figure A1.8 and the **Serendipity** elements described in Figure A1.9.

The Lagrangian elements are so called because the correct number of nodes and terms from the Pascal triangle are used to define a Lagrange interpolation of any desired degree. In this sense, the triangular elements seen above could also be classed as 'Lagrangian'. The Serendipity elements are similar to the Lagrangian elements except that they have no internal nodes. Their name derives from the method of their discovery – by accident.

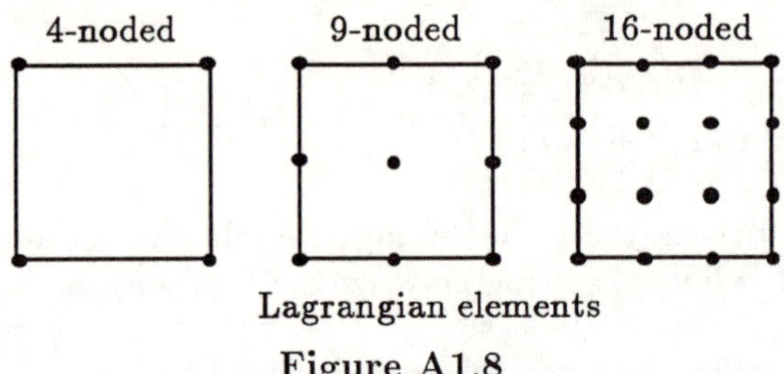

Lagrangian elements

Figure A1.8

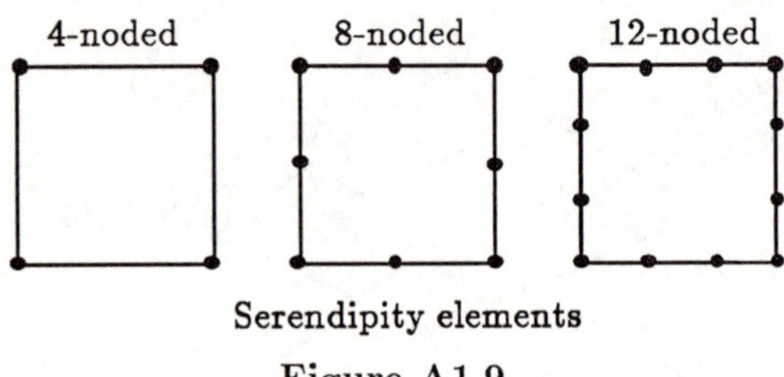

Serendipity elements

Figure A1.9

The terms in the approximation $\tilde{u}(x,y)$ for each of these elements are obtained from the Pascal Triangle as shown in Figure A1.10. Once the precise Pascal Triangle terms are chosen the determinant method can be used to construct the shape functions. For example, in a 9-noded Lagrangian element we assume within the element

$$\tilde{u}(x,y) = a_1 + a_2x + a_3y + a_4xy + a_5x^2 + a_6x^2y + a_7x^2y^2 + a_8xy^2 + a_9y^2$$

leading to

$$G = \{1, x, y, xy, x^2, x^2y, x^2y^2, xy^2, y^2\}$$

$$H_i = G(x = x_i, y = y_i) \quad i = 1, 2, \ldots, 9$$

Similarly for an eight-noded Serendipity element we assume within the element

$$\tilde{u}(x,y) = a_1 + a_2x + a_3y + a_4xy + a_5x^2 + a_6x^2y + a_7xy^2 + a_8y^2$$

leading to

$$G = \{1, x, y, xy, x^2, x^2y, xy^2, y^2\}$$

$$H_i = G(x = x_i, y = y_i) \quad i = 1, 2 \ldots, 8$$

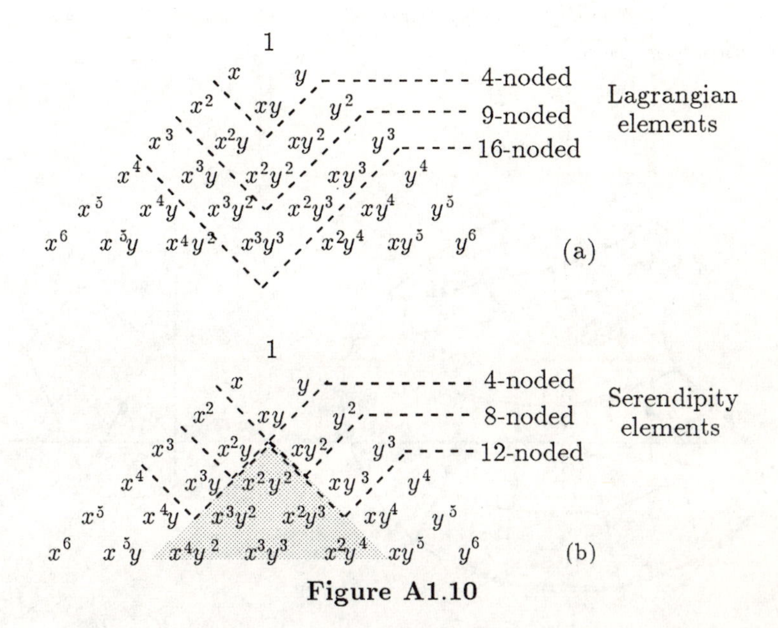

Figure A1.10

In general finite element use the Serendipity elements are more popular because similar levels of accuracy to the corresponding Lagrangian elements can be achieved using fewer nodes per element. Both the Lagrangian and Serendipity elements give rise to a function $\tilde{u}(x, y)$ which is continuous across element boundaries.

The determinant method may be described as a 'formal' method since we do not obtain the explicit form for the shape functions unless we evaluate the determinants that arise. Obviously, for determinants of order five or greater this is somewhat laborious. Fortunately the efficient numerical evaluation of determinants is not the problem that it was once thought to be. Indeed one of the authors has written a Fortran77 program which computes shape functions automatically (Ward, 1987).

• Irregular (transition-type) elements

The two-dimensional elements considered so far have been such that all element edges have equal numbers of nodes. For example in a 10-noded triangular element each side has 4 nodes while in an 8-noded Serendipity element each side has 3 nodes. Using imprecise language one may say that the more nodes per element the more 'accurate' will be the approximation that is obtained. In most finite element modelling we use the same type of element throughout the domain. However, occasionally it may be appropriate to use different elements in different parts of the domain. We might use higher-order elements to obtain a better approximation to the unknown function in that part of the domain where it is thought likely to vary rapidly. Thus a situation such as that shown in Figure A1.11(a) needs to be considered.

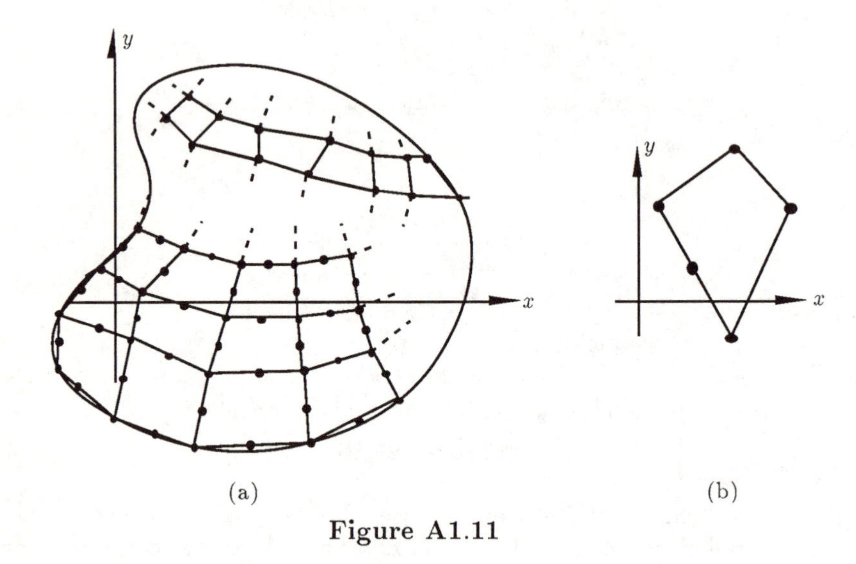

(a) (b)

Figure A1.11

Here four-noded elements are being used in part of the domain and eight-noded elements in the remainder. We need a special **transition-type** element (Figure A1.11(b)) to match one to the other. The determinant method may be applied to determine shape functions for such transition-type elements as we shall demonstrate in the following examples.

Example A1.3

Determine shape functions for the five-noded triangular element shown in Figure A1.12.

Solution

We begin by choosing an expansion from the Pascal Triangle, Figure A1.6, appropriate to an element with 6 nodes. We look for the side with the largest number of nodes and choose the corresponding regular element. Thus within the five-noded element we assume

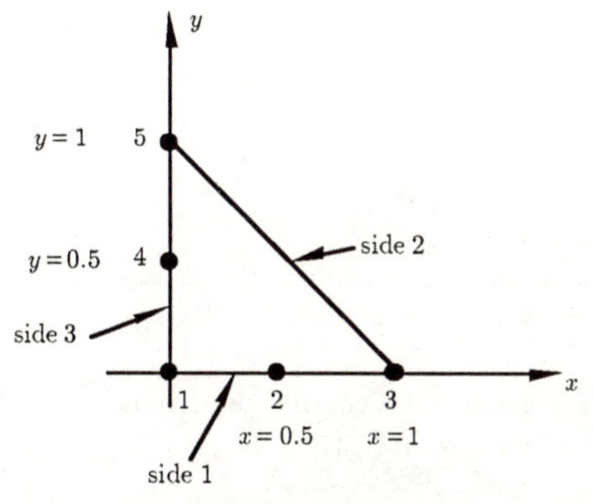

Figure A1.12

$$\tilde{u}(x,y) = a_1 + a_2x + a_3y + a_4x^2 + a_5xy + a_6y^2 \qquad \textbf{(A1.20)}$$

We see that there are six parameters but only five nodes. We must now constrain this expansion so that on the side with two nodes (side 2) the variation of $\tilde{u}$ is linear.

Now on side 2, $x + y = 1$ so that

$$\begin{aligned}\tilde{u}(x, y = 1 - x) &= a_1 + a_2x + a_3(1-x) + a_4x^2 + a_5x(1-x) \\ &\qquad + a_6(1-x)^2 \\ &= a_1 + a_6 + a_3 + x(a_2 - a_3 + a_5 - 2a_6) + x^2(a_4 - a_5 + a_6)\end{aligned}$$

This will be linear only if the quadratic term is absent, that is, if

$$a_4 - a_5 + a_6 = 0 \qquad \textbf{(A1.21)}$$

Therefore (A1.20) becomes (if we eliminate, for example, a_6)

$$\begin{aligned}\tilde{u}(x,y) &= a_1 + a_2x + a_3y + a_4x^2 + a_5xy + (-a_4 + a_5)y^2 \\ &= a_1 + a_2x + a_3y + a_4(x^2 - y^2) + a_5(xy + y^2)\end{aligned}$$

Clearly we now have five parameters to determine from the information at the five nodes. The determinant method will supply the shape functions if we choose

$$G = \{1, x, y, x^2 - y^2, xy + y^2\}$$

$$H_i = H(x = x_i, y = y_i) \quad i = 1, 2, 3, 4, 5$$

(It is irrelevant which of the three parameters is eliminated from (A1.21). The same shape functions are obtained in each case.) We leave it as an exercise for the reader to show that the actual shape functions are:

$$S_1 = 1 - 3x - 3y + 2(x+y)^2$$
$$S_2 = 4x(1 - x - y)$$
$$S_3 = x(-1 + 2(x+y))$$
$$S_4 = y(-1 + 2(x+y))$$
$$S_5 = 4y(1 - x - y)$$

Example A1.4

Determine the shape functions for the five-noded irregular Serendipity element shown in Figure A1.13

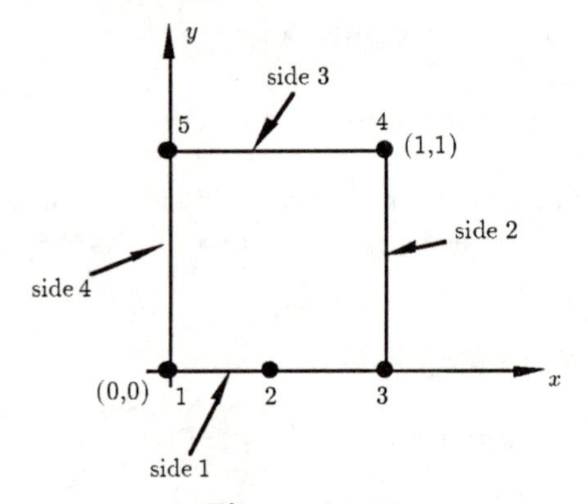

Figure A1.13

Solution

We choose terms appropriate to an eight-noded regular Serendipity element from the Pascal Triangle (see Figure A1.10(b)).

$$\tilde{u}(x,y) = a_1 + a_2x + a_3y + a_4xy + a_5x^2 + a_6x^2y + a_7xy^2 + a_8y^2 \qquad \textbf{(A1.22)}$$

To eliminate three of these parameters we demand linear behaviour on sides 2, 3 and 4.

- side 2 $\qquad x = 1$

$$\tilde{u}(1,y) = a_1 + a_2 + a_3y + a_4y + a_5 + a_6y + a_7y^2 + a_8y^2$$

We require

$$a_7 + a_8 = 0$$

for linear behaviour.

- side 3 $\qquad y = 1$

$$\tilde{u}(x,1) = a_1 + a_2x + a_3 + a_4x + a_5x^2 + a_6x^2 + a_7x + a_8$$

We require

$$a_5 + a_6 = 0$$

• side 4 $\quad x = 0$

$$\tilde{u}(0, y) = a_1 + a_3 y + a_8 y^2$$

We require

$$a_8 = 0$$

Solving the three constraint equations we have $a_7 = 0$, $a_8 = 0$ and $a_6 = -a_5$. Thus (A1.22) becomes

$$\tilde{u}(x, y) = a_1 + a_2 x + a_3 y + a_4 xy + a_5(x^2 - x^2 y)$$

leading to

$$G = \{1, x, y, xy, x^2(1-y)\}$$

$$H_i = G(x = x_i, y = y_i) \quad i = 1, 2, 3, 4, 5$$

So, for example,

$$D_{2G} = \begin{vmatrix} 1 & 0 & 0 & 0 & 0 \\ 1 & x & y & xy & x^2(1-y) \\ 1 & 1 & 0 & 0 & 1 \\ 1 & 1 & 1 & 1 & 0 \\ 1 & 0 & 1 & 0 & 0 \end{vmatrix} = \begin{vmatrix} x & y & xy & x^2(1-y) \\ 1 & 0 & 0 & 1 \\ 1 & 1 & 1 & 0 \\ 0 & 1 & 0 & 0 \end{vmatrix}$$

$$= \begin{vmatrix} x & xy & x^2(1-y) \\ 1 & 0 & 1 \\ 1 & 1 & 0 \end{vmatrix}$$

$$= x(1-y)(x-1)$$

A simple calculation shows that $D = -1/4$ so the corresponding shape function is

$$S_2(x, y) = \frac{D_{2G}}{D} = 4x(1-y)(1-x)$$

The remaining shape functions are similarly obtained and are

$$S_3 = x(1-y)(2x-1) \qquad S_4 = xy \qquad S_5 = y(1-x)$$
$$S_1 = (1-y)(1-3x+2x^2)$$

A1.6 Hermitian elements

The elements considered so far in this appendix have all been characterised by the fact that to each node there corresponds a single unknown function value. We say that such elements have **one degree of freedom** per node and they have, as we have seen in detail in this text, many applications to the determination of approximate solutions to second-order boundary value problems. Of course, it is this type of boundary value problem that has, at most, one essential boundary condition defined at each node residing on the boundary.

However, we have seen in Chapter 10 that fourth-order boundary value problems can have up to two essential boundary conditions imposed on a boundary node. Single-degree of freedom elements are of no use for problems of this kind. What is required is an element in which there are two unknowns at each node – the value of the function and of its derivative. The construction of such elements is considered in this section.

The simplest element of this type is a one-dimensional two-noded element. At each node the approximate function $\tilde{u}(x)$ and its derivative $\tilde{u}'(x)$ are specified. The lowest order polynomial able to model such a function is one with four parameters viz. a cubic. That is we shall assume within the element [*c*] (see Figure A1.14).

$$\tilde{u}(x) = a_1 + a_2\, x + a_3\, x^2 + a_4\, x^3 \qquad \textbf{(A1.23)}$$

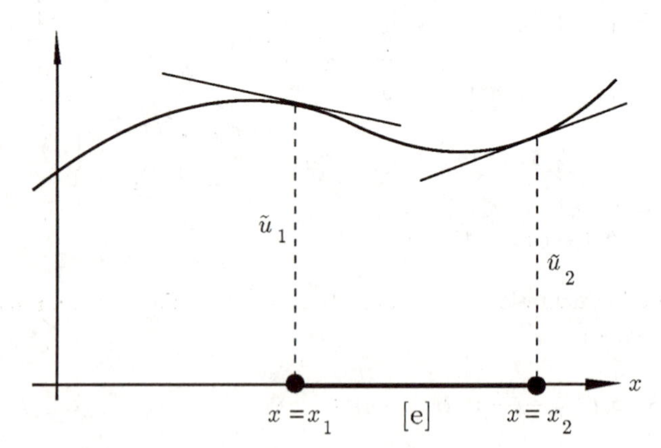

Figure A1.14

The parameters a_i, $i = 1, 2, 3, 4$, are chosen such that

$$\tilde{u}(x = x_i) = \tilde{u}_i \quad \text{and} \quad \frac{d\tilde{u}}{dx}(x = x_i) = \tilde{u}'_i \quad i = 1, 2$$

That is,

$$\begin{aligned}
\tilde{u}_1 &= a_1 + a_2\,x_1 + a_3\,x_1^2 + a_4\,x_1^3 \\
\tilde{u}_1' &= a_2 + 2\,a_3\,x_1 + 3\,a_4\,x_1^2 \\
\tilde{u}_2 &= a_1 + a_2\,x_2 + a_3\,x_2^2 + a_4\,x_2^3 \\
\tilde{u}_2' &= a_2 + 2\,a_3\,x_2 + 3\,a_4\,x_2^2
\end{aligned}$$

As usual in finite elements it is desirable to re-express the expansion (A1.23) in the nodal value/shape function form:

$$\tilde{u}(x) = \tilde{u}_1 C_1 + \tilde{u}_1'\,C_2 + \tilde{u}_2\,C_3 + \tilde{u}_2' C_4$$

where C_i, $i = 1, 2, 3, 4$, are the shape functions for this element. (The 'C' notation is chosen to emphasize the cubic nature of the element.) Shape functions which are used for the interpolation of a function and its derivatives are usually known as **Hermite** shape functions, and the corresponding elements as **Hermitian** elements. Using similar arguments to those outlined earlier it can be shown that the shape functions satisfy the following 'order-properties':

$$\begin{aligned}
&C_1 + C_3 = 1 \\
&x_1\,C_1 + C_2 + x_2\,C_3 + C_4 = x \\
&x_1^2\,C_1 + 2\,x_1\,C_2 + x_2^2\,C_3 + 2\,x_2\,C_4 = x^2 \\
&x_1^3\,C_1 + 3\,x_1^2\,C_2 + x_3^3\,C_3 + 3\,x_2^2\,C_4 = x^3
\end{aligned} \qquad \textbf{(A1.24)}$$

together with the 'interpolation-properties'

$$\begin{aligned}
&C_1 = \begin{cases} 1 & \text{at node 1} \\ 0 & \text{at node 2} \end{cases} \qquad C_3 = \begin{cases} 0 & \text{at node 1} \\ 1 & \text{at node 2} \end{cases} \\
&C_i = 0 \quad \text{at each node} \quad i = 2, 4 \\
&\frac{dC_2}{dx} = \begin{cases} 1 & \text{at node 1} \\ 0 & \text{at node 2} \end{cases} \qquad \frac{dC_4}{dx} = \begin{cases} 0 & \text{at node 1} \\ 1 & \text{at node 2} \end{cases} \\
&\frac{dC_i}{dx} = 0 \quad \text{at each node} \quad i = 1, 3
\end{aligned} \qquad \textbf{(A1.25)}$$

The conditions on the derivative ensure that $\tilde{u}'(x)$ is correctly interpolated at each node.

The determinant method can again be applied to determine the shape functions if we define the following vectors:

$$G = \{1, x, x^2, x^3\}$$

$$H_i = G(x = x_i), \qquad \frac{dH_i}{dx} = \{0, 1, 2\,x_i, 3\,x_i^2\} \quad i = 1, 2$$

Also, if we define

$$D = \det \left[H_1, \frac{dH_1}{dx}, H_2, \frac{dH_2}{dx} \right]^T$$

with D_{iG} defined as before, then the shape functions are given by

$$C_i = \frac{D_{iG}}{D} \qquad i = 1, 2, 3, 4$$

The determinant identity (A1.14) shows, for this choice of G, H_i and D, that the order-properties (A1.24) are satisfied and confirms that the polynomial expansion of $\tilde{u}(x)$ contains the terms $1, x, x^2$ and x^3. Also, with these choices, the interpolation-properties (A1.25) are easily shown to be satisfied.

As an example, the first Hermitian shape function for a standard element with nodes at $x = \pm 1$ is

$$C_1 = \frac{1}{D} \begin{vmatrix} 1 & x & x^2 & x^3 \\ 0 & 1 & -2 & 3 \\ 1 & 1 & 1 & 1 \\ 0 & 1 & 2 & 3 \end{vmatrix} = \frac{1}{D}[8 - 12x + 4x^3]$$

where

$$D = \begin{vmatrix} 1 & -1 & 1 & -1 \\ 0 & 1 & -2 & 3 \\ 1 & 1 & 1 & 1 \\ 0 & 1 & 2 & 3 \end{vmatrix} = 16$$

Hence

$$C_1 = \frac{1}{4}[2 - 3x + x^3]$$

In the same way we find:

$$C_2 = \tfrac{1}{4}(1-x-x^2+x^3) \qquad C_3 = \tfrac{1}{4}(2+3x-x^3) \qquad C_4 = \tfrac{1}{4}(-1-x+x^2+x^3)$$

- **Two-dimensional Hermitian elements**

In two-dimensional problems we are interested in constructing elements in which there are three unknowns at each node – the function $\tilde{u}(x, y)$ and its two partial derivatives $\partial\tilde{u}/\partial x$ and $\partial\tilde{u}/\partial y$. Such elements are applicable to the solution of fourth-order boundary value problems in two dimensions, such as those involving the Biharmonic equation.

For two-dimensional Hermitian elements we normally use a four-noded quadrilateral together with a Serendipity-type polynomial expansion. This is because the eight-noded Serendipity element of Figure A1.9, although popular in solving two-dimensional problems, is impractical in this case as 24 shape functions would be required. The four-noded quadrilateral is more appropriate as only 12 shape functions are required.

For such elements we continue to use a Serendipity-type polynomial expansion. For example, in a four-noded element with three degrees of freedom per node we interpolate $\tilde{u}$, $\partial\tilde{u}/\partial x$, $\partial\tilde{u}/\partial y$ exactly at each node and use the polynomial variation:

$$\begin{aligned}\tilde{u}(x,y) = a_0 + a_1\,x + a_2\,y + a_3\,xy + a_4\,x^2 + a_5\,x^2y \\ + a_6\,x^3 + a_7\,x^3\,y + a_8\,y^2 + a_9\,y^2x + a_{10}\,y^3 + a_{11}\,y^3x\end{aligned}$$

which can be written in nodal value/shape function form as

$$\tilde{u}(x,y) = \sum_{i=1}^{12} \tilde{h}_i\,S_i\,(x,y)$$

where

$$\tilde{h}_1 = \tilde{u}_1, \quad \tilde{h}_2 = \frac{\partial\tilde{u}_1}{\partial x}, \quad \tilde{h}_3 = \frac{\partial\tilde{u}_1}{\partial y}, \quad \ldots, \tilde{h}_{12} = \frac{\partial\tilde{u}_4}{\partial y}$$

Using an obvious notation the shape functions are given by the formula

$$S_i = \frac{D_{iG}}{D} \qquad i = 1,2,\ldots,12$$

where

$$D = \det\left[H_1, \frac{\partial H_1}{\partial x}, \frac{\partial H_1}{\partial y}, \ldots, H_4, \frac{\partial H_4}{\partial x}, \frac{\partial H_4}{\partial y}\right]^T$$

and $G = \{1, x, y, \ldots, y^3, y^3x\}$ with $H = G(x = x_i, y = y_i)$.

It is clear from what has been said earlier that these shape functions have all the properties we expect.

For a standard four-noded element we find

$$S_1 = \frac{1}{8}\{(1-x)(1-y)^2(2+y) - x(1-x^2)(1-y)\}$$

$$S_2 = \frac{1}{8}(1-y)(1+x)(1-x)^2$$

$$S_3 = \frac{1}{8}(1-x)(1+y)(1-y)^2$$

$$S_4 = \frac{1}{8}\{(1-y)(1+x)^2(2-x) - y(1-y^2)(1+x)\}$$

$$S_5 = -\frac{1}{8}(1-y)(1-x)(1+x)^2$$
$$S_6 = \frac{1}{8}(1+y)(1+x)(1-y)^2$$
$$S_7 = \frac{1}{8}\{(1+y)(1+x)^2(2-x) + y(1-y^2)(1+x)\}$$
$$S_8 = -\frac{1}{8}(1-x)(1+x)^2(1+y)$$
$$S_9 = -\frac{1}{8}(1-y)(1+y)^2(1+x)$$
$$S_{10} = \frac{1}{8}\{(1-x)(1+y)^2(2-y) - x(1-x^2)(1+y)\}$$
$$S_{11} = \frac{1}{8}(1+x)(1-x)^2(1+y)$$
$$S_{12} = -\frac{1}{8}(1-y)(1+y)^2(1-x)$$

Of course the evaluation of the 12×12 determinants which arise in this calculation has to be carried out on a computer.

Hermitian elements are rather specialized, being applicable to the solution of fourth-order boundary value problems. We have seen in Chapter 10, that more familiar, one-degree of freedom per node elements can be used if the fourth-order boundary value problem is reformulated appropriately. Consequently unless there is a specific interest in approximating the derivatives of a function it is perhaps advisable to avoid the use of Hermitian elements.

Outline Solutions to Odd-numbered Exercises

Chapter 2

2.1 (i) The single Galerkin equation is

$$\int_0^1 \left\{ \frac{d^2\tilde{u}}{dx^2} + \tilde{u} + x \right\} x(1-x)\, dx = 0 \quad \text{where} \quad \tilde{u} = a_1 x(1-x)$$

whence $a_1 = \frac{5}{18}$ and $\tilde{u}(x) = \frac{5}{18}x(1-x)$.

(ii) The two Galerkin equations are

$$\int_0^1 \left\{ \frac{d^2\tilde{u}}{dx^2} + \tilde{u} + x \right\} x(1-x)\, dx = 0 \quad \int_0^1 \left\{ \frac{d^2\tilde{u}}{dx^2} + \tilde{u} + x \right\} x^2(1-x)\, dx = 0$$

where $\dfrac{d^2\tilde{u}}{dx^2} = -2a_1 + 2a_2(1-3x)$. Substitution and evaluation of the two integrals gives

$$18a_1 + 9a_2 = 5 \qquad 63a_1 + 52a_2 = 21$$

and hence $a_1 = \dfrac{71}{369}$, $a_2 = \dfrac{7}{41}$ and $\tilde{u}(x) = x(1-x)[\dfrac{71}{369} + \dfrac{7}{41}x]$

Exact solution: $u(x) = \dfrac{\sin x}{\sin 1} - x$

2.3 (i) $\theta_0(x) = 150$ (satisfies both b.c.'s)
$\theta_1(x) = x(x-4)$ (satisfies homogeneous form of b.c.'s)
(ii) Standard Galerkin gives $a_1 = 32.3$

2.5 The given trial solutions satisfy only the essential boundary conditions $u(0) = 0$ and not the suppressible condition $u'(1) = 0$. However if we integrate the first term of the standard Galerkin equations

$$\int_0^1 \left\{ -\frac{d^2\tilde{u}}{dx^2} - \frac{1}{1+x} \right\} \theta_i(x)\, dx = 0 \qquad i = 1, 2, \ldots$$

by parts we obtain

$$\left[-\frac{d\tilde{u}}{dx}\theta_i \right]_0^1 + \int_0^1 \frac{d\tilde{u}}{dx}\frac{d\theta_i}{dx}\, dx$$

But $\theta_i(0) = 0$ (essential (homogeneous) b.c. at $x = 0$) and $\dfrac{d\tilde{u}}{dx} = 0$ at $x = 1$ (suppressible b.c.) so

$$\int_0^1 \left\{ \frac{d\tilde{u}}{dx}\frac{d\theta_i}{dx} - \frac{\theta_i}{1+x} \right\} dx = 0$$

where $\tilde{u}(x)$ need only satisfy $\tilde{u}(0) = 0$. Clearly the trial solutions (i) and (ii) are now acceptable.

(i) $\theta_i(x) = x$ gives $\int_0^1 \left\{ a_1 - \frac{x}{1+x} \right\} dx = 0$ giving $a_1 = 1 - \ln 2 = 0.3069$

(ii) $\theta_1(x) = x \quad \theta_2(x) = x^2$ gives modified Galerkin equations

$$\int_0^1 \left\{ (a_1 + 2a_2 x) - \frac{x}{1+x} \right\} dx = 0 \qquad \int_0^1 \left\{ (a_1 + 2a_2 x)2x - \frac{x^2}{1+x} \right\} dx = 0$$

whence $a_1 + a_2 = 0.3069 \qquad 3a_1 + 4a_2 = 0.5794$ and so $a_1 = 0.6482,\ a_2 = -0.3413$. The exact solution is $u(x) = x(1 + \ln 2) - (x+1)\ln(x+1)$

2.7 (a) Modified Galerkin equations are

$$\int_1^3 \left\{ -\frac{d\tilde{u}}{dx}\frac{d\theta_i}{dx} - 2\theta_i \right\} dx + 6\theta_i(3) = 0 \qquad i = 1, 2, \ldots$$

(b)(ii) The two modified Galerkin equations give

$$-2a_1 + a_2 = 1 \qquad -a_1 + a_2 = 5 \qquad \text{whence } a_1 = 4,\ a_2 = 9$$

Note : This problem will be discussed more fully in Chapter 4

2.9 The standard Galerkin method gives the equations

$$a_1(-20 + 5\lambda) + a_2(-10 + 2\lambda) = 0$$

$$a_1(-10 + 2\lambda) + a_2(-8 + \lambda) = 0$$

from which non-trivial solutions are obtained if

$$\lambda^2 - 20\lambda + 60 = 0 \quad \text{or} \quad \lambda = 3.675,\ 16.32$$

2.11 The standard Galerkin equations are

$$\int_a^b \left\{ -\frac{d}{dx}(p(x)\frac{d\tilde{u}}{dx}) + q(x)\tilde{u} - r(x) \right\} \theta_i(x)\, dx = 0 \quad i = 1, 2, \ldots, n$$

Integrating the first term by parts and using $\theta_i(a) = \theta_i(b) = 0$ (because of the given essential boundary conditions) gives the modified Galerkin equations

$$\int_a^b \left\{ p(x)\frac{d\tilde{u}}{dx}\frac{d\theta_i}{dx} + q(x)\tilde{u}\theta_i(x) - r(x)\theta_i(x) \right\} dx = 0$$

But, from (2.2), $\frac{d\tilde{u}}{dx} = \frac{d\theta_0}{dx} + \sum_{j=1}^{n} a_j \frac{d\theta_j}{dx}$. Substituting and rearranging

$$\sum_{j=1}^{n} a_j \int_a^b [p(x)\frac{d\theta_i}{dx}\frac{d\theta_j}{dx} + q(x)\theta_i(x)\theta_j(x)]\,dx = \int_a^b r(x)\theta_i(x)\,dx$$

$$-\int_a^b [p(x)\frac{d\theta_0}{dx}\frac{d\theta_i}{dx} + q(x)\theta_0(x)\theta_i(x)]\,dx \quad i = 1, 2, \ldots, n$$

or $\sum_{j=1}^{n} K_{ij}a_j = F_i$ or $[K]\{a\} = \{F\}$ in matrix form

Chapter 3

3.1 We have

$$\tilde{u}(-1) = \tilde{u}_1 = \alpha_1 + \alpha_2(-1) \qquad \tilde{u}(+1) = \tilde{u}_2 = \alpha_1 + \alpha_2(+1)$$

Solving for α_1, α_2, substituting in the expression for $\tilde{u}(\xi)$ and collecting terms

$$\tilde{u}(\xi) = \frac{\tilde{u}_1}{2}(1-\xi) + \frac{\tilde{u}_2}{2}(1+\xi)$$

and hence $L_1(\xi)$ and $L_2(\xi)$ as in (3.8)

3.3 In all cases we transform to the standard element using

$$x = L_1(\xi)x_1 + L_2(\xi)x_2 = x_1 + \frac{\ell}{2}(1+\xi)$$

(i) $\displaystyle\int_{x_1}^{x_2} L_1^4\,dx = \frac{1}{16}\int_{-1}^{1}(1-\xi)^4\frac{\ell}{2}\,d\xi = \frac{\ell}{5}$

$$\int_{x_1}^{x_2} L_1^2L_2^2\,dx = \frac{1}{16}\int_{-1}^{1}(1-\xi)^2(1+\xi)^2\frac{\ell}{2}\,d\xi = \frac{\ell}{30} \qquad \int_{x_1}^{x_2} L_2^4\,dx = \frac{\ell}{5}$$

Hence the required matrix is $\dfrac{\ell}{30}\begin{bmatrix} 6 & 1 \\ 1 & 6 \end{bmatrix}$

(ii) $\displaystyle\int_{x_1}^{x_2} xL_1\,dx = \frac{1}{2}\int_{-1}^{1}(1-\xi)\left(\frac{\ell}{2}\xi + x_1 + \frac{\ell}{2}\right)\frac{\ell}{2}\,d\xi = \frac{\ell}{2}\left(x_1 + \frac{\ell}{3}\right)$

$$\int_{x_1}^{x_2} xL_2\,dx = \frac{\ell}{2}\left(x_1 + \frac{2\ell}{3}\right)$$

so the required matrix is $\dfrac{\ell}{2}\left\{\begin{matrix} x_1 + \dfrac{\ell}{3} \\ x_1 + \dfrac{2\ell}{3} \end{matrix}\right\}$

3.5 $\tilde{u}^{[1]}(x) = \left(\dfrac{x_2 - x}{\ell^{[1]}}\right)\tilde{u}_1 + \left(\dfrac{x - x_1}{\ell^{[1]}}\right)\tilde{u}_2$ so $\tilde{u}^{[1]}(x_2) = \tilde{u}_2$

$\tilde{u}^{[2]}(x) = \left(\dfrac{x_3 - x}{\ell^{[2]}}\right)\tilde{u}_2 + \left(\dfrac{x - x_2}{\ell^{[2]}}\right)\tilde{u}_3$ so $\tilde{u}^{[2]}(x_2) = \tilde{u}_2$

$$\frac{d\tilde{u}^{[1]}}{dx} = \frac{\tilde{u}_2 - \tilde{u}_1}{\ell^{[1]}} \qquad \frac{d\tilde{u}^{[2]}}{dx} = \frac{\tilde{u}_3 - \tilde{u}_2}{\ell^{[2]}}$$

3.7 All results follow readily from the definitions of the $Q_i(\xi)$

3.9 Transform to the standard element so that

$$\int_{[e]} Q_i(x)\,dx = \frac{\ell}{2}\int_{-1}^{1} Q_i(\xi)\,d\xi. \qquad \text{Results follow readily}$$

3.11 In all cases we can map to the standard element using the linear transformation $x = x_1 + \frac{\ell}{2}(1+\xi)$ since in each case $x_2 = (x_1 + x_3)/2$. We then use

$\tilde{u}(\xi) = \sum_{i=1}^{3} Q_i(\xi)\tilde{u}_i$ to find $\tilde{u}$ at the specified point.

(i) $\tilde{u}(x = 0.25) = \tilde{u}(\xi = -0.75) = 17.41$ (ii) $\tilde{u}(x = 3.10) = \tilde{u}(\xi = 0.3) = 4.229$
(iii) $\tilde{u}(x = 1.25) = \tilde{u}(\xi = -0.5) = 1.0625$

3.13 (i) Mapping to standard element $-1 \leq \xi \leq 1$ gives

$$\int_1^2 x^5\,dx = \int_{-1}^{1} (\xi + 3)^5\,d\xi = \int_{-1}^{1} h(\xi)\,d\xi \quad \text{(say)}$$

Using Table 3.3 for the sample points ξ_i and weighting factors w_i for the 3-point formula

$$\int_{-1}^{1} h(\xi)\,d\xi = 0.4738 + 3.3750 + 6.6512 = 10.5 \quad \text{(exact, as expected)}$$

(ii) $\displaystyle\int_0^2 \frac{dx}{\sqrt{x}} = \int_{-1}^{1} \frac{d\xi}{\sqrt{(1+\xi)}} = 2.4761$ (3-pt formula). Exact value is 2.8284.
Trying 5-pt formula gives 2.6044 which is still only moderately accurate. The problem is the singularity in the integrand at $x = 0$.

Chapter 4

4.1 Proceed as in Section 4.2 or equivalently Section 4.3

(i) $\tilde{u}_2 = \tilde{u}(1) = -\frac{1}{4}$
(ii) The equations for the unknown nodal values are

$$\begin{bmatrix} 4 & -2 & 0 \\ -2 & 4 & -2 \\ 0 & -2 & 4 \end{bmatrix} \begin{Bmatrix} \tilde{u}_2 \\ \tilde{u}_3 \\ \tilde{u}_4 \end{Bmatrix} = \begin{Bmatrix} \frac{5}{2} \\ -\frac{3}{2} \\ -\frac{1}{2} \end{Bmatrix}$$

giving $\tilde{u}_2 = \tilde{u}(\frac{1}{2}) = \frac{1}{2}$, $\tilde{u}_3 = \tilde{u}(1) = -\frac{1}{4}$, $\tilde{u}_4 = \tilde{u}(\frac{3}{2}) = -\frac{1}{4}$

(iii) Again we find $\tilde{u}_2 = \tilde{u}(1) = -\frac{1}{4}$

Also $\tilde{u} = (6\xi^2 - 3\xi - 1)/4$ where $x = 1+\xi$. This agrees with the exact solution: $\tilde{u} = (6x^2 - 15x + 8)/4$.

4.3 The equations for the unknown nodal values are, after inserting the essential boundary conditions and using symmetry,

$$\begin{bmatrix} 3.2 & -2 \\ -4 & 4 \end{bmatrix} \begin{Bmatrix} \tilde{u}_2 \\ \tilde{u}_3 \end{Bmatrix} = \begin{Bmatrix} -2 \\ -2 \end{Bmatrix}$$

so $\tilde{u}_2 = \tilde{u}(200) = -2.5$ cm, $\tilde{u}_3 = \tilde{u}(400) = -3.0$ cm and $\tilde{u}_4 = -2.5$ cm (by symmetry).

4.5 (a) (i) Only unknown nodal value is $\tilde{u}_2$:

$$4\tilde{u}_2 = \frac{14000}{96} \quad \text{giving } \tilde{u}_2 = 36.46$$

(ii) Again only $\tilde{u}_2$ is unknown. We find:

$$\frac{16}{3}\tilde{u}_2 = 200 \quad \text{giving } \tilde{u}_2 = 37.5$$

(b) (i) Final reduced system of equations is

$$\begin{bmatrix} 4 & -2 \\ -2 & 2 \end{bmatrix} \begin{Bmatrix} \tilde{u}_2 \\ \tilde{u}_3 \end{Bmatrix} = \begin{Bmatrix} 145.8 \\ 178.1 \end{Bmatrix}$$

giving $\tilde{u}_2 = 162.0 \quad \tilde{u}_3 = 251.0$

(ii) Final reduced system of equations is

$$\begin{bmatrix} 16 & -8 \\ -8 & 7 \end{bmatrix} \begin{Bmatrix} \tilde{u}_2 \\ \tilde{u}_3 \end{Bmatrix} = \begin{Bmatrix} 600 \\ 453 \end{Bmatrix}$$

giving $\tilde{u}_2 = 163 \quad \tilde{u}_3 = 251.0$

4.7 The final system of equations is

$$\begin{bmatrix} 104 & -55 \\ -55 & 104 \end{bmatrix} \begin{Bmatrix} \tilde{u}_2 \\ \tilde{u}_3 \end{Bmatrix} = \begin{Bmatrix} -0.7777 \\ 2.7778 \end{Bmatrix}$$

giving $\tilde{u}_2 = -0.0299, \ \tilde{u}_3 = -0.0426$
(ii) The final system of equations is

$$\begin{bmatrix} 104 & -55 & 0 \\ -55 & 104 & -55 \\ 0 & -55 & 52 \end{bmatrix} \begin{Bmatrix} \tilde{u}_2 \\ \tilde{u}_3 \\ \tilde{u}_4 \end{Bmatrix} = \begin{Bmatrix} -0.777 \\ -2.777 \\ 15.611 \end{Bmatrix}$$

giving $\tilde{u}_2 = 0.4134, \ \tilde{u}_3 = 0.7958, \ \tilde{u}_4 = 1.142$

(iii) The final system of equations is

$$\begin{bmatrix} 52 & -55 & 0 & 0 \\ -55 & 104 & -55 & 0 \\ 0 & -55 & 104 & -55 \\ 0 & 0 & -55 & 52 \end{bmatrix} \begin{Bmatrix} \tilde{u}_1 \\ \tilde{u}_2 \\ \tilde{u}_3 \\ \tilde{u}_4 \end{Bmatrix} = \begin{Bmatrix} -18.055 \\ -0.777 \\ -2.777 \\ 21.612 \end{Bmatrix}$$

giving $\tilde{u}_1 = -0.561,\ \tilde{u}_2 = -0.206,\ \tilde{u}_3 = 0.189,\ \tilde{u}_4 = 0.616$

4.9 The final system of equations is

$$\begin{bmatrix} 2 & -1 \\ -1 & 126 \end{bmatrix} \begin{Bmatrix} \tilde{u}_2 \\ \tilde{u}_3 \end{Bmatrix} = \begin{Bmatrix} 50 \\ 625 \end{Bmatrix}$$

giving $\tilde{u}_2 = 27.59,\ \tilde{u}_3 = 5.179$

(NB Here, and in the following questions, the derivative boundary condition gives an additional contribution to the final row of $[K]$ as well as to the final row of $\{B\}$.)

4.11 (cf. Example 4.1.) The final system of equations is:

$$\begin{bmatrix} 13.333 & -5.667 & 0 & 0 \\ -5.667 & 13.333 & -5.667 & 0 \\ 0 & -5.667 & 13.333 & -5.667 \\ 0 & 0 & -5.667 & 7.066 \end{bmatrix} \begin{Bmatrix} \tilde{u}_2 \\ \tilde{u}_3 \\ \tilde{u}_4 \\ \tilde{u}_5 \end{Bmatrix} = \begin{Bmatrix} 493 \\ 40 \\ 40 \\ 28 \end{Bmatrix}$$

giving $\tilde{u}_2 = 53.9,\ \tilde{u}_3 = 39.9,\ \tilde{u}_4 = 32.8,\ \tilde{u}_5 = 30.3$

4.13 The final system of equations is

$$\begin{bmatrix} 304 & -102 \\ -102 & 133 \end{bmatrix} \begin{Bmatrix} \tilde{u}_2 \\ \tilde{u}_3 \end{Bmatrix} = \begin{Bmatrix} 18280 \\ 340 \end{Bmatrix}$$

giving $\tilde{u}_2 = 82.10,\ \tilde{u}_3 = 65.60$

Chapter 5

5.1 Clearly $\{n\} = \{\sin\alpha, \cos\alpha, 0\}^T$ and so, from (5.3),

$$\{T\} = [\sigma]\{n\} = \{\frac{F}{A}\sin\alpha, 0, 0\}^T$$

The normal component of $\{T\}$ on the plane is $\{T\}^T\{n\} = F\sin^2\alpha/A$. The magnitude of the shear stress is

$$|\{T\} - \{N\}| = |\{F\sin\alpha\cos^2\alpha/A, F\sin^2\alpha\cos\alpha/A, 0\}^T|$$

$$= \frac{F}{A} \sin\alpha \cos\alpha$$

The total force on the plane is $(F \sin\alpha / A)A'$ in the x–direction where A' is the area of the inclined plane, which is equal to $A/\sin\alpha$. Hence the total force on the plane is F.

5.3 $$\{T\} = [\sigma]\{n\} = \begin{bmatrix} 200 & 400 & 300 \\ 400 & 0 & 0 \\ 300 & 0 & -100 \end{bmatrix} \begin{Bmatrix} 1/3 \\ 2/3 \\ 2/3 \end{Bmatrix} = \begin{Bmatrix} 1600/3 \\ 400/3 \\ 100/3 \end{Bmatrix}$$

The normal stress has magnitude $\{T\}^T\{n\} = 288.88$ and from $\{S\} = \{T\} - \{N\}$ we find the shear stress magnitude is 468.9

5.5 Straightforward use of (5.10)

5.7 The known quantities are $u_1 = 0, u_3 = u, G_3 = 0$. The unknowns are G_1, u_2, G_3. The overall system of equations is

$$\begin{bmatrix} k_a & -k_a & 0 \\ -k_a & k_a + k_b & -k_b \\ 0 & -k_b & k_b \end{bmatrix} \begin{Bmatrix} 0 \\ u_2 \\ u \end{Bmatrix} = \begin{Bmatrix} G_1 \\ 0 \\ G_3 \end{Bmatrix}$$

The solution of these equations is

$$u_2 = \frac{k_b u}{k_a + k_b} \qquad G_1 = -\frac{k_a k_b u}{k_a + k_b} \qquad G_3 = \frac{k_a k_b u}{k_a + k_b}$$

5.9 The overall stiffness matrix is

$$\begin{bmatrix} 12 & -12 & 0 & 0 \\ -12 & 28 & -16 & 0 \\ 0 & -16 & 22 & -6 \\ 0 & 0 & -6 & 6 \end{bmatrix} \begin{Bmatrix} 0 \\ u_2 \\ u_3 \\ 0 \end{Bmatrix} = \begin{Bmatrix} G_1 \\ -60 \\ 120 \\ G_4 \end{Bmatrix}$$

giving $u_2 = 1.67$ and $u_3 = 6.67$

5.11 Symmetry implies $u_3 = -u_2, \quad v_3 = v_2, \quad u_5 = 0$. We find

$$u_2 = \frac{P\ell}{AE} \qquad v_2 = \frac{\sqrt{3}}{3} u_2 \qquad v_5 = \frac{3}{4} v_2 - \frac{\sqrt{3}}{4} u_2$$

Chapter 6

6.1 (a) $\tilde{u}(-1,-1) = \alpha_0 - \alpha_1 - \alpha_2 = \tilde{u}_1$

$$\tilde{u}(1,-1) = \alpha_0 + \alpha_1 - \alpha_2 = \tilde{u}_2$$

$$\tilde{u}(-1,1) = \alpha_0 - \alpha_1 + \alpha_2 = \tilde{u}_3$$

Solving for the α_i in terms of the $\tilde{u}_i$ gives

$$\tilde{u} = -\frac{1}{2}(\xi+\eta)\tilde{u}_1 + \frac{1}{2}(1+\xi)\tilde{u}_2 + \frac{1}{2}(1+\eta)\tilde{u}_3$$

(b) For example $T_1(-1,-1) = 1$, $T_1(1,-1) = 0$, $T_1(-1,1) = 0$ immediately from (6.19)

6.3 (i) Using (6.18) we obtain

$$x = 0.06\xi + 0.19 \quad y = 0.025\xi + 0.06\eta + 0.095$$

$$\text{or} \quad \xi = 16.667(x - 0.19) \quad \eta = 16.667y - 6.950x - 0.264$$

(ii) From Q6.1 above, within the element

$$\tilde{u} = -\frac{190}{2}(\xi+\eta) + \frac{160}{2}(1+\xi) + \frac{185}{2}(1+\eta)$$

$$= -15\xi - 2.5\eta + 172.5$$

But $(x,y) = (0.2, 0.06)$ gives $(\xi,\eta) = (0.1667, -0.6540)$ hence $\tilde{u} = 168.4$. Also $\dfrac{\partial\tilde{u}}{\partial x} = -232.6 \quad \dfrac{\partial\tilde{u}}{\partial y} = -41.617$

6.5 For example the side $\xi = -1$, $-1 \le \eta \le 1$ gives, on substitution,

$$x = x_1L_1(-1,\eta) + x_4L_4(-1,\eta) = -\frac{x_1}{2}(\eta-1) + \frac{x_4}{2}(\eta+1)$$

$$y = y_1L_1(-1,\eta) + y_4L_4(-1,\eta) = -\frac{y_1}{2}(\eta-1) + \frac{y_4}{2}(\eta+1)$$

which are the parametric equations of a straight line. Also

$$\eta = -1 \text{ gives } x = x_1,\ y = y_1 \text{ and } \eta = 1 \text{ gives } x = x_4,\ y = y_4$$

so the line joins (x_1, y_1) and (x_4, y_4).

6.7 C_2: map to standard linear element using $\xi = \dfrac{x}{4}$ or $x = 4\xi$

$$\therefore \quad \int_{C_2} = \int_{-1}^{1} (12\xi)4\,d\xi = 0$$

C_1: choose nodes 1:(4,0) 2:(0,3) 3:(-4,0) so, by (6.8), $x = 4Q_1(\xi) - 4Q_3(\xi)$, $y = 3Q_2(\xi)$ and $\dfrac{dx}{d\xi} = -4$, $\dfrac{dy}{d\xi} = -6\xi$ using (3.16). Hence

$$\int_{C_1} = \int_{-1}^{1} \{(12(Q_1 - Q_3) - 12Q_2)(-4) + (16(Q_1 - Q_3) + 6Q(2)(-6\xi)\}\, d\xi$$

$$= 48 \int_{-1}^{1} (1 + \xi^2)\, d\xi = 128$$

6.9 (a) Nodes $1:(1,0)$ $2:(\cos 22.5^0, \sin 22.5^0)$ $3:(\frac{1}{\sqrt{2}}, \frac{1}{\sqrt{2}})$ so

$$x = Q_1(\xi) + 0.9239Q_2(\xi) + 0.7071Q_3(\xi)$$

$$y = 0.3289Q_2(\xi) + 0.7071Q_3(\xi)$$

$$\frac{dx}{d\xi} = -0.1407\xi - 0.1465 \qquad \frac{dy}{d\xi} = -0.0583\xi + 0.3536$$

Proceeding as in Q6.7 gives answer 0.2070. (This is exact because this line integral is path independent.)

(b) Numerical approximation gives 0.9919, exact answer 1.0783.

6.11 Nodes: $1:(1,0)$ $2:(\frac{\sqrt{3}}{2}, \frac{1}{2})$ $3:(\frac{1}{2}, \frac{\sqrt{3}}{2})$ $4:(0,1)$ so from Q6.9

$$x = C_1(\xi) + \frac{\sqrt{3}}{2}C_2(\xi)) + \frac{1}{2}C_3(\xi)$$

$$y = \frac{1}{2}C_2(\xi) + \frac{\sqrt{3}}{2}C_3(\xi)) + C_4(\xi)$$

then proceeding as in Q6.7 gives exact answer -1. (To be expected as this is a path independent line integral.)

6.13 Transforming to the standard square

$$\mathrm{I} = \int_{-1}^{1}\int_{-1}^{1} e^{(\xi+3)/2}\frac{\pi}{16}(\xi + 3)\, d\xi\, d\eta$$

We obtain $\mathrm{I} \approx 11.606674$. (The exact answer, which can be obtained using polar coordinates, is $\pi\, e^2/2$ or 11.606702.)

6.15 We only require global nodes 1,2,5,3 from Q6.14 which we will take as corresponding to local nodes 1,2,3,4 respectively. Using (6.22) and (6.23) we map to the standard square using

$$x = \frac{\sqrt{2}}{2}L_2(\xi,\eta) + 0.76537L_3(\xi,\eta)$$

$$y = \frac{\sqrt{2}}{2}L_2(\xi,\eta) + 1.84776L_3(\xi,\eta) + L_4(\xi,\eta)$$

From these transformation equations the Jacobian can be determined as an expression in ξ and η. Then the expressions $(x+y)$ and J can be determined at each of the four Gaussian sample points (ξ_i, η_i), $i = 1,2,3,4$. We find

$$(x+y)_{(\xi_1,\eta_1)} = 1.76666 \qquad J_{(\xi_1,\eta_1)} = \begin{vmatrix} 0.35971 & 0.36841 \\ 0.00615 & 0.51486 \end{vmatrix} = 0.18293$$

$$(x+y)_{(\xi_2,\eta_1)} = 1.35983 \qquad J_{(\xi_2,\eta_1)} = \begin{vmatrix} 0.35971 & 0.36841 \\ 0.02297 & 0.55543 \end{vmatrix} = 0.19133$$

$$(x+y)_{(\xi_1,\eta_2)} = 1.12069 \qquad J_{(\xi_1,\eta_2)} = \begin{vmatrix} 0.22709 & 0.25959 \\ 0.00615 & 0.51486 \end{vmatrix} = 0.11533$$

$$(x+y)_{(\xi_2,\eta_2)} = 2.02776 \qquad J_{(\xi_2,\eta_2)} = \begin{vmatrix} 0.22709 & 0.25959 \\ 0.02297 & 0.55543 \end{vmatrix} = 0.12017$$

leading to

$$\iint_{[1]} (x+y)\, dx\, dy = 0.95491$$

which compares with a value of 0.97463 obtained when two triangles were used.

Chapter 7

7.1 Take $\tilde{u} = a_1\theta_1(x) = a_1(x^2 - a^2)(y^2 - b^2)$ then

$$\frac{\partial^2 \tilde{u}}{\partial x^2} = 2a_1(y^2 - b^2) \qquad \frac{\partial^2 \tilde{u}}{\partial y^2} = 2a_1(x^2 - a^2)$$

Hence

$$\iint_A \left[2a_1(y^2 - b^2) + 2a_1(x^2 - a^2) + 2\right](x^2 - a^2)(y^2 - b^2)\, dx\, dy = 0$$

Integrating and solving we obtain $a_1 = \dfrac{5}{4(a^2 + b^2)}$

7.3 The modified Galerkin equations are, for a two-parameter solution,

$$\iint_A \left[-\left\{ \frac{\partial \theta_i}{\partial x}\frac{\partial \tilde{u}}{\partial x} + \frac{\partial \theta_i}{\partial y}\frac{\partial \tilde{u}}{\partial y} \right\} + \theta_i \right] dxdy$$

$$+ \oint_\Gamma \theta_i \left\{ -\frac{\partial \tilde{u}}{\partial y}dx + \frac{\partial \tilde{u}}{\partial x}dy \right\} = 0 \qquad i = 1,2$$

But $\theta_i = 0$ on $x = 0$ and on $y = 0$. Also

$$\frac{\partial \tilde{u}}{\partial n} = \left(\hat{\mathbf{i}}\frac{\partial \tilde{u}}{\partial x} + \hat{\mathbf{j}}\frac{\partial \tilde{u}}{\partial y} \right) \bullet \left(\frac{\hat{\mathbf{i}} + \hat{\mathbf{j}}}{\sqrt{2}} \right) \qquad \text{on } x + y - 1 = 0 \text{ (say C)}$$

that is $\quad \dfrac{\partial \tilde{u}}{\partial x} + \dfrac{\partial \tilde{u}}{\partial y} = 0$

But $$\int_C \theta_i \left\{-\frac{\partial\tilde{u}}{\partial y}dx + \frac{\partial\tilde{u}}{\partial x}dy\right\} = \int_1^0 \theta_i \left\{-\frac{\partial\tilde{u}}{\partial y} - \frac{\partial\tilde{u}}{\partial x}\right\} dx = 0$$

so the line integral in the Galerkin equations is zero. We require each θ_i to satisfy only the essential boundary conditions so choose $\theta_1 = xy$, $\theta_2 = xy(x+y)$ as instructed. Substituting into the modified Galerkin equations and integrating gives two equations

$$4a_1 + 4a_2 - 1 = 0 \qquad 15a_1 + 16a_2 - 3 = 0$$

with solution $a_1 = 1$, $a_2 = -0.75$.

7.7 Use (7.33) for the elements of $[K]$ and the first of (7.34) for the elements of $\{F\}$. Transform to the standard square using (7.12b) which gives

$$x = x_1 + b(1+\xi) \qquad y = y_1 + a(1+\eta) \qquad J = ab = \frac{1}{4}A$$

Replacing R_i by L_i (as in (7.13)) we find

$$K_{ij}^{[e]} = \frac{a}{b}\int_{-1}^{1}\int_{-1}^{1} \frac{\partial L_i}{\partial \xi}\frac{\partial L_j}{\partial \xi} d\xi\, d\eta + \frac{b}{a}\int_{-1}^{1}\int_{-1}^{1} \frac{\partial L_i}{\partial \eta}\frac{\partial L_j}{\partial \eta} d\xi\, d\eta$$

and $$F_i^{[e]} = hab\int_{-1}^{1}\int_{-1}^{1} L_i\, d\xi\, d\eta$$

The integrals and final result then follow using (7.20).

7.9 (*Note*: for Q7.9 and Q7.10 that the suppressible boundary condition $\dfrac{\partial u}{\partial n} = 0$ on the line of symmetry ensures that $\{B\} = 0$ in each case.)

The final reduced system of equations is

$$\begin{bmatrix} 1 & -1 & 0 \\ -1 & 4 & -2 \\ 0 & -2 & 4 \end{bmatrix} \begin{Bmatrix} \tilde{u}_1 \\ \tilde{u}_2 \\ \tilde{u}_4 \end{Bmatrix} = \begin{Bmatrix} 58.12 \\ 174.36 \\ 174.36 \end{Bmatrix}$$

giving $\tilde{u}_1 = 218.0 \quad \tilde{u}_2 = 159.8 \quad \tilde{u}_4 = 123.5$

7.11 The final system of equations is

$$\begin{bmatrix} 4 & -1 & -1 & -2 \\ -1 & 8 & -2 & -2 \\ -1 & -2 & 8 & -2 \\ -2 & -2 & -2 & 16 \end{bmatrix} \begin{Bmatrix} \tilde{u}_1 \\ \tilde{u}_2 \\ \tilde{u}_4 \\ \tilde{u}_5 \end{Bmatrix} = \begin{Bmatrix} 261.5 \\ 523 \\ 523 \\ 1046 \end{Bmatrix}$$

or, since $\tilde{u}_2 = \tilde{u}_4$ by symmetry,

$$\begin{bmatrix} 4 & -2 & -2 \\ -2 & 6 & -2 \\ -2 & -4 & 16 \end{bmatrix} \begin{Bmatrix} \tilde{u}_1 \\ \tilde{u}_2 \\ \tilde{u}_5 \end{Bmatrix} = \begin{Bmatrix} 261.5 \\ 523 \\ 1046 \end{Bmatrix}$$

giving $\tilde{u}_1 = 216.7 \quad \tilde{u}_2 = 168.2 \quad \tilde{u}_5 = 134.5$

7.13 Use of the element matrix (7.48) and that derived in Q7.12(i) followed by assembly gives the system $[K]\{\tilde{u}\} = \{0\}$ where

$$[K] = \frac{1}{200}\begin{bmatrix} 250 & 75 & 0 & -75 & -250 \\ 75 & 250 & -75 & 0 & -250 \\ 0 & -75 & 250 & 75 & -250 \\ -75 & 0 & 75 & 250 & -250 \\ -250 & -250 & -250 & -250 & 1000 \end{bmatrix} + \frac{50\omega^2 c^2}{12}\begin{bmatrix} 4 & 1 & 0 & 1 & 2 \\ 1 & 4 & 1 & 0 & 2 \\ 0 & 1 & 4 & 1 & 2 \\ 1 & 0 & 1 & 4 & 2 \\ 2 & 2 & 2 & 2 & 8 \end{bmatrix}$$

(The matrix $\{B\} = \{0\}$ here because of the given suppressible boundary conditions.)

Chapter 8

8.1 (a) The corresponding boundary value problem is

$$u'' + u + x^2 = 0 \qquad u(0) = u(1) = 0$$

giving $u = -2\cos x + \dfrac{2\cos 1 - 1}{\sin 1}\sin x - x^2 + 2$

(b) The corresponding boundary value problem is

$$u'' + u + x^2 = 0 \qquad u(0) = 0 \quad u'(1) = 0$$

giving $u = -2\cos x + \dfrac{2 - 2\sin 1}{\cos 1}\sin x - x^2 + 2$

8.5 (a) $J[\phi] = \displaystyle\iint_D \left\{ \left(\frac{\partial \phi}{\partial x}\right)^2 + \left(\frac{\partial \phi}{\partial y}\right)^2 - 2\phi \right\} \, dx\, dy$

(b) $J[\phi] = \displaystyle\iint_D \left\{ -\left(\frac{\partial \phi}{\partial x}\right)^2 - \left(\frac{\partial \phi}{\partial y}\right)^2 + 2\phi \right\} \, dx\, dy + \int_{\Gamma_2} (-\phi^2)\, ds$

8.9 Using the given trial solution and the boundary conditions we obtain $\tilde{u}(x) = x(1-x)[a_1 + a_2 x]$ where $a_1 = \alpha_1$ and $a_2 = (\alpha_1 + \alpha_2)$. Hence $\theta_1(x) = x(1-x)$ and $\theta_2(x) = x^2(1-x)$. Also $\theta_0(x) \equiv 0$ (homogeneous b.c.'s here). The components of the (symmetric) matrix $[K]$ are (up to a constant factor), using $p(x) = -1$, $q(x) = 1$, $r(x) = e^x$, and the expressions in Q8.8

$$K_{11} = -\frac{3}{10} \quad K_{12} = K_{21} = -\frac{3}{20} \quad K_{22} = -\frac{13}{105}$$

$$F_1 = 3 - e \quad F_2 = 3e - 8$$

so the system $[K]\{a\} = \{F\}$ is

$$\begin{bmatrix} -\dfrac{3}{10} & -\dfrac{3}{20} \\ -\dfrac{3}{20} & -\dfrac{13}{105} \end{bmatrix} \begin{Bmatrix} a_1 \\ a_2 \end{Bmatrix} = \begin{Bmatrix} 3 - e \\ 3e - 8 \end{Bmatrix}$$

whence $a_1 = -0.7958$, $a_2 = -0.2866$.

Chapter 10

10.1 From first principles consider $\phi(x) = u(x) + \varepsilon\eta(x)$ where $\eta(0) = \eta(\ell) = 0$. Then

$$J[\phi] = \int_0^\ell \left\{ EI(u'' + \varepsilon\eta'')^2 - 2q(u + \varepsilon\eta) \right\} dx$$

$$= J[u] + 2\varepsilon \int_0^\ell \left\{ EIu''\eta'' - q\eta \right\} dx + \varepsilon^2 \int_0^\ell EI(\eta'')^2 \, dx$$

Clearly, following the usual variational process

$$\left.\frac{dJ}{d\varepsilon}\right|_{\varepsilon=0} \quad \text{leading to} \quad \int_0^\ell \left\{ EIu''\eta'' - 2q\eta \right\} dx = 0$$

integrating the first term in the integral by parts twice and incorporating $\eta(0) = \eta(\ell) = 0$ we have

$$\Bigg[EIu''\eta' \Bigg]_0^\ell + \int_0^\ell \left\{ \frac{d^2}{dx^2}(EIu'') - q \right\} \eta \, dx = 0$$

leading to

$$\frac{d^2}{dx^2}(EIu'') - q = 0 \qquad 0 \le x \le \ell$$

$$\left. \mathrm{EI}u'' \right|_{x=\ell} = \left. \mathrm{EI}u'' \right|_{x=0} = 0 \qquad u(0) = u(\ell) = 0$$

The first two boundary conditions are suppressible, the other two are essential

10.3 To obtain a 'feel' for the calculation try the single parameter case first; that is take $a_2 = 0$. The integral $\int_1^2 r^n \, dr = (2^{n+1} - 1)/(n + 1)$ will be required for each value of n from $n = 2$ to 13. The single equation for a_1 is found to be

$$11551.8a_1 - 6.612 = 0 \qquad \text{giving} \quad a_1 = 0.000572$$

The exact value at $r = 1.5$ is found to be 0.00273 whereas the value obtained from this one-parameter approximation is 0.002612.

10.5 We find

$$J[\phi] = \int_0^{\ell} (\phi'')^2\, dx - 2P\phi(\frac{\ell}{2})$$

The final reduced system is

$$\frac{1}{\ell^3}\begin{bmatrix} 192 & 0 \\ 0 & 16\ell^2 \end{bmatrix}\begin{Bmatrix} \tilde{u}_2 \\ \tilde{u}_2' \end{Bmatrix} = \begin{Bmatrix} P \\ 0 \end{Bmatrix}$$

giving $\tilde{u}_2 = P\ell^3/192 \quad \tilde{u}_2' = 0$.

References and Further Reading

Collatz L. (1966). *The Numerical Treatment of Differential Equations* 3rd edn. Berlin: Springer–Verlag.

Ward J.P. (1987). A determinant formalism for shape functions. *Comm. Applied Numerical Methods*, **3**, 129–39.

General text on numerical methods

Gerald C.F. & Wheatley P.O. (1989). *Applied Numerical Analysis* 4th edn. Reading MA: Addison-Wesley.

Introductory finite element texts whose level of treatment is comparable with the present text

Akin J.E. (1986). *Finite Element Analysis for Undergraduates.* New York: Academic Press.

Burnett D.S. (1987). *Finite Element Analysis.* Reading MA: Addison-Wesley.

Segerlind L.J. (1984). *Applied Finite Element Analysis* 2nd edn. New York: John Wiley.

Finite element texts for further reading at a more advanced level and/or with more extensive coverage of applications

Rao S.S. (1989). *The Finite Element Method in Engineering* 2nd edn. Oxford: Pergamon Press.

Reddy J.N. (1984). *An Introduction to the Finite Element Method.* New York: McGraw-Hill.

Stasa F.L. (1985). *Applied Finite Element Analysis for Engineers.* New York: Holt, Rinehart & Winston.

Zienkiewicz O.C. and Taylor R.L. (1989). *The Finite Element Method. Vol. 1. Basic Formulation and LinearProblems* 4th edn. New York: McGraw-Hill.

Texts mainly dealing with the computer implementation of the method

Akin J.E. (1982). *Application and Implementation of Finite Element Methods.* New York: Academic Press.

Hinton E. & Owen D.R.J. (1977). *Programming the Finite Element Method.* New York: Academic Press.

Index